AN ENGINEER'S GUIDE TO MATHEMATICA®

AN ENGINEER'S GUIDE TO MATHEMATICA®

Edward B. Magrab
University of Maryland, USA

WILEY

This edition first published 2014

Registered office
John Wiley & Sons Ltd, The Atrium, Southern Gate, Chichester, West Sussex, PO19 8SQ, United Kingdom

For details of our global editorial offices, for customer services and for information about how to apply for permission to reuse the copyright material in this book please see our website at www.wiley.com.

Library of Congress Cataloging-in-Publication Data applied for.

ISBN: 9781118821268

Set in 10/12pt Times by Aptara Inc., New Delhi, India
Printed and bound in Malaysia by Vivar Printing Sdn Bhd

1 2014

For
June Coleman Magrab

Contents

Preface

The primary goal of this book is to help the reader attain the skills to create Mathematica programs that obtain symbolic and numerical solutions to a wide range of engineering topics, and to display the numerical results with annotated graphics.

Some of the features that make the most recent versions of Mathematica a powerful tool for solving a wide range of engineering applications are their recent introduction of new or expanded capabilities in differential equations, controls, signal processing, optimization, and statistics. These capabilities, coupled with its seamless integration of symbolic manipulations, engineering units, numerical calculations, and its diverse interactive graphics, provide engineers with another effective means of obtaining solutions to engineering problems.

The level of the book assumes that the reader has some fluency in engineering mathematics, can employ the engineering approach to problem solving, and has some experience in using mathematical models to predict the response of elements, devices, and systems. It should be suitable for undergraduate and graduate engineering students and for practicing engineers.

The book can be used in several ways: (1) to learn Mathematica; (2) as a companion to engineering texts; and (3) as a reference for obtaining numerical and symbolic solutions to a wide range of engineering topics involving ordinary and partial differential equations, optimization, eigenvalue determination, statistics, and so on.

The following aids have been used to make it easier to navigate the book's material. Different fonts are used to make the Mathematica commands and the computer code distinguishable from text. In addition, since Greek letters and subscripts can be used in variable names, almost all programs have been coded to match the equations being programmed, thereby making portions of the code more readable. In the first chapter, the use of templates is illustrated so that one can easily create variables with Greek letters and with subscripts. Lastly, since Mathematica is fundamentally different from computer languages usually employed by engineers, the introductory material attempts to make this transition as smooth as possible.

In many of the chapters, tables are used extensively to illustrate families of commands and the effects that different options have on their output. From these tables, the reader can determine at a glance which command and which options can be used to satisfy the current objective. The order of the material is introduced is such a way that the complexity of the examples can be increased as one progresses through the chapters. Thus, the examples range from the ordinary to the challenging. Many of the examples are taken from a wide range of engineering topics. To supplement the material presented in this book, many specific references are made throughout the text to Mathematica's *Documentation Center*, which provide numerous guides and tutorials on topical collections of commands.

The book has two interrelated parts. The first part consists of seven chapters, which introduce the fundamentals of Mathematica's syntax and a subset of commands useful in solving engineering problems. The second part makes extensive use of the material in these seven chapters to show how, in a straightforward manner, one can obtain numerical solutions in a wide range of engineering specialties: vibrations, fluid mechanics and aerodynamics, heat transfer, controls and signal processing, optimization, structures, and engineering statistics. In this part of the book, the vast majority of the solutions are presented as interactive graphics from which one can explore the results parametrically.

In Chapter 1, the basic syntax of Mathematica is introduced and it is shown how to intermingle symbolic and numerical calculations, how to use elementary mathematical functions and constants, and how to create and manipulate complex numbers. Several notational programming constructs are both illustrated and tabulated and examples are given on how to attach physical units to numerical and symbolic quantities. The basic structure of the notebook interface and its customization are presented. In addition, the various templates that can be used to simplify the integration of Greek letters, superscripts and subscripts, and other mathematical symbols into the programming process, and the commands that represent many basic mathematical functions and mathematical constants are illustrated.

In Chapter 2, the commands that can be used to create lists are discussed in detail and their special construction to form vectors and matrices composed of numerical and/or symbolic elements that are commonly employed to obtain solutions engineering applications are introduced. The use of vectors and matrices is discussed in two distinctly different types of applications: to perform operations on an element-by-element basis or to use them as entities in linear algebra operations.

In Chapter 3, ways to create functions, exercise program control, and perform repetitive operations are discussed. The concept of local and global variables is introduced and its implications with respect to programming are illustrated.

In Chapter 4, two types of symbolic manipulations are illustrated. The first is the simplification and manipulation of symbolic expressions to attain a compact form of the result. The second is to perform a mathematical operation on a symbolic expression. The mathematical operations considered are: differentiation, integration, limit, solutions to ordinary and partial differential equations, power series expansion, and the Laplace transform.

In Chapter 5, several Mathematica functions that have a wide range of uses in obtaining numerical solutions to engineering applications are presented: integration, solution to linear and nonlinear ordinary and partial differential equations, solution of equations, determination of the roots of transcendental equations, determining the minimum or maximum of a function, fitting curves and functions to data, and obtaining the discrete Fourier transform.

In Chapter 6, a broad range of 2D and 3D plotting functions are introduced and illustrated using numerous tables and examples from engineering topics. It is shown how to display discrete data values and values obtained from analytical expressions in different ways; that is, by displaying them using logarithmic compression, in polar coordinates, as contours, or as surfaces. The emphasis is on the ways that the basic figure can be modified, enhanced, and individualized to improve its visual impact by using color, inset figures and text, figure titles, axes labels, curve labels, legends, combining figures, filled plot regions, and tooltips.

In Chapter 7, the creation and implementation of interactive graphics and animations are introduced and discussed in detail and illustrated with many examples. The control devices that are considered are the slider/animator, slider, 2D sliders, radio buttons, setter buttons, popup menus, locators, angular gauges, and horizontal gauges.

In Chapter 8, the response of single and two degree-of-freedom systems and thin elastic beams are determined when they are subject to various loadings, damping, initial conditions, boundary conditions, and nonlinearities.

In Chapter 9, the commands used to determine the mean, median, root mean square, variance, and quartile of discrete data are presented and the display of these data using histograms and whisker plots are illustrated. It is shown how to display the results from a regression analysis using a probability plot, a plot of the residuals, and confidence bands. The ways to perform an analysis of variance (ANOVA) and to setup and analyze factorial designs are introduced with examples.

In Chapter 10, the modeling and analysis of control systems using transfer function models and state-space models are presented. It is shown how to connect system components to form closed-loop systems and to determine their time-domain response. Examples are given to show how to optimize a system's response with a PID controller and any of its special cases using different criteria. The creation and use of different models of high-pass, low-pass, band-pass, and band-stop filters are presented and the effects of different types of windows on the short-time Fourier transform are illustrated. The spectral analyses of sinusoidal signals in the presence of noise are presented using root mean square averaging and using vector averaging.

In Chapter 11, several topics in heat transfer and fluid mechanics are examined numerically and interactive environments are developed to explore the characteristics of the different systems. The general topic areas include: conduction, convection, and radiation heat transfer, and internal and external flows.

Edward B. Magrab
Bethesda, MD
USA
October, 2013

Table of Engineering Applications

Part I
Introduction

1

Mathematica® Environment and Basic Syntax

1.1 Introduction

Mathematica is a programming language that integrates, through its notebook interface, symbolic and numerical computations, visualization, documentation, and dynamic interactivity. It provides access to a large collection of such diverse and continually updated and expanded data sets as geometric shapes, a searchable dictionary, and individual country attributes. It also permits one to simultaneously program with different programming paradigms, such as procedural, functional, rule-based, and pattern-based. Its interface has a real-time input semantics evaluator that uses styling and coloring to provide immediate visual feedback on such coding aspects as function names, variable selection, and argument structures. Many of the Mathematica functions used for computation and visualization contain a fair amount of high-level automation so that the user has to interact minimally with their inner workings. If desired, many aspects of the automation procedures can be bypassed and specific choices can be selected.

In this book, we shall employ a subset of Mathematica's library of functions and use them to obtain solutions to a variety of engineering applications. It will be found as one becomes more confident with Mathematica that it is most effectively used interactively. In later chapters, emphasis will be placed on displaying the results as dynamically interactive graphical displays so that real-time parametric investigations can be performed.

In this chapter, we shall introduce the fundamental syntax of Mathematica. In Chapters 2 to 7, we shall introduce additional syntax and illustrate its usage. We start by stating that all variables by default are symbols and global in nature, and unless specifically restricted or cleared, are always available in all open notebooks until Mathematica is closed. Also, because Mathematica treats all variables initially as symbolic entities, any undefined symbol appearing in an expression (that is, any variable appearing on the right-hand side of an equal sign) is perfectly acceptable and will not produce an error message. However, depending on how the expression is used, subsequent operations may not perform as expected depending on the intent for this variable.

An Engineer's Guide to Mathematica®, First Edition. Edward B. Magrab.

Companion Website: www.wiley.com/go/magrab

In addition to the functions that are an integral part of Mathematica, each version of Mathematica comes with what are called standard extra packages that provide specific additional functionality. Frequently, the capabilities of these packages become an integral part of Mathematica. What the names of these packages are and a brief description of what they do can be obtained by entering *Standard Extra Packages* into the search area of the *Documentation Center Window*, which is found in the *Help* menu. Each package is loaded by using the **`Needs`** function. One such case is illustrated in Example 4.11.

1.2 Selecting Notebook Characteristics

Interaction with Mathematica occurs through its notebook interface. As we shall be concerned primarily with presenting graphically solutions to engineering analyses, our discussion will be directed to one type of use of the notebook: entering, manipulating, and numerically evaluating equations typically encountered in engineering.

Upon opening Mathematica, the window shown in Figure 1.1 appears on the computer screen. Since virtually all types of mathematical symbols can appear in Mathematica expressions, it is beneficial to also have its *Special Characters* palette open. As indicated in Figure 1.2, the letters and symbols are accessed by selecting *Palettes* from the Mathematica menu strip and then choosing *Special Characters*. These operations produce the windows shown in Figure 1.2.

To increase or decrease the font size of the characters displayed in the notebook, *Window* from the Mathematica menu strip is selected, then *Magnification* is chosen, and the amount of magnification (or reduction) is clicked. These operations are illustrated in Figure 1.3. As shall be discussed in what follows, various types of expression delimiters are used in constructing expressions: parentheses, brackets, and braces. When nested expressions are employed and various combinations of these delimiters are used, one frequently needs to verify that these delimiters are grouped as intended. A tool that performs this check by highlighting the region that appears between the delimiter selected and its closing delimiter is accessed from the *Edit* menu and then by clicking on *Check Balance*, as shown in Figure 1.4. In Mathematica 9, the placement of the cursor adjacent to either an opening or closing delimiter will highlight them in green. This is a very valuable editing tool; however, it can be disabled by going to *Preferences* in the *Mathematica* menu strip, selecting *Interface*, and then deselecting *Enable dynamic*

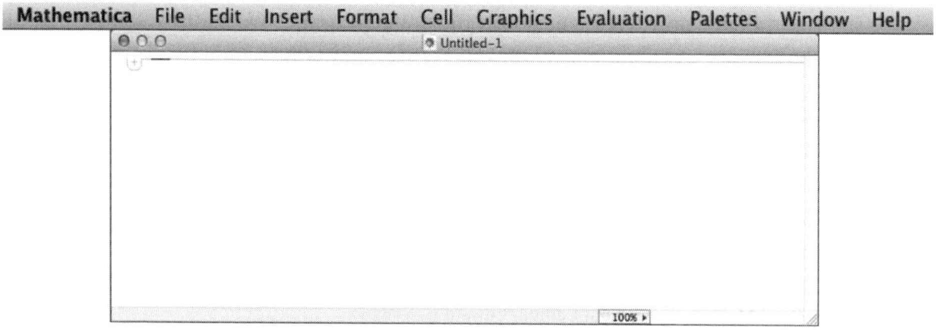

Figure 1.1 Window appearing upon opening Mathematica

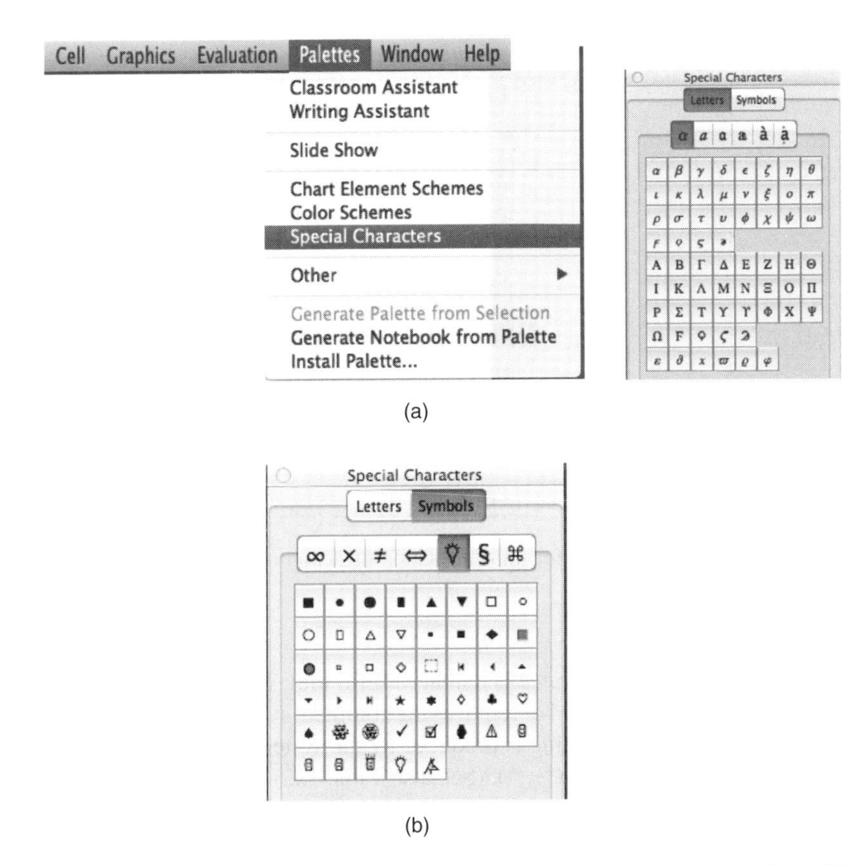

Figure 1.2 (a) Opening the *Special Characters* window to select various alphabet symbols; (b) Accessing various types of symbols; shown here are shapes that can be used as plot markers

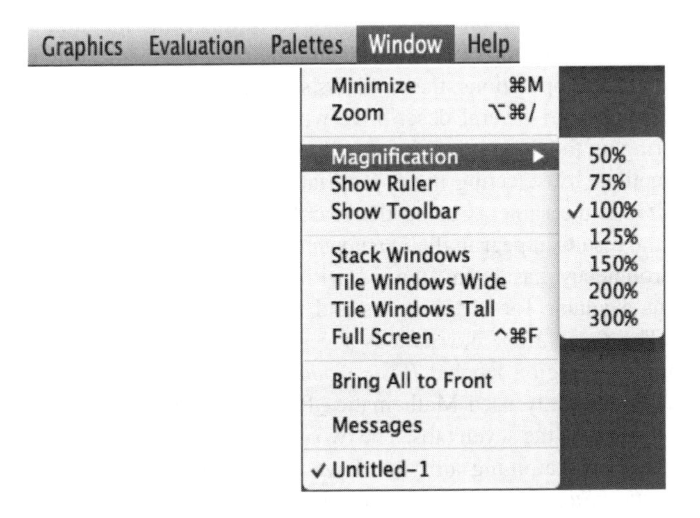

Figure 1.3 Setting the notebook font size

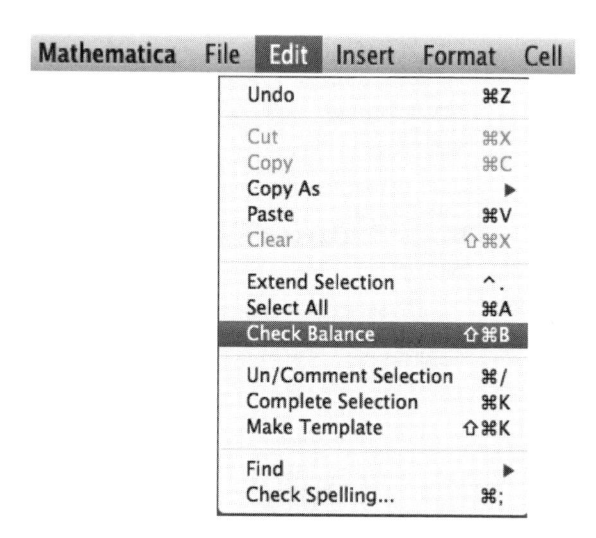

Figure 1.4 Selecting *Check Balance* for implementation of delimiter region identification for (…), […], and {…}

highlighting. Just below *Check Balance* is another useful tool: *Un/Comment Selection*. This feature comments out text highlighted or removes the comment symbols if the selected text had been commented out. The commenting is produced by the system by placing the highlighted text between the asterisks of the set (*…*). (See also Table 1.2.)

Since Mathematica has such a large selection of functions to choose from and since the arguments and their individual form and purpose vary, one should keep the *Documentation Center* window and/or the *Function Navigator* window open for easy access to descriptions of these functions. The *Documentation Center* window is accessed by selecting *Help* from the Mathematica menu strip and then selecting *Documentation Center*. The *Function Navigator* is accessed either by selecting *Function Navigator* from this same menu or by selecting the fourth icon from the left at the top of the *Documentation Center*'s menu strip, which is labeled F[...]. Performing these operations, the windows shown in Figure 1.5 are obtained. Entering either the function name or several descriptive words in the *Documentation Center* search entry area will bring up the appropriate information. In the *Function Navigator*, one will see the candidate functions by selecting the appropriate topic. Using the search function in the *Function Navigator* is the same as using the search function in the *Documentation Center* window; that is, the results appear in the *Documentation Center* window.

After some proficiency has been attained with Mathematica, one can also access the types of functions available for certain tasks and what their arguments are from the *Basic Math Assistant*. The *Basic Math Assistant* is accessed from the *Palettes* menu as shown in Figure 1.6. Visiting the region labeled *Basic Commands*, one can find what arguments are required for many commonly used Mathematica functions. The functions are grouped into seven areas as indicated by the seven tabs. The two rightmost tabs refer to plotting commands. There are two other programming aids that have been added in Mathematica 9. They are the *Next Computation Suggestions Bar* and the *Context-Sensitive Input Assistant*; these are discussed in Section 1.3.

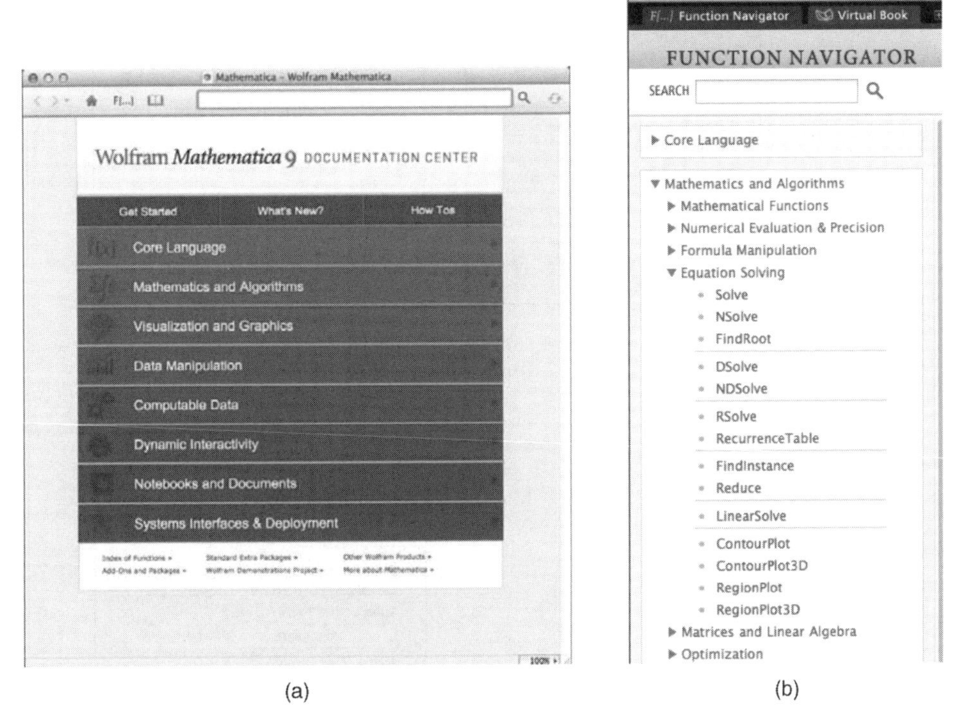

Figure 1.5 (a) *Documentation Center* window and (b) *Function Navigator* window

The *Documentation Center* window also provides access to tutorials on various topics concerning the usage of classes of functions and also has a page that summarizes a collection of functions that can be applied to solve specific topics. Listed in Table 1.1 are selected search entries that can be used as a starting point in determining what is available in Mathematica for obtaining solutions to a particular topic or class of problems. In addition, entering *tutorial/VirtualBookOverview* in the *Documentation Center* search box provides a table of contents to a "how to" introduction to the Mathematica language and contains a very large number of examples illustrating the options available for a specific function.

Lastly, the appearance of the code and the numerical results displayed in the notebook can be altered by selecting *Preferences* in the *Edit* menu. In the *Preferences* window, the *Appearance* tab is chosen and the appropriate tab is selected. For example, the default value of the number of decimal digits to be displayed is 6. To change this value, one goes to the *Numbers* tab and then to the *Formatting* tab. In the box associated with *Displayed precision*, the desired integer value it entered.

Creating New Notebooks or Opening Existing Notebooks

To create a new notebook, one clicks on *File* on the Mathematica menu strip and selects *New* and then *Notebook*. A new notebook window will appear. To open an existing notebook, one clicks on *File* on the Mathematica menu strip and selects *Open* or *Open Recent*. Selecting

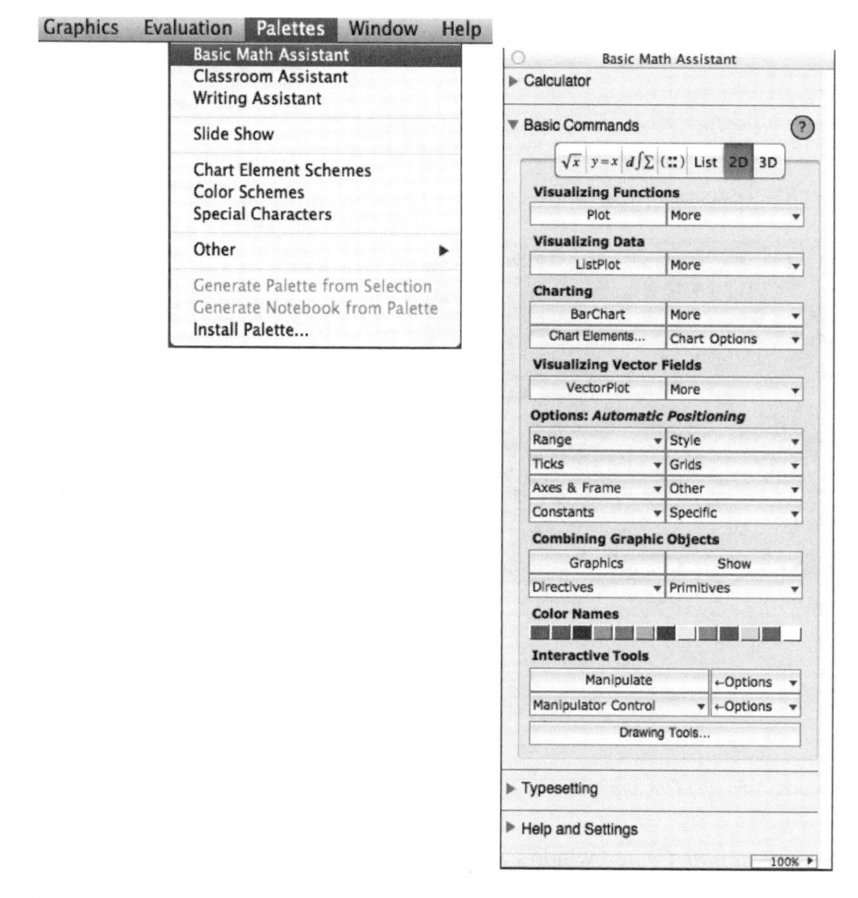

Figure 1.6 Opening the *Basic Math Assistant* window to access the 2D palette of plotting commands

Open will bring up a file directory window, whereas *Open Recent* will bring up a short list of the most recently used notebooks.

Saving Notebooks

To save a notebook that was created during a Mathematica session, one clicks on *File* on the Mathematica menu strip and selects *Save As...* . This brings up a file directory from which an appropriate directory is selected and a notebook name is entered. This procedure is also used for renaming an existing notebook. For an existing notebook that has been modified and the existing notebook name is to remain the same, one clicks on *File* on the Mathematica menu strip and selects *Save*.

1.3 Notebook Cells

To execute an expression or a series of expressions, one has two ways to do it. To execute each expression separately, one types the expression and then simultaneously depresses *Shift*

Table 1.1 Selected topical search entries for the *Documentation Center*

Topic	Search Entry
Trigonometric and inverse trigonometric functions	guide/TrigonometricFunctions
Hyperbolic and inverse hyperbolic functions	guide/HyperbolicFunctions
Special functions	guide/SpecialFunctions
Statistics	guide/DescriptiveStatistics guide/FunctionsUsedInStatistics
Minimum, maximum, optimization, curve fitting, least squares	guide/Optimization
Differentiation and integration	guide/Calculus tutorial/Differentiation tutorial/Integration
Differential equations, roots of polynomials, and roots of transcendental functions	guide/DifferentialEquations guide/EquationSolving tutorial/SolvingEquations tutorial/DSolveOverview
Matrices, vectors, and linear algebra	guide/MatricesAndLinearAlgebra
Fourier and Laplace transforms	guide/IntegralTransforms
Interactive graphical output	Manipulate guide/DynamicVisualization tutorial/IntroductionToManipulate
Lists	guide/ListManipulation
Plotting: 2D and 3D	guide/VisualizationAndGraphicsOverview guide/FunctionVisualization guide/DataVisualization guide/DynamicVisualization guide/PlottingOptions guide/Legends guide/Gauges
Listing of all Mathematica functions	guide/AlphabeticalListing (or click on the *Index of Functions* label at the bottom left of the *Documentation Center* window)
Mathematica's syntax	guide/Syntax
Function creation	tutorial/DefiningFunctions
Program debugging and speed	guide/TuningAndDebugging
Manipulation of symbolic expressions	tutorial/PuttingExpressionsIntoDifferentForms
Controls	guide/ControlSystems
Signal processing	guide/SignalProcessing
Units and units conversion	tutorial/UnitsOverview
Export graphics	tutorial/ExportingGraphicsAndSounds

and *Enter*. The system response appears directly below. When one wants to execute a series of expressions after all the expressions have been entered, each expression is typed on a separate line and after each expression has been typed it is followed by *Enter*. When the collection of expressions is to be executed, the last expression entered is followed by simultaneously depressing *Shift* and *Enter*. Each expression in this group of expressions is executed in the order that they appear and the results from each expression (if not followed by a semicolon) appear directly after the last expression entered.

In the first case, the single expression constitutes an individual cell and is so indicated by a closing bracket that appears at the rightmost edge of the notebook window. The system response also appears in its own cell. However, these two individual cells are part of another cell that is composed of these two individual cells. This is illustrated in Figure 1.7a. In the process of obtaining these cells, Mathematica provided two programming aids automatically. The first is the *Context-Sensitive Input Assistant*, which appeared after the first two letters of `Sin` were typed. As shown in Figure 1.7b, a short list of common Mathematica commands appears that can be expanded to all appropriate Mathematica commands that begin with `Si` by clicking on the double downward facing arrows. Additional information regarding the *Context-Sensitive Input Assistant* can be found in the *Documentation Center* using the entry *tutorial/UsingTheInputAssistant*.

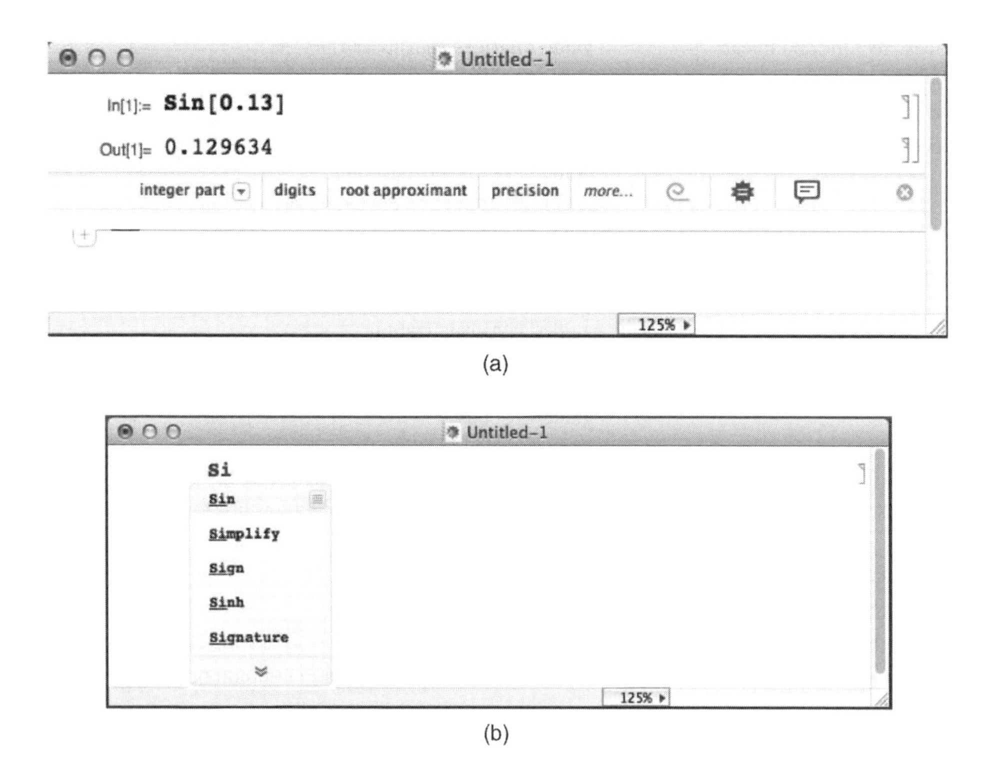

Figure 1.7 (a) Cell delimiters, which appear on the right-hand edge of the window and the *Next Computation Suggestions Bar*; (b) the *Context-Sensitive Input Assistant*, which appeared after the two letters `Si` were typed

Note: The *Context-Sensitive Input Assistant* can be disabled by selecting *Preferences* in the *Mathematica* menu. In the *Preferences* window, the *Interface* tab is chosen and then the check mark adjacent to *Enable autocompletion with a popup …* is removed.

After the execution of a line of code and the display of the result, there appears on a separate line a system-provided set of choices. This line is called the *Next Computation Suggestions Bar*. It can be suppressed for this calculation by clicking on the encircled × that appears at its right edge. The *Next Computation Suggestions Bar* remains suppressed for all subsequent program executions until reactivated; that is, until one clicks on the arrow at the right end of one of the results. It is context dependent, and in this case the system suggests to the user that if additional processing of the result is desired, typical operations could be: **`digital`** – to find one of the various forms the numerical value (floor, ceiling, round, and fractional part in this case); **`digits`** – obtain a list of the digits appearing in the numerical result, and so on. Each choice results in the appearance of another *Next Computation Suggestions Bar*. Depending on the complexity of the result, the *Next Computation Suggestions Bar* will provide appropriate suggestions, such as converting radians to degrees or plotting the result. Additional information about the *Next Computation Suggestions Bar* can be found in the *Documentation Center* by using the entry *guide/WolframPredictiveInterface*.

Note: The *Next Computation Suggestions Bar* can be disabled by selecting *Preferences* in the *Edit* menu. In the *Preferences* window the *Interface* tab is chosen and then the check mark adjacent to *Show Suggestions Bar after last output* is removed.

In Figure 1.8, the case of executing a series of expressions after all the expressions have been entered is shown. In this case, the cursor in placed under and outside of the rightmost cell, which is delineated by the horizontal line. It is seen that the numerical evaluation of each of these trigonometric functions appears in its own cell and corresponds to the order in which they appear. Thus, the first numerical value corresponds to sin(0.13), the second to cos(0.13), and the third to tan(0.13). In addition, it is seen that the three expressions and the three system

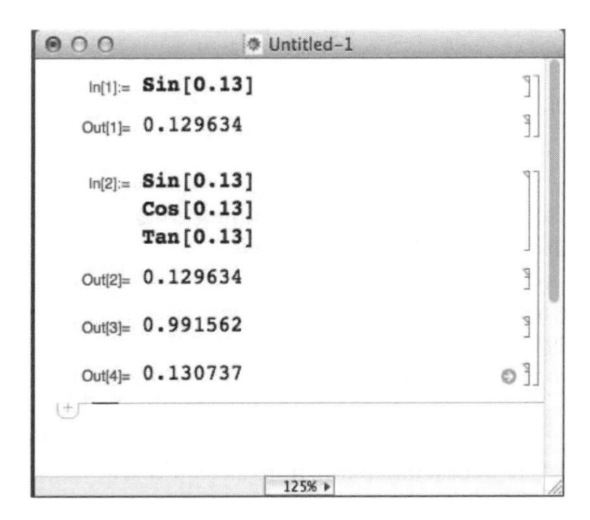

Figure 1.8 Creation of a cell composed of several expressions by using *Enter* following the first two expressions; that is, for **`Sin`** and **`Cos`**, and then *Shift* and *Enter* after **`Tan`** (the *Next Computation Suggestions Bar* has been hidden)

responses reside in a cell that is distinct from the cell of the single computation preceding it. For this case, the *Next Computation Suggestions Bar* has been suppressed.

From Figures 1.7 and 1.8, it is seen that every time *Shift* and *Enter* are simultaneously depressed, the system provides an input identifier, in the first case `In[1]` and in the second case `In[2]`. A similar identifier is created for the output (system response). In the first case, it is `Out[1]` and, in the second case, each cell gets its own identifier: `Out[2]`, `Out[3]`, and `Out[4]`. It is seen that the numerical values of the output identifiers do not have to correspond to the input identifiers. When illustrating how the Mathematica language is used, these input and output identifiers will be omitted.

Aborting a running program

In the event that a calculation appears to be running excessively long, one can abort the calculation by selecting *Evaluation* from the Mathematica menu strip and then choosing *Abort Evaluation*. A calculation is in progress when the thickness of the cell bracket on the rightmost edge of the notebook window increases and appears as a thick black vertical line.

1.4 Delimiters

There are three types of delimiters that are used by Mathematica and each type indicates a very specific set of operational characteristics: ( ) – open/closed parentheses; [ ] – open/closed brackets; and { } – open/closed braces. The parentheses are used to group quantities in mathematical expressions. The brackets are used to delineate the region containing the arguments to all Mathematica and user-created functions and their usage is discussed in Section 3.2. Lastly, the braces are used to delineate the elements of lists, which are defined and discussed in Chapter 2. A summary of these and several other special characters are given in Table 1.2.

1.5 Basic Syntax

1.5.1 Introduction

Mathematica's syntax is different in many respects from traditional programming languages such as Basic, Fortran, and MATLAB® and takes a fair amount of usage to get used to it. In addition, it should be realized that there are often different ways a solution can be coded to obtain one's end results. The "best" way can be judged by such criteria as execution time, code readability, and its number of instructions. The beginner is encouraged to experiment with Mathematica's syntax to see what can be done and how it is done. One's programming sophistication typically increases with continued usage.

The mathematical operators are the traditional ones: + for addition, – for subtraction, / for division, * for multiplication (or a space between variables, which we shall illustrate subsequently), and ^ for exponentiation. In addition, numbers without a decimal point are considered integers and are treated differently from those with a decimal point.

Consider the formula

$$z = \frac{a+b}{c} - \frac{3}{4}ef^2$$

Table 1.2 Special characters and their usage

Character	Name	Usage	Introduced in
`,`	Comma	Separator of elements in a list {..., ...} or arguments of a function [..., ...]	Section 2.2 Section 3.2
`_`	Underscore	Appended to variable name(s) in the argument of a user-defined function and in a few Mathematica functions	Section 3.2.1
`:=`	Colon equal	Delays the evaluation of the expression on the right-hand side of the equal sign	Section 3.2.1
`;`	Semicolon	Suppresses the display of an excuted expression	Section 1.5.1
		Required when employing more than one expression in the argument of certain functions (denoted in Mathematica as a **`CompoundExpression`**)	Section 3.2.1
		Part of the syntax to access elements of lists	Table 2.4
`.`	Period	Decimal point; for expressions with decimal numbers, Mathematica will attempt a numerical evaluation	Section 1.5.1
		Performs matrix/vector product when the entities have the appropriate dimensions	Section 2.4.1
`" "`	Quotation marks	Defines a string expression of all characters appearing within the quotation marks	Section 1.9.1
`%`	Percent	Gives the last result generated; to access **`Out[n]`**, enter **`%n`**	Section 1.5.1
`{ }`	Braces	Defines a list, which is a collection of comma-separated objects	Section 2.2
`[ ]`	Brackets	Defines the region containing the arguments of a function	Section 3.2.1
`( )`	Parentheses	Groups terms in equations	Section 1.5.1
`[[ ]]`	Double brackets	Argument indicates the locations of elements of a list	Section 2.3.3
`(* *)`	Parenthesis asterisk	Nonexecutable comment composed of the characters placed between the asterisks	Section 1.6
`/.`	Slash period	Shorthand notation for **`ReplaceAll`**, which replaces all occurrences of a quantity as specified by a rule construct and is used in the form **`/.a->b`**	Section 1.10

(continued)

Table 1.2 (*Continued*)

Character	Name	Usage	Introduced in
`->`	Hyphen greater than	Indicates a rule construct: `a->b` means that `a` will be transformed to `b`	Section 1.10
`//`	Double slash	Shorthand notation for **`Postfix`**: the instruction following the double slash is often used to specify how an output will be displayed or to simplify an expression or to time the execution of an expression	Section 2.2.1 Table 2.1 Table 2.2
`<>`	Greater than less than	Concatenates string objects appearing on each side of these symbols	Section 1.9.1
`/@`	Slash at symbol	Shorthand notation for **`Map`** and used in the form `f/@h`	Section 3.5.4
`#,#n`	Number sign	Represents the *n*th symbol in a pure function; when $n = 1$, the "1" can be omitted	Section 3.2.2
`##`	Number signs	Represents the sequence of arguments supplied to a pure function	Table 7.1
∈	Element	The Mathematica specification that follows the symbol indicates the domain of the symbol that precedes it.	Table 4.2 Example 4.4
≈	-	Special symbol used in **`NProbability`**	Section 9.2.1

It is entered as

```
z=(a+b)/c-3/4 e f^2
```

Note that we chose to use a space[1] to indicate multiplication instead of the asterisk (*); that is, there is a space between the 4 and `e` and between the `e` and `f`. The system responds with

$$\frac{a+b}{c} - \frac{3\,e\,f^2}{4}$$

Use of the cursor on the above equation would indicate that there are spaces between the `3` and `e` and between the `e` and `f`. Notice that the system has suppressed the display of `z` in the output. However, it is accessible by simply typing in a new cell either `z` or `%` and then *Shift* and *Enter* simultaneously. When either of these operations is performed, the above expression is displayed. Also, it should be noted that the system had no trouble dealing with five undefined variable names. It simply treated them as symbolic quantities and used them accordingly.

[1] In actuality, a number preceding a symbol does not require a space. It is employed here for consistency. However, with a space used, `2 a` means `2×a` and is the same as `a 2`. On the other hand, without a space `2a` is the same as `2×a`, but `a2` is the name of a variable.

On the other hand, if a decimal point was added to either or both of the integers, that is,

```
z=(a+b)/c-3./4 e f^2
```

was typed into the notebook and *Shift* and *Enter* depressed simultaneously, the system's response is

$$\frac{a+b}{c} - 0.75\, e\, f^2$$

The inclusion of the decimal point forces the system to evaluate all numerical values in the expression when possible. Another way to have the system evaluate all integer expressions is with **N**, which is discussed subsequently. It is mentioned that Mathematica makes a distinction between a decimal number and an integer. An integer is a number without a decimal point and a number with a decimal point is labeled internally a real number. Thus, if one searches for real numbers, only those numbers using a decimal point will be identified as such. All integers will be ignored.

The advantages of Mathematica's ability to seamlessly integrate symbolic manipulation and numerical calculations are now illustrated. We shall use the expression given above for **z** except this time it will be preceded by two additional expressions as follows

```
a=9+1.23^3;
e=h/(11. f);
z=(a+b)/c-3./4 e f^2
```

where the semicolons at the end of the first two expressions were used to suppress their output. The system responds with

$$\frac{10.8609+b}{c} - 0.0681818\, f\, h$$

It is seen that Mathematica did the substitutions for the variables **a** and **e**, performed all numerical calculations that it could, and did the algebraic simplification that resulted in the cancellation of one **f**.

1.5.2 Templates: Greek Symbols and Mathematical Notation

Greek symbols and symbols with subscripts can be used as variable names. This has the advantage of making portions of the code more readable. However, it can take a bit longer to write the code because of the additional operations that are required to create these quantities. In this book, we shall use the Greek alphabet and the subscripts when practical so that one can more readily identify the code with the equations that have been programmed.

Greek Symbols

Greek symbols can be used directly for variable names with the use of the *Special Characters* palette. To insert a special character, one places the cursor in the notebook at the location at which the character is to be placed. Then one selects the character from the palette and the selected character will appear at the location selected in the notebook. For example, using the palette to create the expression

```
x=Ω^γ
```

displays

Ω^γ

These expressions can be used in a regular manner; thus, the execution of

```
Ω=7.;
γ=2.;
x=Ω^γ
```

yields `49` and we have used the semicolon (;) to suppress the display of the first two lines of the program.

Mathematical Notation

Mathematica provides a means to easily create program instructions using standard mathematical notation such as exponents, square roots, and integrals. This notation is accessed by using the *Typesetting* (or *Calculator*) portion of the *Basic Math Assistant*, which is selected from the *Palettes* menu, as shown in Figure 1.9. To illustrate the palette's use, we shall repeat the example given in Section 1.5.2. Additional applications of mathematical notation are given in the subsequent chapters. Thus,

```
Ω=7.
γ=2.
x=Ω^γ
```

yields `49`. The last instruction was obtained from clicking on the template superscript symbol (□□).

For another example, consider the cube root of 27. In this case, the use of the *Basic Math Assistant* palette results in

$$x = \sqrt[3]{27}$$

which yields `3`. This instruction was obtained from clicking on the template *n*th root symbol ($\sqrt[\square]{\square}$).

The use of the *Typesetting* portion of the *Basic Math Assistant* is also very useful in annotating the graphical display of results as illustrated in Table 6.8.

One can also use the *Basic Math Assistant* to create variables that contain subscripts and superscripts. Thus, using the *Basic Math Assistant* to create the relation $d_a = e^{b+c}$, we have

$$d_a = E^{b+c}$$

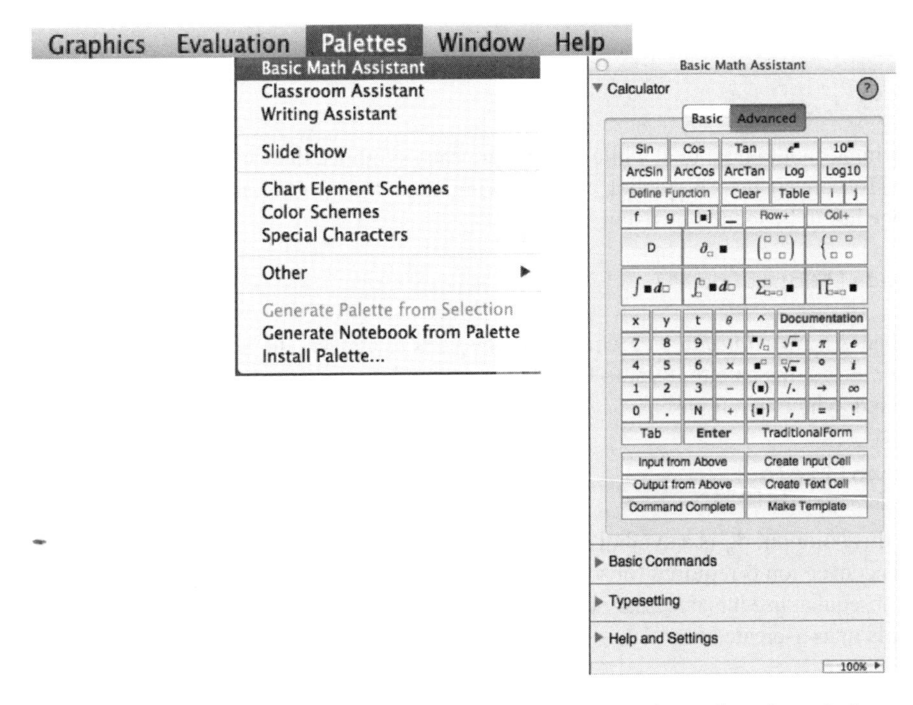

Figure 1.9 Opening the *Basic Math Assistant* window to access advanced mathematical notation templates

which displays

$$\mathtt{e^{b+c}}$$

and

$$\mathtt{f=1+d_a}$$

displays

$$\mathtt{1+e^{b+c}}$$

The symbol `e` is an approximation to the way that Mathematica displays *e*.

A subscript and superscript appearing on the right hand of the equal sign are treated the same. Using the *Basic Math Assistant* template, consider the following

```
a=7;
```

$$\mathtt{b=c^a+d_a}$$

which displays

$c^7 + d_7$

It is important to note that, while **d** and **a** are each symbols, the variable d_a is not a symbol entity. To convert it to a symbol, the following steps have to be taken. First, the *Notation* package has to be loaded by using

```
Needs["Notation`"]
```

This opens a *Notation Palette*, which must be used to convert the subscripted variable to a symbol. From this palette, **Symbolize[□]** is selected. In the square, the subscripted symbol is entered, which in our case is

```
Symbolize[d_a]
```

From this point on, d_a is a symbol.

This conversion is required for several of the templates appearing in the *Typesetting Palette* if in subsequent use it is necessary to treat them as a single symbol. For the use of subscripted symbols in user-created functions, see Section 3.2.1.

1.5.3 Variable Names and Global Variables

User-created names for variables and functions must start with a letter, are case sensitive, and are permanent for the duration of the Mathematica session unless specifically removed or appear in certain Mathematica commands. There does not appear to be a restriction on the number of alphanumeric characters that can be used to create a variable name. It is good practice to remove the variables after one has finished using them and before proceeding further. This removal is done with **Clear** or with **ClearAll**. The arguments of these commands are comma-separated names of the variables to be deleted (cleared). Either of these commands can be used in one of two ways. They can be employed after the completion of a procedure to delete the variable names that were just used or they can be employed before a new procedure to ensure that the variable names to be used do not have another definition.

The naming convention in Mathematica it that *all* Mathematica function names begin with a capital letter and following the last letter of the function name are a pair of open/closed brackets [ ]. Between these brackets, one places expressions, procedures, and lists according to the specifications regarding the usage of that function. Consequently, some care should be exercised when creating variable names and function names. One way to eliminate the possibility of a conflict is to start each variable name with a lower case letter. In any case, do not use the following single capital letters as variable names: **C**, **D**, **E**, **I**, **N**, and **O**.

We shall now show the care that has to be exercised when choosing variable names since, as previously mentioned, all variable names and their respective definitions or assignments remain available until either they are redefined or cleared. Suppose that earlier in the notebook one evaluated the expression

```
a=0.13^2
```

and the system responded with

```
0.0169
```

Then, later in the notebook, one enters the expression

```
z=(a+b)/c-e f^2
```

where it was thought that **a**, **b**, **c**, **e**, and **f** were symbols; that is, no assignment has been given them. However, the system response is

$$\frac{0.0169+b}{c} - e\,f^2$$

since the variable **a** had already been assigned the numerical value shown above. To avoid this type of unintended substitution, one uses **Clear** as follows

```
Clear[a,b,c,e,f]
z=(a+b)/c-e f^2
```

Although it may not always be explicitly shown, it is implicit that for all examples presented in this book the **Clear** *function will have been employed using the appropriate variable names and user-created function names to ensure that this type of unintended substitution doesn't occur.*

One can clear all user-created variable names and function names without specifying them individually by using

```
ClearAll["Global`*"]
```

The "backwards prime" is required and the asterisk (*) indicates that all global names are to be included.

Importance of global variables

Global variables play in important role in the creation and use of functions. It will be seen that when used properly, the fact that variables are by default global variables can simplify the implementation of user-created functions and programming in general. These topics are introduced and discussed in Chapter 3.

1.6 Mathematical Constants

The common mathematical constants $j = i = \sqrt{-1}$, π, ∞, e, $\pi/180$ (conversion from degrees to radians) and $\gamma = 0.57721566$ (Euler's constant) are represented in Mathematica as listed in Table 1.3. Thus, to enter the expression $e^{j\pi/4}$, we type

```
E^(I π/4)
```

Table 1.3 Mathematical constants

Constant	Mathematica function
$i = j = \sqrt{-1}$	`I`
π	`Pi` or[†] π, which is obtained from the *Special Characters* palette under the α tab within the *Letters* tab
∞	`Infinity` or ∞, which is obtained from the *Special Characters* palette under ∞ tab within the *Symbols* tab
e	`E` (or `Exp[1]`)
$\pi/180$	`Degree` (See text for usage)
γ	`EulerGamma` (Euler's constant)

[†] Note: `Pi^2.` and `π^2.` return the same numerical values; however, `Pi^2` and `π^2` return π^2.

and the response is

$$\mathbf{e}^{\frac{i\pi}{4}}$$

where **e** is used to represent the way that Mathematica displays e.

To convert this symbolic expression to a numerical value, we use the Mathematica function

```
N[expr,n]
```

where **expr** is the expression that is to be converted to a numerical value and n is the number of digits of precision that is used during its evaluation of the expression. Thus,

```
N[E^(I π/4)]
```

gives

```
0.707107+0.707107 i
```

where the symbol **i** is an approximation to the way Mathematica displays **I**. By omitting n in **N**, we have settled for machine precision. However, the number of digits displayed is that specified in *Preferences*, which in this case is six. The same result will be obtained without **N** if the integer 4 was replaced by 4.0; that is, by its decimal form. The use of the decimal point to force numerical evaluation of all numbers will be used frequently.

To obtain the previous results using 20 digits of precision, we modify the above expression as

```
N[E^(I π/4),20]
```

which produces

```
0.70710678118654752440+0.70710678118654752440 i
```

To convert an angular value from degrees to radians, the constant **Degree** is used in the following manner. Let the angle be 30°; then to express this angle in radians we use

```
N[Degree 30]
```

which results in

```
0.523599
```

The function **N** is not required if its argument is of the form **Degree 30.0**; that is, if the decimal form is used. However, **Degree** can also be used to append to 30 the degree symbol. Hence, entering

```
Degree 30 (* or 30 Degree *)
```

results in

```
30°
```

We have indicated an equivalent syntax as a comment.

1.7 Complex Numbers

Complex numbers are formed with the appropriate use of **I**. The resulting quantity can be manipulated and parsed by using the functions shown in Table 1.4. For example, let us determine the numerical value of the magnitude and the imaginary part of

$$r = (1 + 2j)^j$$

Since we are interested in numerical values, we express the numerical values in their decimal form and enter

```
r=(1.+2. I)^I
m=Abs[r]
im=Im[r]
```

Table 1.4 Complex number manipulation functions

		Output		
Operation	Mathematica function	$z = 2 + 3j$	$z = 2.0 + 3j$	$z = x + jy$
Real part	`Re[z]`	2	2.	`-Im[y]+Re[x]`
Imaginary part	`Im[z]`	3	3	`Im[x]+Re[y]`
Complex conjugate	`Conjugate[z]`	2-3 i	2.-3 i	`Conjugate[x]- I Conjugate[y]`
Argument	`Arg[z]`	$\text{ArcTan}\left[\frac{3}{2}\right]$	0.982794	`Arg[x+I y]`

The system responds with

```
0.22914+0.23817 i
0.03305
0.23817
```

The first value corresponds to r, the second value to the magnitude of r, and the last value to the imaginary part of r. It is mentioned again that only *Enter* was depressed after each of the first two lines and *Shift* and *Enter* were simultaneously depressed after the third line. These three lines of code reside in one cell and each of the output results resides in its own cell. However, all six lines reside in one cell.

1.8 Elementary, Trigonometric, Hyperbolic, and a Few Special Functions

Elementary Functions

The syntax of several elementary functions is shown in Table 1.5. The functions have been illustrated in this table for real, integer, symbolic, and complex quantities.

Trigonometric Functions

The names for trigonometric and inverse trigonometric functions are listed in Table 1.6. The arguments of these functions can be integers, real numbers, or complex quantities. Thus, to determine the value of the cot($\pi/5$), we enter

```
Cot[π/5]
```

and obtain

$$\sqrt{1+\frac{2}{\sqrt{5}}}$$

On the other hand, if a numerical value was desired, then one would enter

```
Cot[π/5.]
```

and the system would respond with `1.37638`.

If the numerical value of cos($2 + 3j$) were desired, we enter

```
Cos[2.+3 I]
```

and obtain

```
-4.18963-9.10923 i
```

Table 1.5 Elementary functions

Operation	Mathematica function*	Examples of output for explicit forms of z			
		$z = 2.0$	$z = x + jy$	$z = 2 + 3j$	$z = 2.0 + 3j$
$\sqrt{z}$	**Sqrt[z]**	1.41421	$\sqrt{\mathtt{x + i\,y}}$	$\sqrt{\mathtt{2 + 3\,i}}$	1.67415+0.895977 **i**
$\sqrt[3]{x}$	**CubeRoot[z]**	1.25992	$\sqrt[3]{\mathtt{x + i\,y}}$	-	-
e^z	**E^z** or **Exp[z]**	7.38906	$e^{x+i\,y}$	$e^{2+3\,i}$	-7.31511+1.04274 **i**
$\ln_e z$	**Log[z]**	0.693147	Log[x+y i]	Log[2+3 i]	1.28247+0.982794 **i**
$\log_{10} z$	**Log[10,z]** or **Log10[z]**	0.30103	$\frac{\mathtt{Log[x + y\,i]}}{\mathtt{Log[10]}}$	$\frac{\mathtt{Log[2 + 3\,i]}}{\mathtt{Log[10]}}$	0.556972+0.426822 **i**
$\lvert z\rvert$	**Abs[z]**	2.	Abs[x+y i]	$\sqrt{13}$	3.60555
signum(z)	**Sign[z]**	1	Sign[x+y i]	$\frac{2 + 3\,i}{\sqrt{13}}$	0.5547+0.83205 **i**
$z!$	**Factorial[z]** or **z!**	2.	(x+i y)!	(2+i 3)!	-0.44011-0.063637 **i**

* When used symbolically, **Sqrt[z²]** will not return **z** and **Log[z²]** will not return **2 Log[z]**. To obtain these simplifications, **PowerExpand** must be used as shown in Table 4.1.

Table 1.6 Trigonometric and inverse trigonometric functions

Trigonometric function	Mathematica function	Inverse trigonometric function	Mathematica function
sin z	`Sin[z]`	$\sin^{-1} z$	`ArcSin[z]`
cos z	`Cos[z]`	$\cos^{-1} z$	`ArcCos[z]`
tan z	`Tan[z]`	$\tan^{-1} z$ or $\tan^{-1} y/x$	`ArcTan[z]` or `ArcTan[x,y]`
csc z	`Csc[z]`	$\csc^{-1} z$	`ArcCsc[z]`
sec z	`Sec[z]`	$\sec^{-1} z$	`ArcSec[z]`
cot z	`Cot[z]`	$\cot^{-1} z$	`ArcCot[z]`

To obtain the value of a trigonometric function when the angle is given in degrees, we use `Degree`. Thus, if the angle is 35°, then its cosine is found by entering

```
Cos[Degree 35.]
```

which yields

```
0.819152
```

The decimal form of the number is required to produce this result. If the decimal point were not used, the output would have been

```
Cos[35°]
```

For the case when the argument is a complex symbolic quantity $z = x + jy$, refer to Table 4.1.

Hyperbolic Functions

The names for hyperbolic and inverse hyperbolic functions are listed in Table 1.7. The arguments of these functions can be complex quantities. Thus, to determine the value of the cosh(2 + 3j), we enter

```
Cosh[2.+3 I]
```

and obtain

```
-3.72455+0.511823 i
```

For a symbolic complex quantity $z = x + jy$, refer to Table 4.1.

Table 1.7 Hyperbolic and inverse hyperbolic functions

Hyperbolic function	Mathematica function	Inverse hyperbolic function	Mathematica function
$\sinh z$	`Sinh[z]`	$\sinh^{-1} z$	`ArcSinh[z]`
$\cosh z$	`Cosh[z]`	$\cosh^{-1} z$	`ArcCosh[z]`
$\tanh z$	`Tanh[z]`	$\tanh^{-1} z$	`ArcTanh[z]`
$\text{csch}\, z$	`Csch[z]`	$\text{csch}^{-1} z$	`ArcCsch[z]`
$\text{sech}\, z$	`Sech[z]`	$\text{sech}^{-1} z$	`ArcSech[z]`
$\coth z$	`Coth[z]`	$\coth^{-1} z$	`ArcCoth[z]`

Special Mathematical Functions

Mathematica has a large collection of special functions, which can found in the *Documentation Center* window using the search entry *guide/SpecialFunctions*. It will be found that many functions used to obtain solutions to engineering topics are available. The application of several of these special functions will be illustrated in later chapters. Some of the more common functions used in engineering applications can be found in Table 4.4.

1.9 Strings

1.9.1 String Creation: `StringJoin[]` *and* `ToString[]`

Any combination of numbers, letters, and special characters that are linked together to form an expression that does not perform any Mathematica operation is denoted a string. There are a large number of commands that can be used to manipulate strings. Our use for them will be limited primarily to creating annotated output for graphics. Therefore, we shall introduce only a few string creation and manipulation commands. Additional enhancements to string expressions such as subscripts and superscripts are presented in Table 6.8.

A string is created by enclosing the characters with quotation marks. Thus,

```
s1="text"
```

creates a string variable **s1**, which is composed of four characters. To concatenate two or more strings, we use either

```
StringJoin[s1,s2,...]
```

or

```
s1<>s2<>...
```

where **sN** are strings. Thus, if

```
s1="text";
s2="Example of ";
```

then either

```
s2<>s1
```

or

```
StringJoin[s2,s1]
```

yields

```
Example of text
```

One is also able to convert a numerical quantity to a string by using

```
ToString[expr]
```

where **expr** is a numerical value or an expression that leads to a numerical value. Thus, for example, if we would like to evaluate and display in an annotated form the value of

$$\sqrt{|\sin(x^2)|}$$

when $x = 0.35$, the instructions are

```
x=0.35;
p="When x = "<>ToString[x]<>", √|sin(x²)| = "<>
 ToString[√Abs[Sin[x²]]]
```

which displays

```
When x = 0.35, √|sin(x²)| = 0.349562
```

In obtaining this expression, we have used the appropriate *Basic Math Assistant* templates in the expression for **p**.

1.9.2 Labeled Output: Print[], NumberForm[], EngineeringForm[], *and* TraditionalForm[]

An improved way of displaying output is to combine the string conversion commands with **Print**, which is implemented with

```
Print[expr1,expr2,...]
```

where **exprN** is a constant, a symbolic expression, the numerical value of a computed variable, a string, or a Mathematica object. To print multiple lines, one can use multiple **Print** commands or one **Print** command and the **Row** command, which is discussed in Figure 6.2 and Table 6.8.

For example, the results of the previous example using complex numbers is modified as follows

```
r=(1+2 I)^I;
Print["r = ",r]
Print["|r| = ",N[Abs[r]],"  Im[r] = ",N[Im[r]]]
```

which yields

```
r=(1+2 i)^i
|r|=0.3305  Im[r]=0.23817
```

The number of digits that are displayed and their form can be controlled by using **NumberForm** or **EngineeringForm**. These commands, respectively, are given by

```
NumberForm[val,n]
EngineeringForm[val,n]
```

where **val** is the numerical quantity to be displayed with **n** digits. To illustrate the usage of these commands, we shall display $|r|$ in the above example with 10 digits using both formats. Then,

```
r=Abs[(1+2. I)^I];
Print["|r| (Number form) = ",NumberForm[r,10]]
Print["|r| (Engrg Form) = ",EngineeringForm[r,10]]
```

which displays

```
|r| (Number form) = 0.3304999676
|r| (Engrg Form) = 330.4999676×10^-3
```

The functions **NumberForm** and **EngineeringForm** are frequently used to format numerical output when annotating graphics.

The function that generates traditional mathematical notation is

```
TraditionalForm[expr]
```

where **expr** is a mathematical expression using Mathematica's syntax. For example, if

```
expre=BesselJ[n,x] Sin[x];
```

then

```
Print["y = ",TraditionalForm[expre]]
```

displays

```
y = sin(x) J_n(x)
```

Table 1.8 Decimal-to-integer conversion functions

Method	Mathematica function	x (Argument)	y (Output)
Closest integer to x	`y=Round[x]`	`-0.6`	`-1`
		`2.7`	`3`
		`2.49-2.51 I`	`2-3`
Smallest integer $\geq x$	`y=Ceiling[x]`	`-0.6`	`0`
		`2.7`	`3`
		`2.49-2.51 I`	`3-2`
Greatest integer $\leq x$	`y=Floor[x]`	`-0.6`	`-1`
		`2.7`	`2`
		`2.49-2.51 I`	`2-3`
Integer part of x	`y=IntegerPart[x]`	`-0.6`	`0`
		`2.7`	`2`
		`2.49-2.51 I`	`2-2`
Replace a real number $< 10^{-10}$ with integer 0	`Chop[x]`	`1.2 10^(-15)+4.5 I`	`4.5 I`

1.10 Conversions, Relational Operators, and Transformation Rule

Decimal-to-Integer Conversion

There are several ways that one can round a decimal value to an integer depending on the rounding criterion. These methods are summarized in Table 1.8.

Relational Operators

Several relational operators are listed in Table 1.9. These operators are typically used in such functions as **`If`**, **`Which`**, **`While`**, **`Solve`**, **`DSolve`**, and **`Simplify`**.

Substitution: Transformation Rule

Mathematica has a very specific manner in which one or more expressions can be substituted into another expression. The procedure that is introduced here is used throughout

Table 1.9 Relational and logical operators

Mathematical symbol	Mathematica symbol
$=$	`==`
$>$	`>`
$\geq$	`>=`
$<$	`<`
$\leq$	`<=`
$\neq$	`!=`
And	`&&`
Or	`||`

Mathematica's implementation and is one of the primary ways in which one gains access to the expressions and results from many of its built-in functions. It will be illustrated here using some simple examples and then again in later sections with other applications.

There are two distinct sets of symbols that will now be defined. The first is /. (slash period), which is shorthand for replace all (**ReplaceAll**). The second notation is **a** -> **b** (hyphen greater than), which is a rule construct that means that **a** will be transformed to **b**. Then the statement

```
v/.a->b
```

instructs Mathematica to perform the following: in the expression **v**, everywhere **a** occurs replace it with **b**. The quantity **b** can be a number, a symbol, or an expression. If **a** does not appear in **v**, then nothing happens. Furthermore, several substitutions can be made sequentially in the order that they appear. Thus, an extension of the previous expression could be[2]

```
v/.a->b/.c->Sqrt[e+f]
```

For an illustration of the use of this construct and its effect, we return to the previous example given by

```
a=9+1.23^3;
e=h/(11. f);
z=(a+b)/c-3./4 e f^2
```

which is one way that substitution can be performed. The same results can be obtained as follows

```
z=(a+b)/c-3./4 e f^2;
z/.a->9+1.23^3/.e->h/(11. f)
```

which results in

$$\frac{10.8609+b}{c} - 0.0681818\ f\ h$$

and is the same result that was obtained previously.

Finding Built-in Function Options

Many built-in functions have options that can be changed. The options that can be changed and that are currently being used by a built-in function are determined by typing and executing

```
Options[FunctionName]
```

[2] This transformation can also be performed using a list, which is discussed in Chapter 2, as follows:

```
v/.{a->b,c->Sqrt[e+f]}
```

which displays all options, their option names, and their current selections for the function **FunctionName**. Another way to determine these options is to enter **FunctionName** into the *Documentation Center* search box. As an example, we shall see what options are available for **NDSolveValue**, which performs a numerical evaluation of differential equations. Then,

```
Options[NDSolveValue]
```

displays

```
{AccuracyGoal->Automatic, Compiled->Automatic,
DependentVariables->Automatic, EvaluationMonitor->None,
InterpolationOrder->Automatic, MaxStepFraction->1/10,
MaxSteps->10000, MaxStepSize->Automatic, Method->Automatic,
NormFunction->Automatic, PrecisionGoal->Automatic,
SolveDelayed->Automatic, StartingStepSize->Automatic,
StepMonitor->None, WorkingPrecision->MachinePrecision}
```

1.11 Engineering Units and Unit Conversions: Quantity[] and UnitConvert[]

With the introduction of Mathematica 9, the use of units has become an integral part of Mathematica[3]. In this section, we shall introduce its basic use.

The function that attaches units to a numerical or symbolic quantity is

```
xu=Quantity[mag,unit]
```

where **mag** is the magnitude of the quantity with units **unit**, which is the unit designation presented as a string; that is, it is contained within quotation marks. Common engineering units are listed in Table 1.10. In this table, the middle column gives the unambiguous unit designations that are used by Mathematica. These are the ones that are used in this section. The units in the first column are those that are commonly used and are internally converted by **Quantity** to those in the second column. The arguments of **Quantity** can be accessed with

```
mag=QuantityMagnitude[xu]
unit=QuantityUnit[xu]
```

For example, if the length of x is 6.5 m, then

```
xu=Quantity[6.5,"Meters"] (* or Quantity[6.5,"m"] *)
```

displays

```
6.5 m
```

[3] Additional information can be found by entering *tutorial/UnitsOverview* in the *Documentation Center* search window.

Table 1.10 Common engineering units for `Quantity[]`

```
p=QuantityUnit[Quantity[x,str]]
q=Quantity[18.,ToString[p]]
```

str	p	q
Length/Volume		
"m"	Meters	18. m
"cm"	Centimeters	18. cm
"mm"	Millimeters	18. mm
"nm"	Nanometers	18. nm
"Ang"	Angstroms	18. Å
"ft"	Feet	18. ft
"in"	Inches	18. in
"mi"	Miles	18. mi
"L"	Liters	18. L
"gal"	Gallons	18. gal
Force/weight/pressure		
"kg"	Kilograms	18. kg
"N"	Newtons	18. N
"g"	Grams	18. g
"Pa"	Pascals	18. Pa
"lb"	Pounds	18. lb
Temperature		
"K"	KelvinsDifference	18. K
"Kelvins"	Kelvins	18. K
"C"	DegreesCelsiusDifference	18. °C
"Celsius"	Celsius	18. °C
"F"	DegreesFahrenheitDifference	18. °F
"Fahrenheit"	Fahrenheit	18. °F
Energy/power		
"W"	Watts	18. W
"J"	Joules	18. J
"btu"	BritishThermalUnitsIT	18. BTU_{IT}
Electrical		
"A"	Amperes	18. A
"H"	Henries	18. H
"ohm"	Ohms	18. Ω
"farad"	Farads	18. F
"coulomb"	Coulombs	18. C
"V"	Volts	18. V
"T"	Teslas	18. T
Time/frequency		
"s"	Seconds	18. s
"ms"	Milliseconds	18. ms
"micros"	Microseconds	18. μs
"Hz"	Hertz	18. Hz
"GHz"	Gigahertz	18. GHz
"hr"	Hours	18. h

To determine the magnitude and the units of x, one uses

```
mag=QuantityMagnitude[xu]
unit=QuantityUnit[xu]
```

which displays, respectively,

```
6.5
Meters
```

To perform a calculation with units, consider the following determination of the average velocity of a device that travels 5 meters in 2 seconds

```
v=Quantity[5.,"Meters"]/Quantity[2,"Seconds"]
```

which displays

```
2.5 m/s
```

To illustrate the usage of **Quantity** further, consider the following quantities: e_{33} (C·m^{-2}), c_{33} (N·m^{-2}), and ε_{33} (F·m^{-1}). We shall show that the following quantity is nondimensional

$$k = \frac{e_{33}^2}{c_{33}\varepsilon_{33}}$$

Thus,

```
k=Quantity[e33,"Coulombs"/"Meters"^2]^2/
  (Quantity[c33,"Newtons"/"Meters"^2]*
  Quantity[Subscript[ε33,"Farads"/"Meters"])
```

which gives

$$\frac{e_{33}^2}{c_{33}\varepsilon_{33}}$$

Since there are no units appearing in this result, k has no dimensions.

The conversion from one system of units to another is done with

```
UnitConversion[xunit,unit]
```

where **xunit** has been created by **Quantity** and **unit** is a string representing the desired unit(s) to which **xunit** is to be converted.

For example, to convert a temperature of, say, 60 degrees Fahrenheit to degrees Celsius, we use

```
UnitConvert[Quantity[60.,"Fahrenheit"],"Celsius"]
```

to obtain

```
15.5556 °C
```

To covert the temperature to degrees Kelvin, we use

```
UnitConvert[Quantity[60.,"Fahrenheit"],"Kelvins"]
```

to obtain

```
288.706 K
```

To convert 55 miles/hr to meters/second, we use

```
UnitConvert[Quantity[55.,"Miles"/"Hours"],
 "Meters"/"Seconds"]
```

to obtain

```
24.5872 m/s
```

To convert 20 lb·ft^{-2} to its SI equivalent, we use

```
UnitConvert[Quantity[20.,"Pounds"/"Feet"^2],"SI"]
```

to obtain

```
97.6486 kg/m^2
```

Other useful unit conversion commands are **CommonUnits**, which converts the various units to compatible dimensions, and **UnitSimplify**, which attempts to simplify the specified units.

1.12 Creation of CDF Documents and Documents in Other Formats

The notebook environment can also be used to create documents as if the notebook were a word processor. This is done by using the *Writing Assistant* found in the *Palette* menu to create cells that accept text and cells that accept equations in addition to those cells that provide the usual Mathematica computational capabilities. This palette is shown in Figure 1.10. The final document can then be saved in different formats by selecting *Save As* from the *File* menu and then choosing the desired format from the *Format* selections at the bottom of the window.

One format that provides a unique capability is Mathematica's own CDF (computable document format) document creator, which permits one to create electronic documents that can contain interactive graphics. The creation of interactive graphics is performed by **Manipulate**, which is discussed in detail in Chapter 7. When the file is opened by the CDF reader, the document has all the characteristics of a regular noneditable file, except that now any graphic entities created by **Manipulate** retain their interactivity. There are several ways to convert a notebook to a CDF file. One such method, after the notebook has been completed,

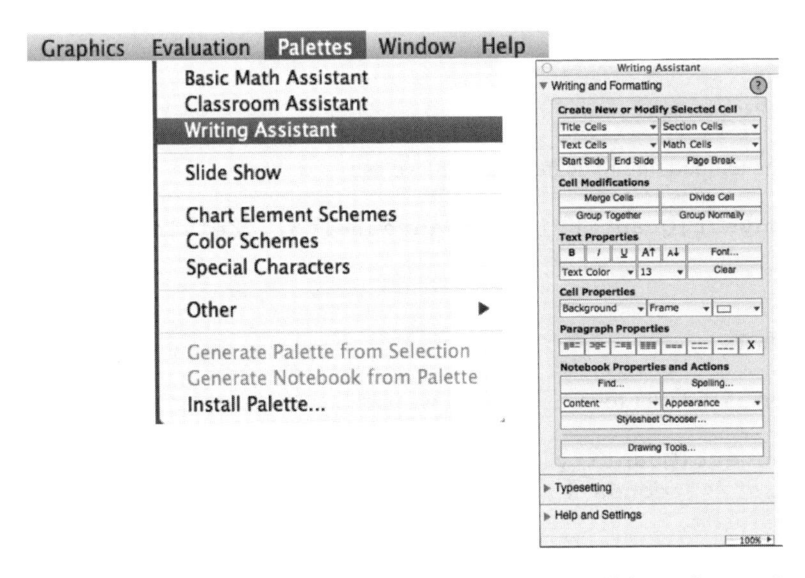

Figure 1.10 Opening the *Writing Assistant* window to access cell formatting templates

is to select *CDF Export* from the *File* menu and then select *Standalone*. This opens a window in which the user selects a file name for the converted notebook and a directory in which it is to reside and then clicks *Continue*. The file is created and the window is exited.

1.13 Functions Introduced in Chapter 1

In addition to the functions listed in Tables 1.4 to 1.8, the additional functions introduced in this chapter are listed in Table 1.11.

Table 1.11 Summary of additional commands introduced in Chapter 1

Command	Usage
`Clear`	Clears from memory specified user-created definitions
`ClearAll["Global`*"]`	Clears from memory all user-created definitions
`EngineeringForm`	Prints real numbers in engineering notation with specified precision
`N`	Gives the numerical value of an expression with specified precision
`NumberForm`	Prints to a specified precision real numbers in an expression
`Options`	Gives the options available for a specified function and their current values
`Print`	Prints an expression
`Quantity`	Appends units to a number or symbol
`StringJoin`	Concatenates a set of string objects in the order that they appear
`Symbolize`	Convert subscripted variable for internal use as single symbol
`ToString`	Converts an expression to a string
`UnitConvert`	Converts a quantity from one system of units to another

Exercises

Sections 1.5 and 1.8

1.1 For $b = 1.2$, $a = 1.5b$, and $\nu = 0.3$, determine κ when

$$\kappa = \frac{(40 + 37\nu)b^4 + (16 + 10\nu)a^2b^2 + \nu a^4}{12(1 + \nu)b^2(3b^2 + a^2)}$$

1.2 For $\gamma_1 = 0.6$, $\gamma_2 = 0.4$, and $m_o = 0.71$, determine Ω_c when

$$\Omega_c = \frac{1}{\sqrt{\gamma_1^3\gamma_2^3\left(m_o + \alpha + \beta\right)}}$$

and

$$\alpha = \gamma_1\left[\frac{\left(3\gamma_1 + \gamma_2\right)^2}{28\gamma_2^2} + \frac{9}{20\gamma_2^2} - \frac{3\gamma_1 + \gamma_2}{4\gamma_2^2}\right]$$

$$\beta = \gamma_2\left[\frac{\left(3\gamma_2 + \gamma_1\right)^2}{28\gamma_1^2} + \frac{9}{20\gamma_1^2} - \frac{3\gamma_2 + \gamma_1}{4\gamma_1^2}\right]$$

1.3 For $n = 5$, determine the value of c when

$$c = \frac{M\left(1 - M^2\right)\sin\alpha}{\left(1 + M^2 - 2M\cos\alpha\right)^2}$$

and

$$\cos\alpha = \sqrt{\left(\frac{1 + M^2}{4M}\right)^2 + 2} - \left(\frac{1 + M^2}{4M}\right)$$

$$M = \frac{1}{\sin(\pi/n)}$$

1.4 For $\gamma = 60°$, $\alpha = 35°$, and $n = 4/3$, find the value of d when

$$d = \alpha - \gamma + \sin^{-1}\left[n\sin\left(\gamma - \sin^{-1}\left\{\frac{\sin\alpha}{n}\right\}\right)\right]$$

1.5 For $x = 0.45$, determine the value of k when

$$k = \frac{1.2}{x}\left[\sqrt{16x^2 + 1} + \frac{1}{4x}\ln\left(\sqrt{16x^2 + 1} + 4x\right)\right]^{-2/3}$$

1.6 For $\varepsilon = 0.00025$, $\mathrm{Re} = 8 \times 10^5$, and $D = 0.3$, determine f when

$$f = \left\{ \left(\frac{64}{\mathrm{Re}}\right)^8 + 9.5 \left[\ln\left(\frac{\varepsilon}{3.7D} + \frac{5.74}{\mathrm{Re}^{0.9}}\right) - \left(\frac{2500}{\mathrm{Re}}\right)^6 \right]^{-16} \right\}^{1/8}$$

1.7 For $\lambda = 4.73$, $r = 3$, $\Omega = 300$, and $k = 450$, determine ω_c when

$$\omega_c = \frac{1}{1 - \beta^2} \left[-\Omega\beta^2 + \sqrt{\Omega^2\beta^4 + \left(1 - \beta^2\right)\omega_o^2} \right]$$

and

$$\beta = \lambda r$$
$$\omega_o = kr\lambda^2$$

1.8 For $x = 0.4$ and $y = 0.6$, determine u when

$$u = -\frac{1}{x} \left\{ \ln\left(1 - \sqrt{1 - y}\right) - \sqrt{1 - y} \right\}$$

1.9 For $g = 9.81$, $\theta = 17°$, $x = 45$, and $v_o = 14$, determine z when

$$z = -\frac{gx^2}{2v_o^2 \cos^2\theta} + x \tan\theta$$

1.10 For $M = 0.75$ and $k = 1.4$, determine p/p_o when

$$\frac{p}{p_o} = \left[1 + \{(k - 1)/2)\} M^2\right]^{\frac{-k}{k-1}}$$

1.11 For $r_1 = 0.01$, $\varepsilon_1 = 0.02$, $T_1 = 77$, $r_2 = 0.05$, $\varepsilon_2 = 0.05$, $T_2 = 300$, and $\sigma = 5.67 \times 10^{-8}$, determine q_{12} when

$$q_{12} = \frac{4\pi\sigma r_1^2 \left(T_1^4 - T_2^4\right)}{\dfrac{1}{\varepsilon_1} + \dfrac{1 - \varepsilon_2}{\varepsilon_2}\left(\dfrac{r_1}{r_2}\right)^2}$$

1.12 For $\varphi = \pi/10.0$, determine the value of χ when

$$\chi = 2\tan^{-1}\left\{ \tan\left(\frac{\pi}{4} + \frac{\varphi}{2}\right) \left[\frac{1 - e\sin\varphi}{1 + e\sin\varphi}\right]^{e/2} \right\} - \frac{\pi}{2}$$

1.13 For $B = 2$ (an integer), determine the value of w when

$$w = \frac{1}{2\sqrt{2}}\sqrt{3B^2 + \sqrt{9B^4 + 16B^2}}$$

Convert the result to a decimal number.

1.14 For $\theta_o = 27°$, determine the value of p when

$$p = 2\sin^2\left(\frac{\theta_o}{2}\right)\left\{1 + \frac{2}{\pi}\cot\left(\frac{\theta_o}{2}\right)\ln\left[\tan\left(\frac{\pi}{4} + \frac{\theta_o}{2}\right)\right]\right\}$$

Sections 1.6 and 1.8

1.15 Show numerically using 50 digits of precision that

$$\sin(\pi/15) = \frac{1}{4}\sqrt{7 - \sqrt{5} - \sqrt{30 - 6\sqrt{5}}}$$

Section 1.11

1.16 The Reynolds number is given by

$$\mathrm{Re} = \frac{\rho v L}{\mu}$$

where L (m) is a characteristic length, μ (Pa·s) is the dynamic viscosity, v (m·s^{-1}) is the mean velocity, and ρ (kg·m^{-3}) is the density. Show that Re is a nondimensional quantity.

1.17 A similarity coordinate transform is given by

$$\eta = y\sqrt{\frac{U\rho}{\mu x}}$$

where x (m) and y (m) are coordinates, U (m/s) is a velocity, μ (Pa·s) is the dynamic viscosity, and ρ (kg·m^{-3}) is the density. Show that η is a nondimensional quantity.

1.18 The natural frequency coefficient Ω for a beam is defined as

$$\Omega^4 = \frac{\omega^2 \rho A L^4}{EI}$$

where E (N·m^{-2}) is the Young's modulus, I (m^4) is the moment of inertia, L (m) is the length, A (m^2) is the area, ρ (kg·m^{-3}) is the density, and ω (rad·s^{-1}) is the radian frequency. Show that Ω is a nondimensional quantity.

2

List Creation and Manipulation: Vectors and Matrices

2.1 Introduction

A list is an object that is composed of a collection of other objects: expressions, numbers, arrays, graphical entities, and just about any legitimate Mathematica quantity. A pair of open/closed braces {…} encloses the list. Each element of the list is separated by a comma: {…, …, …}. In addition, each element of a list can itself be a list. There are effectively no size restrictions on the number of elements that a list can contain. In addition, about 40% of Mathematica's functions can operate on lists. In this chapter, we shall concentrate on the construction of lists and on the operations to these lists that are most useful in obtaining solutions to engineering applications. Consequently, it will be seen that a list whose elements are numerical values or variable names and do not contain any other lists can be considered a vector and those lists whose elements are themselves lists of numerical values or variable names meeting certain size requirements can be considered matrices. A summary of several list creation and manipulation commands is shown in Table 2.1 and a summary of several matrix creation and manipulation commands are presented in Tables 2.2 and 2.3. The application of many of these commands will be illustrated in the sections that follow.

2.2 Creating Lists and Vectors

2.2.1 Introduction

A list of numerical values is created by entering

```
c={1.,3.,5.,7.,9.}
```

which is a five-element vector. This particular list of values could have also been created using

```
c=Range[1.,9.,2]
```

The significance of each number in **`Range`** is described in Table 2.1.

An Engineer's Guide to Mathematica®, First Edition. Edward B. Magrab.

Companion Website: www.wiley.com/go/magrab

Table 2.1 Creation and manipulation functions for vector lists (For additional commands, enter *guide/RearrangingAndRestructuringLists* in *Documentation Center*.)

		Usage with **x={a,b,c}**, **y={e,f}**, **w={17,-4.1,3}**	
Description	Mathematica function*	Expression	Output
Add an element to the end of a list and reset list: length of list is increased by one	**AppendTo[list,elem]**	AppendTo[x,d] AppendTo[x,y]	{a,b,c,d} {a,b,c,{e,f}}
Add an element to the beginning of a list and reset list: length of list is increased by one	**PrependTo[list,elem]**	PrependTo[x,d] PrependTo[x,y]	{d,a,b,c} {{e,f},a,b,c}
Inset a new element at the *n*th location in a list: length of list is increased by one	**Insert[list,elem,n]**	Insert[x,d,3] Insert[x,y,3]	{a,b,d,c} {a,b,{e,f},c}
Replace the *n*th element in a list: length of list is unchanged	**ReplacePart[list, n->elem]** or **list[[n]]=elem**	ReplacePart[x,1->d] or x[[1]]=d; x ReplacePart[x,{{1}, {3}}->d]	{d,b,c} {d,b,c} {d,b,d}
Delete the element in the *n*th position: length of list is decreased by one	**Delete[list,n]**	Delete[x,2] Delete[x,{{1},{3}}]	{a,c} {b}
Number of elements in a list	**Length[list]**	Length[x]	3
Sum of the elements of a list	**Total[list]**	Total[x] Total[w]	a+b+c 15.9
Cumulative sum of the elements: length of list is unchanged	**Accumulate[list]**	Accumulate[x] Accumulate[w]	{a,a+b,a+b+c} {17,12.9,15.9}
Take differences in successive elements in a list: length of list is decreased by one‡	**Differences[list]**	Differences[x] Differences[w]	{-a+b,-b+c} {-21.1,7.1}
Place list's elements together	**Row[list]**	Row[x]	abc

Create a list of numerical values starting at x_s and ending at x_e in steps of d_v or by using n equally spaced values	`Range[xs,xe,dv]` or	`Range[1,2.1,0.25]`	`{1.,1.25,1.5,1.75,2.}`
	`Range[xs,xe,(xe-xs)/(n-1)]`	`Range[1,2.1,(2.1-1)/4]`	`{1.,1.275,1.55,1.825,2.1}`
	`Range[n]`	`Range[5]`	`{1,2,3,4,5}`
Delete inner braces within a list	`Flatten[list,level]`	`Flatten[{{e,f},a,b,c}]`	`{e,f,a,b,c}`
		`z={{{e,f},{e,f}}, {{e,f},{e,f}}};`	
		`Flatten[z,1]`	`{{e,f},{e,f},{e,f},{e,f}}`
Sort elements of a list in the order specified	`Sort[list,order]`	`Sort[w,Greater]`	`{17,3,-4.1}`
		`Sort[w,Less]`	`{-4.1,3,17}`
Obtain minimum value of elements	`Min[list]`	`Min[w]`	`-4.1`
Obtain maximum value of elements	`Max[list]`	`Max[w]`	`17`
Locate position of a value or a pattern within a list	`Position[list,pat]`	`Position[x,a]`	`{{1}}`
		`Position[w,3]`	`{{3}}`
		`Position[w,3.]`	`{0}`
Display list as a column of values	`Column[list]` or `list//Column`	`Column[y]` or `y//Column`	`e` `f`
Join two or more lists	`Join[list1,list2,...]`	`Join[x,w]`	`{a,b,c,17,-4.1,3}`
Reverse order of a list	`Reverse[list]`	`Reverse[x]`	`{c,b,a}`
		`Reverse[Join[x,y]]`	`{f,e,c,b,a}`

*`list` = list of m elements that has previously been defined, $m \geq 0$ [`list` = {} is acceptable]; `elem` = new element to be placed in a list; `n` = location in list where new element is to be placed or deleted; `pat` = pattern

‡Refer to the Documentation Center for additional options for `Differences`.

Table 2.2 Creation and manipulation functions for matrices[†]

Description	Mathematica function*	Expression (Usage[+] with `matx={{2.,3.,5.},{7.,11.,13.},{17.,19.,23.}}`, `b={6,1,8}`)	Output
Create an $n \times m$ array of constant values	`ConstantArray[const, {n,m}]`	`ConstantArray[3,{2,3}]` `ConstantArray[3,4]`	`{{3,3,3},{3,3,3}}` `{3,3,3,3}`
Create an $n \times n$ identity matrix	`IdentityMatrix[n]`	`IdentityMatrix[3]`	`{{1,0,0},{0,1,0},{0,0,1}}`
Create an $n \times n$ diagonal matrix	`DiagonalMatrix[list]`	`DiagonalMatrix[b]`	`{{6,0,0},{0,1,0},{0,0,8}}`
Obtain a diagonal of a matrix	`Diagonal[mat]`	`Diagonal[matx]`	`{2.,11.,23.}`
Dimensions (order) of a matrix	`Dimensions[mat]`	`Dimensions[matx]`	`{3,3}`
Transpose of a matrix	`Transpose[mat]`	`Transpose[matx]`	`{{2.,7.,17.},{3.,11.,19.},{5.,13.,23.}`
Determinant of a matrix	`Det[mat]`	`Det[matx]`	`-78.`

Inverse of a matrix	`Inverse[mat]`	`Inverse[matx]`	`{{-0.0769231,-0.333333,0.205128},`
			`{-0.769231,0.5,-0.115385},`
			`{0.692308,-0.166667,-0.0128205}}`

Delete inner braces within a list	`Flatten[mat]`	`Flatten[matx]`	`{2.,3.,5.,7.,11.,13.,17.,19.,23.}`
Display a matrix in standard form	`MatrixForm[mat]` or `mat//MatrixForm`	`MatrixForm[matx]`	$\begin{pmatrix} 2. & 3. & 5. \\ 7. & 11. & 13. \\ 17. & 19. & 23. \end{pmatrix}$
Display a matrix as a table of values, with or without headers	`TableForm[mat, TableHeadings-> {{rows},{columns}}]`[††]	`TableForm[matx, TableHeadings-> {{"row 1","row 2", "row 3"},{"col 1", "col 2","col 3"}}]`	(see below)

Output of the `TableForm` example:

	col 1	col 2	col 3
row 1	2.	3.	3.
row 2	7.	11.	13.
row 3	17.	19.	23.

[†] Two additional array-creation functions, **`Array`** and **`ArrayFlatten`** are discussed in Section 3.2.2.
*`mat` = array of $m \times m$ elements, $m > 1$; `list` = list of m elements; `const` = numerical or symbolic value; and `m` and `n` are integers
[+] Also performs these operations on symbolic quantities
[††] `rows` = comma-separated list of string labels for each row; `columns` = comma-separated list of string labels for each column

Table 2.3 Additional examples of functions operating on matrices

Description	Mathematica function*	Usage with `matx={{2.,3.,5.},{7.,11.,13.},{17.,19.,23.}}` Expression	Output
Find the largest value in an array	`Max[mat]`	`Max[matx]`	`23.`
Find the smallest value in an array	`Min[mat]`	`Min[matx]`	`2.`
Locate the position of a value or a pattern within a matrix	`Position[mat,pat]`	`Position[matx,Min[matx]]` `Position[matx,Max[matx]]`	`{{1,1}}` `{{3,3}}`
Sum the values in the columns of a matrix	`Total[mat]`	`Total[matx]`	`{26.,33.,41.}`
Sum all the elements of a matrix	`Total[Total[mat]]`	`Total[Total[matx]]`	`100.`
Cumulative sum of the elements of each column	`Accumulate[mat]`	`Accumulate[matx]//MatrixForm`	$\begin{pmatrix} 2. & 3. & 5. \\ 9. & 14. & 18. \\ 26. & 33. & 41. \end{pmatrix}$
Square matrix to the *k*th integer power	`MatrixPower[mat,k]`	`MatrixPower[matx,3]// MatrixForm`	$\begin{pmatrix} 3946. & 4920. & 6064. \\ 11456. & 14278. & 17588. \\ 20632. & 25700. & 31654. \end{pmatrix}$
Exponential of a square matrix	`MatrixExp[mat]`	`MatrixExp[matx]//MatrixForm`	$\begin{pmatrix} 0.7687 & 0.9578 & 1.179 \\ 2.23 & 2.778 & 3.422 \\ 4.014 & 5.001 & 6.16 \end{pmatrix} \times 10^{15}$

*`mat` = array of $m \times m$ elements, $m > 1$; `pat` = pattern or value

The list `c` can now be used in many Mathematica functions, as shown in the following examples. It can be used in symbolic exponentiation where

```
x^c
```

results in

$\{x^{1.}, x^{3.}, x^{5.}, x^{7.}, x^{9.}\}$

or as the argument of sine, where

```
Sin[c]
```

results in

```
{0.841471,0.14112,-0.958924,0.656987,0.412118}
```

A more compact way to create these same results is

```
x^Range[1.,9.,2]
```

and

```
Sin[Range[1.,9.,2]]
```

This compact form can be used if `c` is not needed subsequently and/or readability is not an issue.

These results could have been displayed as a column of entities using

```
x^c//Column
```

or

```
Column[x^c]
```

either of which displays

$x^{1.}$
$x^{3.}$
$x^{5.}$
$x^{7.}$
$x^{9.}$

Element-by-element addition, subtraction, multiplication, division, and exponentiation of vector lists of the same length or of arrays of the same order (dimension) are performed automatically using the same notation that is used when these quantities are scalars (a single value) or a symbolic quantity. If their respective lengths or orders are not appropriate for the intended operation, an error message will appear.

To illustrate this, consider the following addition, subtraction, multiplication, division, and exponentiation of the two lists in the program below. The **Print** command is used to identify the output.

```
c={1.,2.,3.};
d={0.1,0.2,0.3};
Print["c+d = ",c+d]
Print["c-d = ",c-d]
Print["c*d = ",c d]
Print["c/d = ",c/d]
Print["c^d = ",c^d]
```

which displays

```
c+d = {1.1,2.2,3.3}
c-d = {0.9,1.8,2.7}
c*d = {0.1,0.4,0.9}
c/d = {10.,10.,10.}
c^d = {1.,1.1487,1.39039}
```

These types of calculations can be extended to include operations involving all elementary functions and many special functions. For example, consider the expression

$$s = J_1(x) + |\sin y|^{\ln x}$$

where $J_1(x)$ is the Bessel function of the first kind of order 1. The evaluation of this expression when $x = \{1.0, 2.0, 3.0\}$ and $y = \{0.1, 0.2, 0.3\}$ is obtained with the following program

```
x={1.,2.,3.};
y={0.1,0.2,0.3};
s=BesselJ[1,x]+Abs[Sin[y]]^Log[x]
```

which gives

```
{1.44005,0.902938,0.601107}
```

2.2.2 *Creating a List with* `Table[]`

A more general way to generate a list of numerical values is with **Table**. The form for **Table** to create a vector is

```
Table[expr,{n,ns,ne,dt}]
```

or

```
Table[expr,{n,{lst}}]
```

where **expr** is a Mathematica expression that can be a function of the index **n**, and in the first form, **n** is a quantity whose value starts at **ns**, ends at **ne**, and is incremented over this range by an amount **dt**. When **dt** is omitted, a value of 1 is used. In the second form, **lst** is a list of values and **n** sequentially assumes each value in the list in the order that they appear.

Hence, to create the previous list for **c**, we enter

```
c=Table[n,{n,1.,9,2}]
```

which gives

```
{1.,3.,5.,7.,9.}
```

To create the symbolic list of exponential values of **c** and the sine of **c**, we use

```
Table[x^c,{c,1.,9.,2.}]
Table[Sin[c],{c,1.,9.,2.}]
```

which produce the previously obtained results.

2.2.3 *Summing Elements of a List:* Total[]

The elements of a list can be summed by using **Total**. Modifying the previous results, we have

```
Total[Table[x^c,{c,1,9,2}]]
Total[Table[Sin[c],{c,1.,9.,2.}]]
```

which yield, respectively,

$$x^1+x^3+x^5+x^7+x^9$$

```
1.09277
```

The two functions, **Table** and **Total**, can be used to evaluate a series, as shown in the following example.

Example 2.1

Evaluating a Fourier Series

Consider the Fourier series expansion of a periodic pulse of duration-to-period ratio d_T over the nondimensional period $-0.5 \le \tau \le 0.5$. Its Fourier series expansion is

$$f(\tau) = d_T \left[1 + 2 \sum_{k=1}^{150} \frac{\sin\left(k\pi d_T\right)}{k\pi d_T} \cos\left(2k\pi\tau\right) \right]$$

where we have assumed that summing 150 terms will give sufficient convergence. We shall plot the results for one period using **Plot**, whose basic form is given in Table 6.1.

If we set $d_T = 0.25$, then the series is summed and plotted as follows

```
kp=π Range[1,150]; dT=0.25;
Plot[dT (1+2 Total[Sin[kp dT]/(kp dT) Cos[2 kp t]]),
  {t,-0.5,0.5}]
```

The results are shown in Figure 2.1.

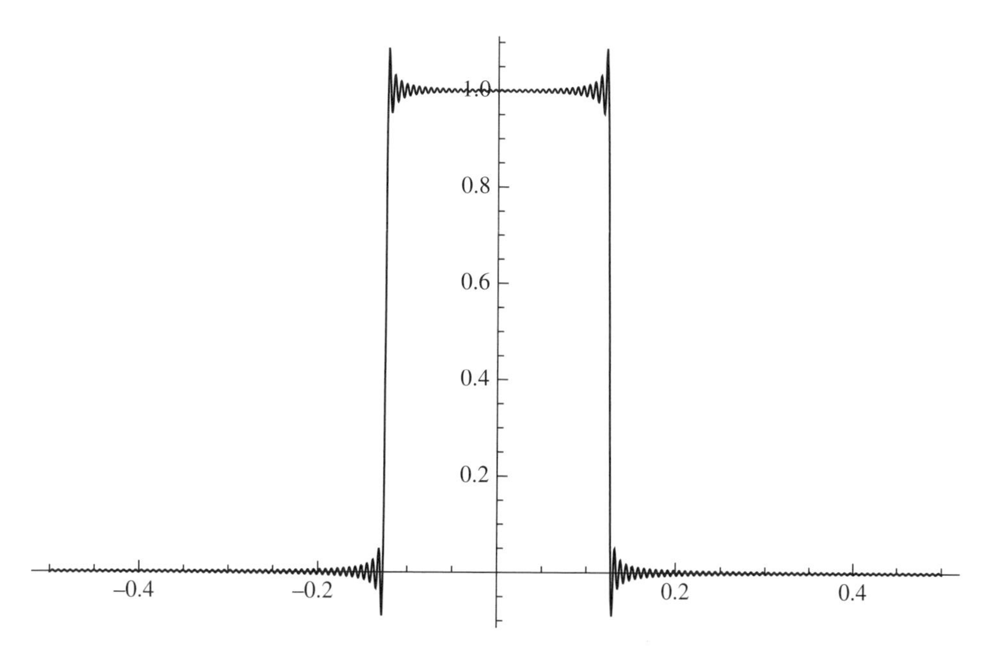

Figure 2.1 Fourier series representation of one period of a periodic pulse of duration $d_T = 0.25$.

2.2.4 *Selecting Elements of a List*

There are several ways to access elements of lists. A few of these are shown in Table 2.4. Additional information can be obtained by entering *Part* in the *Documentation Center* search area. The elements of a list can be accessed by using a pair of double brackets: [[...]]. Thus, to extract the third element from **c**, the syntax that is employed is

```
c=Range[1.,9.,2] (* Creates the list {1.,3.,5.,7.,9.} *)
c3=c[[3]]
```

which yields 5.

To select the first element in a list, one can use either

```
c1=c[[1]]
```

Table 2.4 Ways to access elements of a list or an array

		Usage with `matx={{2,3,5},{7,11,13},{17,19,23}}`, `b={6,1,8,-4}`	
Description	Mathematica function*	Expression	Output
Select an element	-	`b[[3]]` `matx[[2,3]]`	`8` `13`
Select a row	-	`matx[[2]]`	`{7,11,13}`
Select a column	-	`matx[[All,1]]`	`{2,7,17}`
Select a submatrix	-	`matx[[2;;3,1;;2]]`	`{{7,11},{17,19}}`
Select the first element	`First[list]`	`First[b]`	`6`
Select the first row	`First[mat]`	`First[matx]`	`{2,3,5}`
Select the last element	`Last[list]`	`Last[b]`	`-4`
Select the last row	`Last[mat]`	`Last[matx]`	`{17,19,23}`
Take the first *n* elements of a list	`Take[list,n]`	`Take[b,2]`	`{6,1}`
Take the last *n* elements of a list	`Take[list,-n]`	`Take[b,-2]`	`{8,-4}`
Take the *n*th to *k*th elements of a list	`Take[list,{n,k}]`	`Take[b,{2,4}]`	`{1,8,-4}`

*`mat` = array of $m \times m$ elements, $m > 1$; `list` = list of m elements; `n,k` = integer

or

```
c1=First[c]
```

both of which give `1.` To select the last element of the list one can use

```
cLast=c[[5]]
```

when the length of the list is known or when it is unknown either

```
cLast=c[[Length[c]]]
```

or

```
cLast=Last[c]
```

each of which gives `9`.

To access, say, the first, third and last elements of `c`, we can use a list of these locations as follows

```
c[[{1,3,5}]]
```

which displays

```
{1.,5.,9.}
```

Example 2.2

Convergence of a Series

Consider the series

$$s = 1 + 2 \sum_{n=1}^{N\to\infty} \frac{1}{1+(2n)^2} = \frac{\pi}{2}\coth\frac{\pi}{2}$$

We shall sum this series for $N = 50$, 100, 200, and 500 terms and then compare the results to the exact value. A straightforward way is to obtain the evaluations for these four values of N is to use **Table** and **Accumulate** as follows

```
terms=1+Accumulate[2. Table[1/(1+(2 n)^2),{n,1,500}]];
n={50,100,200,500};
p=terms[[n]];
exact=π/2. Coth[π/2.];
pd=100. (exact-p)/exact;
TableForm[{{n,p,pd}},TableHeadings->{None,{"N","s",
  "% diff"}}]
```

which displays

N	s	% diff
50	1.70279	0.578059
100	1.70771	0.290481
200	1.71019	0.145605
500	1.71169	0.058329

2.2.5 *Identifying List Elements Matching a Pattern:* **Position[]**

To determine the locations in a list of all elements that meet a specified value or condition or pattern one uses **Position**, whose form is given by

```
Position[lst,pat]
```

where **lst** is a list and **pat** is a value or a condition or a pattern whose locations in **lst** are being sought. The output of **Position** is a list of locations of those elements that have met **pat**. Another function that determines the values of all those elements that meet a specified criterion is **Select**. This function is introduced in Table 3.2.

To illustrate the use of **Position**, consider the following list of numerical values

```
c={7.,20.,-6.,31.,1.,-8.4,9.3}
```

We shall determine for **c** the location of the maximum and minimum values, respectively, which are found with **Max** and **Min**. Then,

```
c={7.0,20.0,-6.0,31.0,1.0,-8.4,9.3};
px=Position[c,Max[c]]
pm=Position[c,Min[c]]
Print["Maximum value of c = ",c[[px[[1,1]]]],
  " in position ",px[[1,1]]]
Print["Minimum value of c = ",c[[pm[[1,1]]]],
  " in position ",pm[[1,1]]]
```

displays

```
Maximum value of c = 31. in position 4
Minimum value of c = -8.4 in position 6
```

The value of **pat** used in **Position** can also be used to determine if any of the elements meet the general descriptions that are given in Table 3.1: integer, symbol, real, complex, or string. For example, consider the list

```
g={3,4.,-7.6,3+I,31.,4. I,11,"aa" };
```

To determine the locations of any of these five types of list elements, we use **Position** as follows

```
Position[g,_Integer]
Position[g,_Real]
Position[g,_String]
Position[g,_Symbol]
Position[g,_Complex]
```

where in the second argument of **Position**, the leading underscore character is required. Execution of this program yields

```
{{1},{7}}
{{2},{3},{5}}
{{8}}
{{0}}
{{4},{6}}
```

If the actual values of the elements in the list were desired, then **Print** can be used as follows to, say, find the values in **g** that are integers

```
Print["Integers in g:",g[[Flatten[Position[g,_Integer]]]]]
```

which displays

```
Integers in g: {3,11}
```

The command **Flatten** was used to reduce the output of **Position** to a list containing only one set of braces.

2.3 Creating Matrices

2.3.1 Introduction

To create an array of values, the elements of a vector list are themselves replaced by vector lists. For example, to create a (3×3) matrix, we use a three element vector list in which each of the elements is a three-element vector list, as shown below

```
mat={{2.,3.,5.},{7.,11.,13.},{17.,19.,23.}}
```

Thus, the first inner set of braces delineates the three elements of the first row of the matrix; the second set of inner braces the second row; and so on. In general, Mathematica does not require that the number of elements in each list comprising each row have the same length. For our purposes, the length requirement is dictated by how one subsequently uses the array, such as to perform operations using the rules of matrix algebra.

Returning to **mat**, it is noted that the first nine prime numbers have been selected for the array elements. Then, **mat** can be used as previously shown for the vector variable **c**; that is,

```
x^mat
```

results in

$\{\{x^{2.},x^{3.},x^{5.}\},\{x^{7.},x^{11.},x^{13.}\},\{x^{17.},x^{19.},x^{23.}\}\}$

and

```
Sin[mat]
```

results in

```
{{0.909297,0.14112,-0.958924},
 {0.656987,-0.99999,0.420167},
 {-0.961397,0.149877,-0.84622}}
```

The output can be displayed in the more conventional matrix form as

```
mat={{2.,3.,5.},{7.,11.,13.},{17.,19.,23.}}//MatrixForm
```

which results in

$$\begin{pmatrix} 2. & 3. & 5. \\ 7. & 11. & 13. \\ 17. & 19. & 23. \end{pmatrix}$$

It is noted that **mat** is now in display form and, therefore, is no longer available for further mathematical operations. To preserve the computable form of **mat** and still display **mat** in matrix form, the following is used instead

```
mat={{2.,3.,5.},{7.,11.,13.},{17.,19.,23.}};
MatrixForm[mat]
```

As indicated in Table 2.2, the array given by **mat** can be converted to a vector by using **Flatten**; thus,

```
Flatten[mat]
```

yields

```
{2.,3.,5.,7.,11.,13.,17.,19.,23.}
```

A matrix can also be created with the appropriate template from the *Advanced* option of the *Calculator* portion of the *Basic Math Assistant*. The 3×3 matrix is created by starting with the 2×2 template. With the cursor positioned at one of the elements of the bottom row, the *Row+* button is clicked. Then the cursor is positioned to the right of the last column and the *Col+* button is clicked. Hence, these operations are used to obtain

$$\text{mat} = \begin{pmatrix} 2. & 3. & 5. \\ 7. & 11. & 13. \\ 17. & 19. & 23. \end{pmatrix};$$

```
Sin[mat]//MatrixForm
```

which displays

$$\begin{pmatrix} 0.909297 & 0.14112 & -0.958924 \\ 0.656987 & -0.99999 & 0.420167 \\ -0.961397 & 0.149877 & -0.84622 \end{pmatrix}$$

As with vector lists, element-by-element addition, subtraction, multiplication, division, and exponentiation can be performed automatically on arrays of the same order (dimension) using the same notation as if these quantities were scalars (a single value) or a symbolic quantity. If their respective orders are not appropriate for the intended operation, an error message will appear. Illustration of these operations on numerical and symbolic matrices is shown in Table 2.5. In creating **mat** and **mat1** in the table, the appropriate templates from the *Basic Math Assistant* were used.

Table 2.5 Illustration of element-by-element operations with matrices

Numeric	Symbolic
mat = $\begin{pmatrix} 2. & 3. & 5. \\ 7. & 11. & 13. \\ 17. & 19. & 23. \end{pmatrix}$ mat1 = $\begin{pmatrix} 0.2 & 0.3 & 0.5 \\ 0.7 & 1.1 & 1.3 \\ 1.7 & 1.9 & 2.3 \end{pmatrix}$	mats= $\begin{pmatrix} a & b \\ c & d \end{pmatrix}$
mat+mat1→ $\begin{pmatrix} 2.2 & 3.3 & 5.5 \\ 7.7 & 12.1 & 14.3 \\ 18.7 & 20.9 & 25.3 \end{pmatrix}$	mats+mats→ $\begin{pmatrix} 2\,a & 2\,b \\ 2\,c & 2\,d \end{pmatrix}$
mat-mat1→ $\begin{pmatrix} 1.8 & 2.7 & 4.5 \\ 6.3 & 9.9 & 11.7 \\ 15.3 & 17.1 & 20.7 \end{pmatrix}$	mats-mats→ $\begin{pmatrix} 0 & 0 \\ 0 & 0 \end{pmatrix}$
mat mat1→ $\begin{pmatrix} 0.4 & 0.9 & 2.5 \\ 4.9 & 12.1 & 16.9 \\ 28.9 & 36.1 & 52.9 \end{pmatrix}$	mats mats→ $\begin{pmatrix} a^2 & b^2 \\ c^2 & d^2 \end{pmatrix}$
mat/mat1→ $\begin{pmatrix} 10. & 10. & 10. \\ 10. & 10. & 10. \\ 10. & 10. & 10. \end{pmatrix}$	mats/mats→ $\begin{pmatrix} 1 & 1 \\ 1 & 1 \end{pmatrix}$
mat^mat1→ $\begin{pmatrix} 1.1487 & 1.3904 & 2.2361 \\ 3.9045 & 13.981 & 28.063 \\ 123.53 & 268.93 & 1355.1 \end{pmatrix}$	mats^mats→ $\begin{pmatrix} a^a & b^b \\ c^c & d^d \end{pmatrix}$

These types of calculations can be extended to include operations involving all elementary functions and many special functions. For example, if

$$s = J_1(x) + |\sin y|^{\ln x}$$

where

$$x = \begin{pmatrix} 2. & 3. & 5. \\ 7. & 11. & 13. \\ 17. & 19. & 23. \end{pmatrix} \quad y = \begin{pmatrix} 0.2 & 0.3 & 0.5 \\ 0.7 & 1.1 & 1.3 \\ 1.7 & 1.9 & 2.3 \end{pmatrix}$$

then s is determined from the following program

```
x={{2.,3.,5.},{7.,11.,13.},{17.,19.,23.}};
y={{0.2,0.3,0.5},{0.7,1.1,1.3},{1.7,1.9,2.3}};
s=BesselJ[1,x]+Abs[Sin[y]]^Log[x]//MatrixForm
```

which displays the results in matrix notation as

$$\begin{pmatrix} 0.902938 & 0.601107 & -0.0212837 \\ 0.420323 & 0.581887 & 0.838857 \\ 0.878896 & 0.744298 & 0.358987 \end{pmatrix}$$

2.3.2 Matrix Generation Using `Table[]`

As with vector lists, another way in which matrix quantities can be created is with **`Table`**. Since the values of **`mat`** were chosen to be the first 9 prime numbers, they can be generated by **`Prime`**. Then we can use **`Table`** in the following manner to recreate **`mat`**. First, we note that

```
Table[Range[3(m-1)+1,3 m,1],{m,1,3,1}]
```

creates the array

```
{{1,2,3},{4,5,6},{7,8,9}}
```

and that

```
Prime[{1,2,3}]
```

results in the first three prime numbers

```
{2,3,5}
```

Therefore, **`mat`** is obtained with

```
N[Table[Prime[Range[3(m-1)+1,3 m,1]],{m,1,3,1}]]
```

which yields

```
{{2.,3.,5.},{7.,11.,13.},{17.,19.,23.}}
```

In obtaining this result, **`N`** has been used to convert the integer output from **`Prime`** to decimal values.

These same results can be obtained without the use of **`Range`** by using **`Table`** twice as follows

```
N[Table[Table[Prime[m+3(k-1)],{m,1,3,1}],{k,1,3,1}]]
```

The nested **`Table`** functions work as follows: the outer **`Table`** index **`k`** is set to 1 and the index **`m`** of the inner **`Table`** function is incremented through all of its values. Then **`k`** is set to 2 and the inner **`Table`** function is again incremented through all of its values for **`m`**. The process is repeated until all values specified for **`k`** have been used.

The **Table** function can also be used to form the (x, y) coordinate pairs necessary for the plotting of discrete data using the family of list plotting functions such as **ListPlot**, as shown in Table 6.1. We shall illustrate the technique with some artificially created data using a simple function. **ListPlot** requires that an array of n coordinate pairs be of the form $\{\{x_1,y_1\}, \{x_2,y_2\},\ldots,\{x_n,y_n\}\}$. Let us create this array by determining the values of $y = x^2$ for $-1 \leq x \leq 1$ and for x incremented by 0.25. Then,

```
dat=Table[{x,x^2},{x,Range[-1,1,0.25]}]
```

gives

```
{{-1.,1.},{-0.75,0.5625},{-0.5,0.25},{-0.25,0.0625},
 {0.,0.},{0.25,0.0625},{0.5,0.25},{0.75,0.5625},{1.,1.}}
```

Example 2.3

Summing a Double Series

Consider the double series, which appears in the determination of squeeze film damping for a rectangular surface (see Section 8.2.2),

$$S_{r,k}(\beta,\sigma) = \frac{64\sigma^2}{\pi^8} \sum_{n=1,3,5}^{\infty} \sum_{m=1,3,5}^{\infty} \frac{1}{m^2 n^2 \left\{ \left(m^2 + (n/\beta)^2\right)^2 + \sigma^2/\pi^4 \right\}}$$

where σ is the squeeze number and β is an aspect ratio. Both of these quantities are nondimensional parameters. We shall sum the series for $n_{\max} = m_{\max} = 101$ and for $\sigma = 31.0$ and $\beta = 1$. Then, with the use of the *Special Character* palette,

```
σ=31.; β=1.;
srk=64. σ^2/π^8 Total[Table[Total[Table[
 1/(m^2 n^2 ((m^2+(n/β)^2)^2+σ^2/π^4)),{m,1,101,2}]],
 {n,1,101,2}]]
```

which gives **0.4818**.

2.3.3 *Accessing Elements of Arrays*

To access any element of an $m \times m$ array, we again use the double brackets [[ ]], but in an expanded way. For matrices, the first index accesses a row in the array and the second index accesses a column in the array. To access, say, the second row in the matrix, we use

```
mat={{2.,3.,5.},{7.,11.,13.},{17.,19.,23.}}
row2=mat[[2]]
```

which produces

```
{7.,11.,13.}
```

To select the second element of the second row, we use

```
ele22=mat[[2,2]]
```

which yields **11**.

Selection of all the elements of the third column of **mat** is obtained from

```
col=mat[[All,3]]
```

which gives

```
{5.,13.,23.}
```

The methods just illustrated and several other methods are summarized in Table 2.4. Additional ways to access elements of an array can be found in the *Documentation Center* by entering *Part* in the search area.

It should be realized that although most applications in engineering will employ arrays whose rows have an equal number of elements this restriction does not exist in Mathematica. Consider the following valid list and its access

```
w={{1,2,3},{4,5}};
w[[1,All]]
w[[2,All]]
```

which displays

```
{1,2,3}
{4,5}
```

This could also be written as

```
{w1,w2}={{1,2,3},{4,5}};
w1
w2
```

which also displays the previously obtained results.

2.4 Matrix Operations on Vectors and Arrays

2.4.1 Introduction

Matrix multiplication is performed by using a decimal point (.) instead of an asterisk (∗) or a space to indicate multiplication between two quantities. In addition, Mathematica internally decides whether the list of length n should be considered a $(1\times n)$ row vector or an $(n\times 1)$

Table 2.6 Illustration of vector and matrix multiplication

Vectors	Matrices*
`q={e,f};` `w={g,h};`	`m=` $\begin{pmatrix} a & b \\ c & d \end{pmatrix}$`;` `n=` $\begin{pmatrix} o & p \\ r & s \end{pmatrix}$`;`
`q.q` → `e^2+f^2`	`m.q` → `{a e+b f,c e+d f}`
`q.w` → `e g+f h`	`q.m` → `{a e+c f,b e+d f}`
`w.q` → `e g+f h`	`q.m.q` → `e (a e+c f)+f (b e+d f) (* scalar *)`
	`m.n//MatrixForm` → $\begin{pmatrix} ao+br & ap+bs \\ co+dr & cp+ds \end{pmatrix}$
	`n.m//MatrixForm` → $\begin{pmatrix} ao+cp & bo+dp \\ ar+cs & br+ds \end{pmatrix}$

*When multiplying a square matrix **m** by itself, it should be realized that **m.m** ≠ **m^2**. As shown in Table 2.3, to obtain the value of a square matrix **m** to an integer power **k**, one can use **MatrixPower[m,k]**.

column vector when multiplying an $(n\times n)$ matrix with a $(1\times n)$ row vector or with an $(n\times 1)$ column vector. In all other matrix multiplication operations, Mathematica checks that the value of the inner product dimensions are equal; that is, $(m\times n)(n\times k)$ is correct, whereas $(m\times n)(k\times n)$ will produce an error message. The vector and matrix multiplication process is illustrated in Table 2.6. In creating **m** and **n** in the table, the appropriate templates from the *Basic Math Assistant* were used.

2.4.2 *Matrix Inverse and Determinant:* Inverse[] *and* Det[]

The inverse of a matrix is obtained with **Inverse**. Thus,

```
mat={{2.,3.,5.},{7.,11.,13.},{17.,19.,23.}};
matI=Inverse[mat]//MatrixForm
```

yields

$$\begin{pmatrix} -0.07692311 & -0.333333 & 0.205128 \\ -0.769231 & 0.5 & -0.115385 \\ 0.692308 & -0.166667 & -0.0128205 \end{pmatrix}$$

If the determinant of the coefficients of **mat** were zero, then an error message will appear stating that the matrix is singular.

To verify that the inverse is correct, we perform the multiplication

```
mat={{2.,3.,5.},{7.,11.,13.},{17.,19.,23.}};
Chop[mat.Inverse[mat]]//MatrixForm
```

and obtain

$$\begin{pmatrix} 1. & 0 & 0 \\ 0 & 1. & 0 \\ 0 & 0 & 1. \end{pmatrix}$$

which is the desired result: the identity matrix. As indicated in Table 1.8, **Chop** is used to replace approximate real numbers that are close to zero with the integer 0. The default value of 10^{-10} is used by Mathematica to define "close to zero."

The determinant of **mat** is obtained by using **Det**. Then

```
mat={{2.,3.,5.},{7.,11.,13.},{17.,19.,23.}};
matD=Det[mat]
```

gives `-78.`

2.5 Solution of a Linear System of Equations: LinearSolve[]

The solution to a system of linear equations is obtained with

```
LinearSolve[m,b]
```

where **m** is an $n\times n$ matrix and **b** is a list with n elements. If the equation is singular or numerically close to it, an error message will appear stating that the matrix is badly conditioned and that the solution may contain significant errors.

Example 2.4

Solution of a System of Equations

Consider the system of equations

$$2x_1 + 3x_2 + 5x_3 = -1$$

$$7x_1 + 11x_2 + 13x_3 = 7.5$$

$$17x_1 + 19x_2 + 23x_3 = 31$$

which can be written in matrix form as

$$\begin{bmatrix} 2 & 3 & 5 \\ 7 & 11 & 13 \\ 17 & 19 & 23 \end{bmatrix} \begin{Bmatrix} x_1 \\ x_2 \\ x_3 \end{Bmatrix} = \begin{Bmatrix} -1 \\ 7.5 \\ 31 \end{Bmatrix}$$

The solution to this system of equations is obtained from

```
m={{2.,3.,5.},{7.,11.,13.},{17.,19.,23.}};
b={-1,7.5,31};
x123=LinearSolve[m,b]
```

which yields

```
{3.9359,0.942308,-2.33974}
```

Thus, x_1 = `x123[[1]] = 3.9359`, x_2 = `x123[[2]] = 0.942308`, and x_3 = `x123[[3]] = -2.33974`. To verify this result, we perform the following matrix multiplication

```
b=m.x123
```

and obtain

```
{-1.,7.5,31.}
```

It is mentioned that `Solve` could also be used to obtain a solution to this system of equations, as shown in Example 4.2.

2.6 Eigenvalues and Eigenvectors: `EigenSystem[]`

Consider the following equation composed of the $(n \times n)$ matrices $[m]$ and $[k]$, a vector $\{x\}$ of length n, and the free parameter λ

$$[k]\{x\} - \lambda[m]\{x\} = 0$$

The eigenvalues and eigenvectors of this system are determined from

```
{lam,vec}=Eigensystem[{k,m}]
```

where `lam` is a list of eigenvalues and `vec` is a matrix of the corresponding eigenvectors. The correspondence of the values in these lists is shown in Example 2.5.

If `b` is a $(n \times n)$ matrix, then the following statement

```
{lam,vec}=Eigensystem[{b,IdentityMatrix[n]}]
```

gives the same results as

```
{lam,vec}=Eigensystem[{b}]
```

If only the eigenvalues are desired, **lam** can be obtained with

```
lam=Eigenvalues[b]
```

and if only the eigenvectors are desired, **vec** can be obtained with

```
vec=Eigenvectors[b]
```

Example 2.5

Natural Frequencies of a Three Degrees-of-Freedom System

Consider a three-degree-of-freedom system with a mass matrix m (kg) and a stiffness matrix k (N·m^{-1}), which have the following values

$$m = \begin{pmatrix} 0.2 & 0 & 0 \\ 0 & 0.3 & 0 \\ 0 & 0 & 12.0 \end{pmatrix}, \quad k = \begin{pmatrix} 936.0 & -768.0 & 0 \\ -768.0 & 1664.0 & -4000.0 \\ 0 & -4000.0 & 25000.0 \end{pmatrix}$$

The natural frequencies and mode shapes can be determined from the following program.

```
k={{936.,-768.,0},{-768.,1664.,-4000.},{0,-4000.,25000.}};
m=DiagonalMatrix[{0.2,0.3,12.}];
{lam,vec}=Eigensystem[{k,m}]
```

Execution of these commands gives

```
{{8678.84,3165.2,465.958},
 {{-0.692177,0.720808,-0.0364293},
  {0.924348,0.364635,-0.112347},
  {-0.66586,-0.730719,-0.150598}}}
```

where

```
lam={8678.84,3165.2,465.958}
```

and

```
vec={{-0.692177,0.720808,-0.0364293},
   {0.924348,0.364635,-0.112347},
   {-0.66586,-0.730719,-0.150598}}}
```

The first row of **vec** is the eigenvector that corresponds to the eigenvalue in the first element of **lam**, the second row of **vec** corresponds to the eigenvalue in the second element of **lam**, and so on. Therefore, the first element of the first row of **vec** corresponds to the value of x_1, the second element to x_2, and the third element to x_3.

Table 2.7 Summary of commands introduced in Chapter 2

Command	Usage
`Eigensystem`	Obtains the eigenvalues and eigenvectors of a square matrix
`LinearSolve`	Solves a system of linear equations
`Prime`	Finds the *n*th prime number
`Table`	Generates vectors and matrices of arbitrary size

A more traditional way of presenting these results is obtained with

```
MatrixForm[Transpose[vec]]
```

which displays

$$\begin{pmatrix} -0.692177 & 0.924348 & -0.66586 \\ 0.720808 & 0.364635 & -0.730719 \\ -0.0364293 & -0.112347 & -0.150598 \end{pmatrix}$$

In this form, the first column contains the amplitudes of $\{x\}$ that correspond to $\lambda = 8678.84$, the second column to those corresponding to $\lambda = 3165.2$, and the third column to those corresponding to $\lambda = 465.958$.

2.7 Functions Introduced in Chapter 2

In addition to the functions listed in Tables 2.1 to 2.4, the additional functions introduced in this chapter are listed in Table 2.7.

References

[1] S. Ramanujan, "Modular equations and approximations to π," *Quarterly Journal of Mathematics*, 1914, Vol. 45, pp. 350–372.

[2] A. E. H. Love, *A Treatise on the Mathematical Theory of Elasticity*, 4th edn, Dover, New York, First American edition 1944, p. 495.

[3] W. D. Pilkey, *Formulas for Stress, Strain, and Structural Matrices*, John Wiley & Sons, New York, 1994, pp. 602–3.

Exercises

Section 2.2.3

2.1 Plot the following Fourier series

$$f(\tau) = \frac{\pi}{2} - \frac{4}{\pi}\sum_{n=1}^{200} \frac{1}{(2n-1)^2}\cos((2n-1)\pi\tau) \qquad 0 \le \tau \le 2$$

using `Plot[f,{t,0,2}]`.

2.2 Given

$$\delta = \left[\frac{1}{n} \sum_{i=1}^{n} x_i^{\beta} \right]^{1/\beta}$$

For $\beta = 2.65$ and $x = \{73, 81, 98, 102, 114, 116, 127, 125, 124, 140, 153, 160, 198, 208\}$, determine δ.

2.3 Evaluate the expression

$$u(\xi, \eta) = 4 \sum_{n=1}^{25} \frac{1 - \cos n\pi}{(n\pi)^3} e^{-n\pi\xi} \sin n\pi\eta$$

at the following six locations: $\xi = 0$, 0.1, and 0.2 and $\eta = 0.4$ and 0.5. Present the output in the following tabular form

η	ξ	u(η, ξ)
0.4	0.	0.239997
0.4	0.1	0.177074
0.4	0.2	0.130061
0.5	0.	0.250007
0.5	0.1	0.185091
0.5	0.2	0.136277

2.4 For the following relationship

$$\sum_{m=1,3,5}^{M \to \infty} \frac{1}{n^2 \left(n^2 + \alpha\right)} = \frac{\pi^2}{8\alpha} \left(1 - \frac{2}{\pi\sqrt{\alpha}} \tanh\left(\frac{\pi\sqrt{\alpha}}{2}\right)\right)$$

show that for $\alpha = 0.5$ and $M = 31$ the difference between the exact value and the value obtained from the truncated summation is 5.07488×10^{-6}.

2.5 For the following formula for the estimation of $1/\pi$ [1], show that when $M = 6$, the difference between the exact value and the result from this formula is 4.660×10^{-57}.

$$\frac{1}{\pi} = \frac{\sqrt{8}}{9801} \sum_{m=0}^{M \to \infty} \frac{(4m)!(1103 + 26390m)}{(m!)^4 396^{4m}}$$

2.6 Given the following series

$$g = \sum_{m=1}^{M} \frac{J_1(\alpha\gamma_m)}{\gamma_m^2 J_1^2(\gamma_m)} J_1(\gamma_m) \tanh(\beta\gamma_m)$$

where $J_n(x)$ is the Bessel function of the first kind of order n and γ_n are the solutions to

$$J_0(\gamma_m) = 0$$

The values of γ_n can be obtained with **N[BesselJZero[]]**. Evaluate this series when $\alpha = 0.5$, $\beta = 0.3$, and $M = 12$.

Section 2.3.2

2.7 Evaluate the following series when $x_1 = 0.4$, $y_1 = 0.65$, $x_2 = 0.3$, and $y_2 = 0.45$

$$s = \sum_{k=1,3,\ldots}^{11} \sum_{m=2,4,\ldots}^{12} \frac{\sin\left((2m+k)x_1/3\right)\sin(ky_1)\sin\left((2m+k)x_2/3\right)\sin(ky_2)}{\left(k^2+km+m^2\right)^2}$$

2.8 Consider the following set of equations

$$\sum_{n}^{M}\sum_{m}^{M} A_{nm}B_{nmlk} - \lambda \sum_{n}^{M}\sum_{m}^{M} A_{nm}C_{nmlk} = 0 \quad l,k = 1,2,\ldots,M$$

where λ is a parameter, A_{nm} are the unknown constants, and B_{nmlk} and C_{nmlk} are known coefficients in terms of specified functions. For a given value of M, create a program that will generate an $M^2 \times M^2$ symbolic matrix in terms of the coefficients B_{nmlk}; that is, when, for example, $M = 3$

$$\begin{pmatrix}
B_{1111} & B_{1211} & B_{1311} & B_{2111} & B_{2211} & B_{2311} & B_{3111} & B_{3211} & B_{3311} \\
B_{1112} & B_{1212} & B_{1312} & B_{2112} & B_{2212} & B_{2312} & B_{3112} & B_{3212} & B_{3312} \\
B_{1113} & B_{1213} & B_{1313} & B_{2113} & B_{2213} & B_{2313} & B_{3113} & B_{3213} & B_{3313} \\
B_{1121} & B_{1221} & B_{1321} & B_{2121} & B_{2221} & B_{2321} & B_{3121} & B_{3221} & B_{3321} \\
B_{1122} & B_{1222} & B_{1322} & B_{2122} & B_{2222} & B_{2322} & B_{3122} & B_{3222} & B_{3322} \\
B_{1123} & B_{1223} & B_{1323} & B_{2123} & B_{2223} & B_{2323} & B_{3123} & B_{3223} & B_{3323} \\
B_{1131} & B_{1231} & B_{1331} & B_{2131} & B_{2231} & B_{2331} & B_{3131} & B_{3231} & B_{3331} \\
B_{1132} & B_{1232} & B_{1332} & B_{2132} & B_{2232} & B_{2332} & B_{3132} & B_{3232} & B_{3332} \\
B_{1133} & B_{1233} & B_{1333} & B_{2133} & B_{2233} & B_{2333} & B_{3133} & B_{3233} & B_{3333}
\end{pmatrix}$$

2.9 An intermediate step in the calculation of the displacement response of a rectangular plate of sides with lengths a and b that is clamped on all four edges requires the solution of the following system of equations [2]

$$\sum_{m=1,3,5}^{M} c_{nm}^{(11)}A_m + c_n^{(12)}B_n = 1$$

$$c_n^{(21)}A_n + \sum_{m=1,3,5}^{M} c_{nm}^{(22)}B_m = \frac{1}{\alpha^4} \qquad n = 1,3,5\ldots$$

where

$$c_{nm}^{(11)} = \frac{n^4 \sin(m\pi/2)}{\left(n^2 + \alpha^2 m^2\right)^2}$$
$$c_{n}^{(12)} = \frac{n\pi\alpha \sin(n\pi/2)\,(n\pi/\alpha + \sinh(n\pi/\alpha)}{16\cosh^2(n\pi/(2\alpha))}$$
$$c_{nm}^{(21)} = \frac{n\pi \sin(n\pi/2)\,(n\pi\alpha + \sinh(n\pi\alpha)}{16\alpha^5\cosh^2(n\pi\alpha/2)}$$
$$c_{n}^{(22)} = \frac{n^4 \sin(m\pi/2)}{\left(\alpha^2 n^2 + m^2\right)^2}$$

and $\alpha = b/a$ is the aspect ratio. Create a symbolic matrix in terms of α as a function of odd M. The matrices for $M = 3$ are given by

$$[C]\{B\} = \{R\}$$

where

$$[C] = \begin{pmatrix} \dfrac{1}{\left(\alpha^2+1\right)^2} & -\dfrac{1}{\left(9\alpha^2+1\right)^2} & \begin{array}{l}\frac{1}{16}\pi\alpha\operatorname{sech}^2\left(\frac{\pi}{2\alpha}\right)\times \\ \left(\sinh\left(\frac{\pi}{\alpha}\right)+\frac{\pi}{\alpha}\right)\end{array} & 0 \\ \dfrac{81}{\left(\alpha^2+9\right)^2} & -\dfrac{81}{\left(9\alpha^2+9\right)^2} & 0 & \begin{array}{l}\frac{1}{16}(-3)\pi\alpha\operatorname{sech}^2\left(\frac{3\pi}{2\alpha}\right)\times \\ \left(\sinh\left(\frac{3\pi}{\alpha}\right)+\frac{3\pi}{\alpha}\right)\end{array} \\ \begin{array}{l}\pi\operatorname{sech}^2\left(\frac{\pi\alpha}{2}\right)\times \\ (\pi\alpha+\sinh(\pi\alpha))/ \\ 16\alpha^5\end{array} & 0 & \dfrac{1}{\left(\alpha^2+1\right)^2} & -\dfrac{1}{\left(\alpha^2+9\right)^2} \\ 0 & \begin{array}{l}-3\pi\operatorname{sech}^2\left(\frac{3\pi\alpha}{2}\right)\times \\ (3\pi\alpha+\sinh(3\pi\alpha))/ \\ 16\alpha^5\end{array} & \dfrac{81}{\left(9\alpha^2+1\right)^2} & -\dfrac{81}{\left(9\alpha^2+9\right)^2} \end{pmatrix}$$

and

$$\{B\} = \begin{Bmatrix} A_1 \\ A_3 \\ B_1 \\ B_3 \end{Bmatrix}, \{R\} = \begin{Bmatrix} 1 \\ 1 \\ \dfrac{1}{\alpha^4} \\ \dfrac{1}{\alpha^4} \end{Bmatrix}$$

Section 2.4

2.10 If

$$\boldsymbol{B} = \begin{bmatrix} 3 & 6 & 6 \\ 6 & 3 & 6 \\ 6 & 6 & 3 \end{bmatrix}$$

show that $\boldsymbol{B}^2 - 12\boldsymbol{B} - 45\boldsymbol{I} = 0$, where $\boldsymbol{I}$ is the identity matrix.

2.11 If

$$k = \begin{bmatrix} 1 & 0 & -1 & 0 \\ 0 & 0 & 0 & 0 \\ -1 & 0 & 1 & 0 \\ 0 & 0 & 0 & 0 \end{bmatrix} \quad T = \begin{bmatrix} \cos\theta & \sin\theta & 0 & 0 \\ -\sin\theta & \cos\theta & 0 & 0 \\ 0 & 0 & \cos\theta & \sin\theta \\ 0 & 0 & -\sin\theta & \cos\theta \end{bmatrix}$$

then determine $\boldsymbol{T'kT}$, where the prime indicates the transpose.

2.12 If

$$C = \begin{bmatrix} \cos\gamma & -\sin\gamma & 0 \\ \sin\gamma & \cos\gamma & 0 \\ 0 & 0 & 1 \end{bmatrix} \quad B = \begin{bmatrix} \cos\beta & 0 & \sin\beta \\ 0 & 1 & 0 \\ -\sin\beta & 0 & \cos\beta \end{bmatrix} \quad A = \begin{bmatrix} 1 & 0 & 0 \\ 0 & \cos\alpha & -\sin\alpha \\ 0 & \sin\alpha & \cos\alpha \end{bmatrix}$$

then determine the matrix product $\boldsymbol{ABC}$.

2.13 The equation of a circle can be determined for three noncolinear points (x_i, y_i), $i = 1, 2, 3$ from

$$\det \begin{vmatrix} x^2 + y^2 & x & y & 1 \\ x_1^2 + y_1^2 & x_1 & y_1 & 1 \\ x_2^2 + y_2^2 & x_2 & y_2 & 1 \\ x_3^2 + y_3^2 & x_3 & y_3 & 1 \end{vmatrix} = 0$$

Determine the equation for the circle that passes through the points (−2,2), (0,0), and (1,1).

Section 2.5

2.14 For the following system of equations

$$\begin{bmatrix} \frac{10}{3} & \frac{5}{3} & \frac{10}{7} \\ \frac{10}{3} & \frac{30}{7} & \frac{50}{9} \\ \frac{20}{9} & 5 & \frac{50}{7} \end{bmatrix} \begin{Bmatrix} x_1 \\ x_2 \\ x_3 \end{Bmatrix} = \begin{Bmatrix} \frac{5}{3} \\ \frac{9}{7} \\ \frac{2}{3} \end{Bmatrix}$$

determine the values for x_i. Verify the solution.

2.15 For the following system of equations

$$\begin{aligned} -6Ri_1 + 4R(i_2 - i_1) &= -V_1 \\ 2R(i_3 - i_2) - 3Ri_2 - 4R(i_2 - i_1) &= -V_2 \\ -Ri_3 - 2R(i_3 - i_2) &= V_3 \end{aligned}$$

determine expressions for i_k.

2.16 Given the following matrix of three equations in the Laplace transform domain

$$\begin{bmatrix} 2s+\tau_o & -s & 0 \\ -s & 2s+\tau_o & -s \\ 0 & -s & s+\tau_o \end{bmatrix} \begin{Bmatrix} \bar{V}_1(s) \\ \bar{V}_2(s) \\ \bar{V}_3(s) \end{Bmatrix} = \begin{Bmatrix} s\bar{U}(s) \\ 0 \\ 0 \end{Bmatrix}$$

where s is the Laplace transform parameter. Determine expressions for $\bar{V}_k(s)$.

Section 2.6

2.17 Determine the eigenvalues of the following matrix

$$\begin{bmatrix} 30 & -\sqrt{6} & -\sqrt{6} \\ -\sqrt{6} & 41 & -15 \\ -\sqrt{6} & -15 & 41 \end{bmatrix}$$

2.18 Determine the largest real eigenvalue of

$$\begin{pmatrix} 0 & 0 & 0 & 0.83 & 0.83 & 0.5 & 0.5 & 0.11 & 0.11 & 0 \\ 0.94 & 0 & 0 & 0 & 0 & 0 & 0 & 0 & 0 & 0 \\ 0 & 0.98 & 0 & 0 & 0 & 0 & 0 & 0 & 0 & 0 \\ 0 & 0 & 0.98 & 0 & 0 & 0 & 0 & 0 & 0 & 0 \\ 0 & 0 & 0 & 0.98 & 0 & 0 & 0 & 0 & 0 & 0 \\ 0 & 0 & 0 & 0 & 0.98 & 0 & 0 & 0 & 0 & 0 \\ 0 & 0 & 0 & 0 & 0 & 0.98 & 0 & 0 & 0 & 0 \\ 0 & 0 & 0 & 0 & 0 & 0 & 0.98 & 0 & 0 & 0 \\ 0 & 0 & 0 & 0 & 0 & 0 & 0 & 0.97 & 0 & 0 \\ 0 & 0 & 0 & 0 & 0 & 0 & 0 & 0 & 0.97 & 0 \end{pmatrix}$$

2.19 Determine the values of ω that satisfy [3]

$$(\boldsymbol{K} - \omega^2 \boldsymbol{M})\{\phi\} = 0$$

where

$$\boldsymbol{K} = \boldsymbol{k}_1 + \boldsymbol{k}_2 + \boldsymbol{k}_3$$
$$\boldsymbol{M} = \boldsymbol{m}_1 + \boldsymbol{m}_2 + \boldsymbol{m}_3$$

and

$$\boldsymbol{m}_1 = \begin{bmatrix} 10.8 & 0.4 & 0 & 0 \\ 0.4 & 12.8 & 0 & 0 \\ 0 & 0 & 0 & 0 \\ 0 & 0 & 0 & 0 \end{bmatrix} \quad \boldsymbol{m}_2 = \begin{bmatrix} 0 & 0 & 0 & 0 \\ 0 & 12.5 & 0.25 & 0 \\ 0 & 0.25 & 5.4 & 0 \\ 0 & 0 & 0 & 0 \end{bmatrix}$$

$$\boldsymbol{m}_3 = \begin{bmatrix} 0 & 0 & 0 & 0 \\ 0 & 41.6 & 0 & -0.4 \\ 0 & 0 & 0 & 0 \\ 0 & -0.4 & 0 & 5.4 \end{bmatrix} \quad \boldsymbol{k}_1 = \begin{bmatrix} 3\times10^4 & -3\times10^4 & 0 & 0 \\ -3\times10^4 & 3\times10^4 & 0 & 0 \\ 0 & 0 & 0 & 0 \\ 0 & 0 & 0 & 0 \end{bmatrix}$$

$$\boldsymbol{k}_2 = \begin{bmatrix} 0 & 0 & 0 & 0 \\ 0 & 5\times10^4 & -5\times10^4 & 0 \\ 0 & -5\times10^4 & 5\times10^4 & 0 \\ 0 & 0 & 0 & 0 \end{bmatrix} \quad \boldsymbol{k}_3 = \begin{bmatrix} 0 & 0 & 0 & 0 \\ 0 & 40\times10^4 & 0 & 20\times10^4 \\ 0 & 0 & 0 & 0 \\ 0 & 20\times10^4 & 0 & 10\times10^4 \end{bmatrix}$$

3

User-Created Functions, Repetitive Operations, and Conditionals

3.1 Introduction

In this chapter, we shall introduce several ways to create functions, exercise program control by using `If` and `Which`, and perform repetitive operations by using `Do`, `While`, `Nest`, and `Map`.

There are several reasons to create functions: avoid duplicate code; limit the effect of changes to specific sections of a program; reduce the apparent complexity of the overall program by making it more readable and manageable; isolate complex operations; and perhaps make debugging and error isolation easier.

3.2 Expressions and Procedures as Functions

3.2.1 Introduction

An expression is any legitimate combination of Mathematica objects such as a mathematical formula, a list, a graphical entity, or a built-in function. Mathematica evaluates each expression in any of several ways; for example, by computing the expression ($1 + 2 \rightarrow 3$), by simplifying it ($a - 4a + 2 \rightarrow 2 - 3a$), or by executing a definition ($r = 7 \rightarrow 7$). During the evaluation process, an attempt is made by Mathematica to reduce expressions to a standard form.

A procedure is a sequence of expressions to be evaluated. When this procedure is used once, it can appear within one cell and be evaluated after all the expressions have been entered or it can be evaluated on an expression-by-expression basis. However, when a procedure will be used more than once or it will be used by a built-in function, it is better to create a special object called a function. There are several ways to create a function: all but one of these ways uses global variables; the one that can create local variables is `Module`. They are all implemented in a similar manner.

An Engineer's Guide to Mathematica®, First Edition. Edward B. Magrab.

Companion Website: www.wiley.com/go/magrab

Table 3.1 Restrictions that can be placed on a function's input variables

		Usage	
Restriction	Function definition*	Expression	Output
Integers only	`fcn[x_Integer]:=1+x^2`	`fcn[3}]`	`10`
		`fcn[3.4]`	`fcn[3.4]`
Symbols only	`fcn[x_Symbol]:=1+x^2`	`fcn[b]`	$1+b^2$
		`fcn[2]`	`fcn[2]`
Lists only	`fcn[x_List]:=1+x^2`	`fcn[{3,5}]`	`{10,26}`
		`fcn[2]`	`fcn[2]`
		`fcn[{2}]`	`{5}`
Real values only	`fcn[x_Real]:=1+x^2`	`fcn[2.]`	`5.`
		`fcn[2]`	`fcn[2]`
		`fcn[2.+3. I]`	`fcn[2.+3. i]`
Complex values only	`fcn[x_Complex]:=1+x^2`	`fcn[2+3 I]`	`-4+12 i`
		`fcn[3 I]`	`-8`
		`fcn[2.]`	`fcn[2.]`
Strings only	`fcn[x_String]:=2 x`	`fcn["a b c"]`	`2 "a b c"`
		`fcn[2.]`	`fcn[2.]`

*If there is more than one argument, a restriction on each argument can be independently specified; for example, **`fcn[x_Integer,y_Complex]:=...`**

When one expression is to be made into a function, a common way to create it is with

```
examfcn[x_,y_, ...]:=expr1
```

where **examfcn** is the function name created by the programmer, the underscore (_) is required for the arguments **x_**, **y_**, ..., and **expr1** is a function of **x**, **y**, ... The colon preceding the equal sign tells Mathematica that **expr1** will be evaluated anew every time that **examfcn** is implemented (used). Lastly, each of the arguments **x_**, **y_**, ..., can be a numerical value or a list of numerical values, a symbolic expression, or a function; either a built-in function or a user-defined function. In addition, as shown in Table 3.1, each input variable can be subject to a condition to ensure that it is a specific type of variable. If that condition is not met, the function is not executed.

When the function is composed of **M** expressions, then the function is created with

```
examfcn[x_,y_, ...]:=(a1=expr1;
  a2=expr2;
  ...
  exprM)
```

where the parentheses and semicolons are required. However, there is no semicolon after **exprM**. The result from the evaluation of the last expression, **exprM**, is the value of the function. If a semicolon were placed there, then the execution of **examfcn** would have no output. In creating the function indicated, each semicolon is followed by *Enter* and depending on the application, either *Enter* or *Shift* and *Enter* simultaneously is used after the closing parenthesis ")". It is assumed that the quantities **a1, a2**, ..., appear in one or more of the expressions following their introduction. The output **exprM** can be a single value **exprM**, a list **{a1,a2,}**, or an array **{{a1,a2, ...},{ ...}, ...}**. Lastly, the variables **a1**, **a2**, ..., are global variables; that is, they are permanently available to all subsequent expressions that use these variable names outside this function definition.

It is also noted that functions created by the user can be treated like any other expression: if appropriate, it can be numerically or symbolically differentiated, integrated, maximized, and so on.

These function-creation capabilities are now illustrated with several examples.

Example 3.1

Function Creation and Usage #1

We shall create a function that evaluates and expands $(1 + ax)^n$ for arbitrary x and n. The expansion of the polynomial is obtained with **Expand**. Then,

```
geqn[x_,n_]:=Expand[(1+a x)^n]
```

creates a function with this capability, where **a** is defined outside the function and is initially a symbol. Then

```
geqn[2,3]
```

gives

$$1+6\ a+12\ a^2+8\ a^3$$

whereas

```
geqn[1+Cos[z],3]
```

yields

$$1+3\ a+3\ a^2+a^3+3\ a\ \mathrm{Cos}[z]+6\ a^2\ \mathrm{Cos}[z]+3\ a^3\ \mathrm{Cos}[z] +3\ a^2\ \mathrm{Cos}[z]^2+3\ a^3\ \mathrm{Cos}[z]^2+a^3\ \mathrm{Cos}[z]^3$$

Furthermore,

```
a=2;
geqn[2,3]
```

yields **125**.

This latter usage illustrates the global nature of **a** in the definition of **geqn**. However, it is important to realize that when **geqn** is used again, the value for **a** remains **a = 2**. To have it revert to a symbol, **Clear[a]** must be used first.

Example 3.2

Function Creation and Usage #2

Consider the evaluation of the following expression

$$f = 1 + at^2 + \sqrt{t}$$

where a is a constant and

$$t = \cos x + a \sinh y^2$$

The function for the evaluation of these quantities is

```
feqn[x_,y_]:=(t=Cos[x]+a Sinh[y^2];
  1+a t^2-Sqrt[t])
```

and **a** is again specified outside the function definition. Then,

```
Print["feqn[2.,3.] = ",feqn[2.,3.]]
Print["t = ",t]
```

gives

$$\text{feqn[2., 3.]} = 1 - \sqrt{-0.416147+4051.54a}+a(-0.416147+4051.54a)^2$$
$$t = -0.416147+4051.54\ a$$

This result again shows the global nature of variables, irrespective of where they are defined.

If one wants to have a choice as to whether a should remain a symbol or be given a numerical value, then the function is redefined as follows

```
feqn[x_,y_,a_]:=(t=Cos[x]+a Sinh[y^2];
  1+a t^2-Sqrt[t])
```

Thus,

```
feqn[2.,3.,b]
```

gives

$$1 - \sqrt{-0.416147+4051.54b}+b(-0.416147+4051.54b)^2$$

and

```
feqn[2.,3.,4.]
```

yields

$$1.05051\times10^9$$

In this case,

```
t
```

displays

```
16205.8
```

Each expression of **feqn** can handle lists so that **x_** and/or **y_** and/or **a_** can be lists. When lists are used, they must be of the same size. Hence,

```
feqn[2.,{3.,4.},{b,c}]
```

gives

$$\left\{1-\sqrt{-0.416147+4051.54b}+b(-0.416147+4051.54b)^2,\right.$$
$$\left.1-\sqrt{-0.416147+4.44306\times 10^6 c}+c(-0.416147+4.44306\times 10^6 c)^2\right\}$$

The first element of this list corresponds to the input values of `x = 2.0, y = 3.0`, and `a = b` and the second element to the input values `x = 2.0, y = 4.0`, and `a = c`.

This particular set of definitions for *f* and *t* is suitable for the use of **With**, which is defined as

```
With[{x=xo,y=y0, ...},expr]
```

where in **expr** all occurrences of **x**, **y**, ... , respectively, are replaced with **xo**, **yo**, ... Thus,

```
feqn[x_,y_,a_]:=With[{t=Cos[x]+a Sinh[y^2]},1+a t^2-Sqrt[t]]
```

In this case, **t** is a local variable and not accessible outside this function.

Example 3.3

Special Syntax for Subscripted Variables

When symbols have been created using **Symbolize** from the *Notation Palette* (recall Section 1.5.2), the arguments of the function definition are different. If it is assumed that a variable g_b is to appear in the function creation statement, then the following steps are taken.

```
Needs["Notation`"]
Symbolize[gb] (* From Notation palette *)
examfcn[gb:_,y_, ...]:=
```

where it is seen that a colon (:) precedes the underscore character (_). On the right-hand side of the equal sign, g_b is used in the regular manner. **Note**: Neither **Symbolize** nor the colon is used in this context if the subscript b is a positive integer; that is, for example, if $g_b \rightarrow g_2$.

As an example, we shall create a function that evaluates $1 + q_m^3/\sqrt{y}$. and evaluate it when $q_m = 1 + c$ and $y = 4$. Then,

```
Needs["Notation`"]
Symbolize[qm] (* From Notation palette *)
fcn[qm:_,y_]:=1+qm^3/Sqrt[y]
fcn[1+c,4]
```

displays

$$1+\frac{1}{2}(c+1)^3$$

3.2.2 *Pure Function:* **Function[]**

Another syntax that can be used to create a function is with what is called a pure function. It is formally created with **Function**. However, in much of the material in the *Documentation Center* it appears in terms of a shorthand syntax that is best introduced by example. Although the pure function has all the attributes discussed above, its implementation is a little different.

Consider the function $f(x,y) = x^2 + ay^3/x$, where a is a constant. Using the previous syntax, a function representing f is given by

```
f[x_,y_]:=x^2+a y^3/x
```

Thus, to determine its value at $x = 2.0$ and $y = 3.0$, we use

```
f[2.,3.]
```

and find that

```
4.+13.5 a
```

A pure function on the other hand can be represented either of two ways. The first way is with

```
Function[{x,y, ...},expr]
```

where **{x,y, ...}** is a list of the independent variables and **expr** is an expression that is a function of the independent variables. Then, using this notation to create $f(x,y)$, we have

```
ff=Function[{x,y},x^2+a y^3/x]
```

Then, as before, to determine its value at $x = 2.0$ and $y = 3.0$, we use

```
ff[2.,3.]
```

and find again that

```
4.+13.5 a
```

The second way to represent this function as a pure function is to employ the following syntax

```
g=#1^2+a #2^3/#1 &
```

In this expression, **#1** represents one variable, x in this case, and **#2** represents a second variable, y in this case. If there were only one variable, then **#1** can be replaced with **#**. The expression is completed by ending it with a space after the last character followed by the **&**. In both cases, **a** is a global variable. In addition, **g** is shorthand for **g[var1,var2]** and although it cannot be explicitly written as such when created, it is employed in that way, as shown below.

The execution of the expression for **g** gives

$$\#1^2+\frac{a\#2^3}{\#1}\ \&$$

Then, to evaluate it, we use

```
g[2.,3.]
```

which yields the same as before; namely, **4.+13.5 a**.

Again the evaluation of the pure function could be delayed in the same manner as the regular function by using the "**:=**" syntax; that is, **g** could have been written as

```
g:=#1^2+a #2^3/#1 &
```

Another way to employ the pure function is to provide the values for **#1** and **#2** immediately following the creation of the pure function as follows

```
g=#1^2+a #2^3/#1 &[2.,3.]
```

The execution of this expression displays the result previously obtained with **g[2.,3.]**.

The use of a pure function also has utility in such operations as incrementing subscripts of symbolic variables and in placing units on numerical values. We shall illustrate this with several examples, each of which uses a pure function and **Prefix**, whose shorthand notation is **/@** (recall Table 1.2).

To create a list of the integers 1 to 5 and their units, say acceleration, we use **Quantity** as follows

```
Quantity[#,"Meters"/"Seconds"^2] &/@Range[5]
```

which yields

$\{1\ m/s^2, 2\ m/s^2, 3\ m/s^2, 4\ m/s^2, 5\ m/s^2\}$

As a second example, a list of a symbolic variable that is subscripted from 1 to 5 will be created. The statement to do this operation is, with the use of the *Basic Math Assistant*,

```
a# &/@Range[5]
```

which results in

$\{a_1, a_2, a_3, a_4, a_5\}$

As a third example, we shall create a 3×3 symbolic array containing the appropriate subscripts for each element. This task is most easily performed with the use of **Array**, which generates an ($n \times m$) array of f_{nm} elements. Its syntax is

```
Array[f,{n,m}]
```

where, in general, f is a function of n and m. Then the program is

```
Array[aToString[#1]<>ToString[#2] &,{3,3}]//MatrixForm
```

which displays

$$\begin{pmatrix} a_{11} & a_{12} & a_{13} \\ a_{21} & a_{22} & a_{23} \\ a_{31} & a_{32} & a_{33} \end{pmatrix}$$

For the last example, we shall create a 4×4 array whose elements are m^n and display the result in matrix form. The value of m corresponds to the row position and n to the column position. The statement to perform these operations is

```
Array[#1^#2 &,{4,4}]//MatrixForm
```

which gives

$$\begin{pmatrix} 1 & 1 & 1 & 1 \\ 2 & 4 & 8 & 16 \\ 3 & 9 & 27 & 81 \\ 4 & 16 & 64 & 256 \end{pmatrix}$$

Example 3.4

Array Creation

A pure function, **Array**, and **ArrayFlatten** will be used to produce the following arrays whose size is $3N \times 3N$.

$$\begin{pmatrix} 0 & 1 & 1 \\ 1 & 0 & 1 \\ 1 & 1 & 0 \end{pmatrix}, \begin{pmatrix} 0 & 1 & 1 & 0 & 2 & 2 \\ 1 & 0 & 1 & 2 & 0 & 2 \\ 1 & 1 & 0 & 2 & 2 & 0 \\ 0 & 3 & 3 & 0 & 4 & 4 \\ 3 & 0 & 3 & 4 & 0 & 4 \\ 3 & 3 & 0 & 4 & 4 & 0 \end{pmatrix}, \begin{pmatrix} 0 & 1 & 1 & 0 & 2 & 2 & 0 & 3 & 3 \\ 1 & 0 & 1 & 2 & 0 & 2 & 3 & 0 & 3 \\ 1 & 1 & 0 & 2 & 2 & 0 & 3 & 3 & 0 \\ 0 & 4 & 4 & 0 & 5 & 5 & 0 & 6 & 6 \\ 4 & 0 & 4 & 5 & 0 & 5 & 6 & 0 & 6 \\ 4 & 4 & 0 & 5 & 5 & 0 & 6 & 6 & 0 \\ 0 & 7 & 7 & 0 & 8 & 8 & 0 & 9 & 9 \\ 7 & 0 & 7 & 8 & 0 & 8 & 9 & 0 & 9 \\ 7 & 7 & 0 & 8 & 8 & 0 & 9 & 9 & 0 \end{pmatrix}$$

The first matrix is for $N = 1$, the second is for $N = 2$, and the third is for $N = 3$, and so on. The form for **Array** is given above; the form of **ArrayFlatten** is

```
ArrayFlatten[array,r]
```

where **array** is the array to be altered and **r** is the levels in **array** to be altered.

The submatrices in the above arrays are created using

```
ConstantArray[nN (m-1)+n,{3,3}]-
  DiagonalMatrix[{nN (m-1)+n,nN (m-1)+n,nN (m-1)+n}]
```

where **nN** = N, **n** is the location of the column of the submatrix, and **m** is the location of the row of the submatrix. We convert the above expression into a pure function and use it in **Array** as follows for $N = 3$

```
nN=3;
Array[ConstantArray[nN (#1-1)+#2,{3,3}]-
  DiagonalMatrix[{nN (#1-1)+#2,nN (#1-1)+
    #2,nN (#1-1)+#2}] &,{nN,nN}]//MatrixForm
```

which gives

$$\begin{pmatrix} \begin{pmatrix} 0 & 1 & 1 \\ 1 & 0 & 1 \\ 1 & 1 & 0 \end{pmatrix} & \begin{pmatrix} 0 & 2 & 2 \\ 2 & 0 & 2 \\ 2 & 2 & 0 \end{pmatrix} & \begin{pmatrix} 0 & 3 & 3 \\ 3 & 0 & 3 \\ 3 & 3 & 0 \end{pmatrix} \\ \begin{pmatrix} 0 & 4 & 4 \\ 4 & 0 & 4 \\ 4 & 4 & 0 \end{pmatrix} & \begin{pmatrix} 0 & 5 & 5 \\ 5 & 0 & 5 \\ 5 & 5 & 0 \end{pmatrix} & \begin{pmatrix} 0 & 6 & 6 \\ 6 & 0 & 6 \\ 6 & 6 & 0 \end{pmatrix} \\ \begin{pmatrix} 0 & 7 & 7 \\ 7 & 0 & 7 \\ 7 & 7 & 0 \end{pmatrix} & \begin{pmatrix} 0 & 8 & 8 \\ 8 & 0 & 8 \\ 8 & 8 & 0 \end{pmatrix} & \begin{pmatrix} 0 & 9 & 9 \\ 9 & 0 & 9 \\ 9 & 9 & 0 \end{pmatrix} \end{pmatrix}$$

We see that this result is not yet in the desired form. To obtain the desired form, we use **`ArrayFlatten`** as follows

```
nN=3;
ArrayFlatten[Array[ConstantArray[nN (#1-1)+#2,{3,3}]-
  DiagonalMatrix[{nN (#1-1)+#2,nN (#1-1)+
    #2,nN (#1-1)+#2}] &,{nN,nN}],2]//MatrixForm
```

which gives the desired result.

3.2.3 `Module[]`

A third way that one can create a function is to use **`Module`**, which allows one to create local variables. The general form for **`Module`** is

```
heqn[xx_,yy_, ...]:=Module[
  {x=xx,y=yy, ...,a1,a2, ...,b1=val1,b2=val2, ...},proced]
```

where **`x`**, **`y`**, **`a1`**, **`a1`**, … , **`b1`**, **`b2`**, … , are local variables and **`proced`** is a procedure that is a function of the local variables and possibly other variables not placed between the preceding pair of braces {…}; these other variables will be global variables. The syntax **`b1=val1`**, **`b1=val1`**, … , additionally assigns the value (or a symbolic expression) given by **`val1`** to **`b1`**, the value of **`val2`** to **`b2`**, … . In addition, the form of the procedure follows the rules given previously for **`examfcn`** that contained **`M`** expressions; that is, a semicolon is placed at the end of each expression except the last one. It is mentioned again that, in general, **`exprM`** can be a list or an array.

We now illustrate the use of **`Module`** with the following example.

Example 3.5

Natural Frequency Coefficient of a Two Degrees-of-Freedom System

The nondimensional natural-frequency coefficients of a two degrees-of-freedom system are given by

$$\Omega_{1,2} = \sqrt{\frac{1}{2}\left[a_1 \mp \sqrt{a_1^2 - 4a_2}\right]}$$

where

$$a_1 = 1 + \omega_r^2 \left(1 + m_r\right)$$
$$a_2 = \omega_r^2$$

and ω_r and m_r are nondimensional parameters.

To determine these frequency coefficients, we create the following function using **Module**

```
om12[mrr_,wrr_]=Module[{mr=mrr,wr=wrr,a1,a2},
  a1=1+wr^2(1+mr); a2=wr^2;
    {Sqrt[0.5(a1-Sqrt[a1^2-4 a2])],
      Sqrt[0.5(a1+Sqrt[a1^2-4 a2])]}];
```

In this construction, **mr**, **wr**, **a1**, and **a2** are local variables. Also, since the last statement is a list with two elements, the execution of **om12** will result in its output being a two-element list. The first element will correspond to the value of Ω_1 and the second element to Ω_2.

Then,

```
om=om12[0.4,1.2]
```

yields

```
{0.771187,1.55604}
```

and therefore, Ω_1 = **om[[1]] = 0.771187** and Ω_2 = **om[[2]] = 1.55604**.

Since we know that the output of this function is a two-element list, another way that this function can be used is

```
{om1,om2}=om12[0.4,1.2]
```

which again gives

```
{0.771187,1.55604}
```

However, this time Ω_1 = **om1 = 0.771187** and Ω_2 = **om2 = 1.55604**. This form may be more convenient if these results are used in further computations.

To illustrate the effects of declaring certain variables local, we revise **om12** by removing the requirement that **a1** and **a2** be local variables. Then,

```
om12[mrr_,wrr_]=Module[{mr=mrr,wr=wrr},
  a1=1+wr^2(1+mr);
  a2=wr^2;
  {Sqrt[0.5(a1-Sqrt[a1^2-4 a2])],
    Sqrt[0.5(a1+Sqrt[a1^2-4 a2])]}];
```

Hence,

```
{om1,om2}=om12[0.4,1.2]
```

still gives

```
{0.771187,1.55604}
```

However, entering

```
{a1,a2}
```

yields

```
{1+(1+mrr) wrr^2,wrr^2}
```

Thus, if one were to use **a1** and/or **a2** in subsequent calculations, these are the expressions that would be used.

3.3 Find Elements of a List that Meet a Criterion: Select[]

Pure functions are frequently needed in certain Mathematica functions. One such function that uses pure functions is **Select**, which provides the means to select only those elements of a list that meet certain criteria. The definition of **Select** is

```
Select[list,crit]
```

where **list** is a vector list and **crit** is the criterion by which the elements of **list** will be evaluated. The argument **crit** is typically a pure function that operates on each element of **list**. Examples of some selection criteria are shown in Table 3.2. The output of **Select** is a list of the elements that satisfy the criteria in the precedence that they appear in the list.

Table 3.2 Examples of the use of `Select[d,crit]`

Selection criterion	Criterion specification*	Usage with `d={1.0,20,2.0,5.0,-7,6.0,-10.0}`	
		Expression	Output
All values < *a*	`#<value &`	`Select[d,#<5 &]`	`{1.,2.,-7,-10.}`
All values > *a*	`#>value &`	`Select[d,#>5 &]`	`{20,6.}`
All values in the range *a* to *b*	`value1<=#<= value2 &`	`Select[d,0<=#<7 &]`	`{1.,2.,5.,6.}`
All even numbers	`EvenQ`	`Select[d,EvenQ]`	`{20,2.,6.,-10.}`
All odd numbers	`OddQ`	`Select[d,OddQ]`	`{1.,5.,-7}`
All positive numbers	`Positive`	`Select[d,Positive]`	`{1.,20,2.,5.,6.}`
All negative numbers	`Negative`	`Select[d,Negative]`	`{-7,-10.}`
All integers	`IntegerQ`	`Select[d,IntegerQ]`	`{20,-7}`
All numbers whose magnitudes are less than a specified value	`Abs[#]<value &`	`Select[d,Abs[#]<10 &]`	`{1.,2.,5.,-7,6.}`

* **value** = number that is followed by a space and an **&**

In addition, **EvenQ**, **OddQ**, **Positive**, and **Negative** given in Table 3.2 can be used individually as commands to determine the specific properties of the elements of a list. **IntegerQ** can be used on a single term only. **EvenQ** and **OddQ** are limited to integers only. The output of these commands is either **True** or **False**.

To illustrate their usage, consider the following example

```
d={6,-6,7,-7};
Print["Even? ",EvenQ[d]]
Print["Odd? ",OddQ[d]]
Print["Positive? ",Positive[d]]
Print["Negative? ",Negative[d]]
```

the output of which is

```
Even? {True,True,False,False}
Odd? {False,False,True,True}
Positive? {True,False,True,False}
Negative? {False,True,False,True}
```

Example 3.6

Usage of `Select[]`

Consider the expression

$$d = \frac{1}{2}\sum_{k}^{K}\frac{1}{k}$$

We shall determine the smallest value of K and the corresponding value of d for which $d > 1$ and for which $d > 2$. Based on the commands introduced so far, we shall do the following. We let $K = 50$ and use **`Table`** to create a list of the first 50 terms of d as a function of k. The elements of this list are denoted d_k. Next, we create a list of cumulative sums using **`Accumulate`**. The elements of this list are denoted a_k. We then use **`Select`** on the elements a_k to find the elements that are, respectively, less than 1 and less than 2. From the length of each of these lists plus 1, the location of the first element whose value exceeds the specified values is obtained. This location is used to select the appropriate d_K.

The program is

```
ak=0.5 Accumulate[Table[1/k,{k,1,50}]];
Table[len=Length[Select[ak,#<m &]]+1;
  Print["When K = ",len,", d = ",ak[[len]]],{m,1,2}];
```

which yields

```
When K = 4, d = 1.04167
When K = 31, d = 2.01362
```

3.4 Conditionals

In this section, we shall introduce several conditional functions and in Section 3.6 we shall give several examples of how they can be applied.

3.4.1 `If[]`

The test of a conditional or set of conditionals, denoted **`cond`**, is obtained with

```
r=If[cond,tr,fa] (* or If[cond,r=tr,r=fa] *)
```

where **`tr`** is selected if **`cond`** is true and **`r=tr`**, and **`fa`** is selected if **`cond`** is false and **`r=fa`**. In general, **`tr`** and **`fa`** can be a constant, an expression, or a procedure. When **`tr`** is a procedure, then each expression in the procedure is followed by a semicolon except the last expression. The last expression of **`tr`** is followed by a comma if **`fa`** is present, otherwise there is no punctuation.

3.4.2 **Which[]**

When a series of many conditionals **cond1**, **cond2**, … , **condM** have to be tested, one uses **Which**, whose form is

```
r=Which[cond1,tr1,cond2,tr2, ... ,condM,trM]
```

Here, each **CondK**, **K** = **1**, **2**, … , **M** is tested from left to right until a **True** is encountered at which point no more conditionals are tested. If **True** is encountered at **K = L**, then **r=trL**. In general, **trK** can be a constant, an expression, or a procedure. When **trK** is a procedure, then each expression in the procedure is followed by a semicolon except the last expression. The last expression is followed by a comma when **K** < **M**; when **K** = **M**, there is no punctuation.

3.5 Repetitive Operations

We shall introduce several functions that perform repetitive operations and in Section 3.6 we shall give several examples of how they can be applied.

3.5.1 **Do[]**

A specified number of repetitive operations are performed by **Do**. It is similar to **Table**, except that instead of forming a list it performs only an evaluation. The form for **Do** is

```
Do[expr,{n,ns,ne,dn}]
```

or

```
Do[expr,{n,{lst}}]
```

where **expr** is an expression or a procedure that is evaluated repetitively over the range of **n** and can be a function of the index **n**. The value of **n** is a quantity whose value starts at **ns**, ends at **ne**, and is incremented over this range by an amount **dn** or **n** can have the values of the elements appearing in **{lst}**. When **expr** is a procedure composed of **M** expressions **expr1**, **expr2**, … , **exprM**, then each expression is followed by a semicolon except **exprM**, which is followed by a comma.

3.5.2 **While[]**

The function **While** performs an indeterminate number of repetitive operations until a specified condition **cond** is met. The form for **While** is

```
While[cond,expr]
```

where **expr** is an expression or a procedure. When **expr** is a procedure composed of **M** expressions **expr1**, **expr2**, ..., **exprM**, then each expression is followed by a semicolon except **exprM**, which has no punctuation.

3.5.3 Nest[]

The function **Nest** applies **n** times an operation **f** to an expression **expr**. Its form is

```
Nest[f,expr,n]
```

The operation **f** is typically expressed as a pure function.

A companion function called **NestList** performs in the same way that **Nest** does except that it presents in a list all the intermediate results; the first term in the list is **expr** and the last result is that given by **Nest**. A second related function is **NestWhile**, in which **n** in **Nest** is replaced by a test such that **Nest** continues until the test is satisfied. The test is typically expressed as a pure function. **Nest** can be used as a compact way to evaluate recurrence relations of the type $y_{n+1} = g(y_n)$.

3.5.4 Map[]

The function **Map** applies **f** to each element in an expression **expr**. Its form is

```
Map[f,expr] (* Alternate form: f/@expr *)
```

This function is best understood by illustrating its capabilities. For example, to create a table of the evaluation of several elementary functions of, say, the number 2 and label these functions, we use **Map** as follows

```
Map[{#[2],#[2.]} &,{Sin,Cos,Tan,Cot,Exp,Sqrt}]//Column
```

which displays

```
{Sin[2], 0.909297}
{Cos[2], -0.416147}
{Tan[2], -2.18504}
{Cot[2], -0.457658}
{e², 7.38906}
{√2, 1.41421}
```

In using **Map** this way, we have taken advantage of how Mathematica treats integers and decimal quantities: with integers no numerical evaluation is performed and with decimal numbers it is.

For another example, we shall add the same pair of elements $\{x,y\}$ to the square of each respective component of the list $\{\{z_1,z_2\},\{z_3,z_4\},\ldots\}$; that is, we shall create the list $\{\{z_1^2+x,z_2^2+y\},\{z_3^2+x,z_4^2+y\},\ldots\}$. Thus,

```
c={{1,2},{3,4},{5,6}};
Map[#^2+{x,y} &,c]
```

gives

```
{{1+x,4+y},{9+x,16+y},{25+x,36+y}}
```

3.6 Examples of Repetitive Operations and Conditionals

Example 3.7

Recurrence Relation #1

The following expression

$$y_{n+1}=y_n\left(1+x_{n+1}\right)^2-2^{n+1}x_{n+1}\quad n=0,1,2,\ldots$$

where

$$x_{n+1}=\frac{1-\sqrt{1-x_n^2}}{1+\sqrt{1-x_n^2}}$$

converges to $1/\pi$ when $x_0=1/\sqrt{2}$ and $y_0=1/2$.

We shall use **While** to determine the value of n when the difference $|1/\pi-y_{n+1}|$ is less than 10^{-15}. We shall print the intermediate values of y_n. The program is

```
x=1/Sqrt[2.]; y=0.5; n=0;
While[Abs[1/π-y]>10^(-15),
  x=(1-Sqrt[1-x^2])/(1+Sqrt[1-x^2]);
  y=y (1+x)^2-2^(n+1) x;
  n=n+1;
  Print["n = ",n-1," y"ToString[n]," = ",NumberForm[y,16]]];
```

Execution of this program displays

```
n = 0 y1 = 0.3431457505076199
n = 1 y2 = 0.3184125985146642
n = 2 y3 = 0.3183098869311611
n = 3 y4 = 0.3183098861837902
```

Example 3.8

Interval halving

We shall determine the value of x that satisfies $f(x) = 0$ when $f(x)$ is a transcendental function. The method that shall be employed is called interval halving, which is an effective but inefficient method. The method works as follows. The independent variable x is given a starting value $x = x_{start}$ and the sign of $f(x_{start})$ is determined. The variable x is then incremented by an amount $dx = \Delta$ and the sign of $f(x_{start}+\Delta)$ is determined. The signs of these two values are compared. If the signs are the same, then x is again incremented by Δ and the sign of $f(x_{start}+2\Delta)$ is evaluated and compared to that of $f(x_{start})$. If the signs are different, then the current value of x is decremented by half the interval size; that is, $dx = \Delta/2$ and the sign of $f(x_{start}+\Delta/2)$ is compared to $f(x_{start})$. If the sign of $f(x_{start}+\Delta/2)$ is the same as $f(x_{start})$ then we set $dx = dx/2 = \Delta/4$ and we determine the sign of $f(x_{start}+3\Delta/4)$. If the sign is different, then we determine the sign of $f(x_{start}+\Delta/4)$. This process is repeated until $dx/x <$ tolerance, which is typically on the order of 10^{-6} or smaller.

To illustrate this technique, we shall determine the lowest root of $\cos(x)$, which is $\pi/2 = 1.5707963268$. Selecting a tolerance of 10^{-6}, the program is

```
x=0.05;  dx=0.5;   tol=10^(-6);
s1=Sign[Cos[x]];
While[dx/x>tol,
  If[s1≠Sign[Cos[x+dx]],dx=dx/2.,x=x+dx]]
NumberForm[{x,Cos[x]},8]
```

which displays

```
{1.5707958,5.0465134×10^-7}
```

Example 3.9

Recurrence Relation #2

Given the following relation

$$\begin{aligned} x_{n+1} &= x_n/2 \quad &&\text{if } x_n \text{ is an even positive integer} \\ &= 3x_n + 1 \quad &&\text{if } x_n \text{ is an odd positive integer} \end{aligned}$$

where $n = 1, 2, 3, \ldots$. Let x_1 be an arbitrary positive integer. This recurrence process appears to always converge at some value of $n + 1 = N$ such that $x_N = 1$. The value of N cannot be predicted as a function of x_1. We shall print all the values of x_n for the values of $x_1 = 3, 19,$

and 23. The determination of whether a number is even is obtained by using **EvenQ** (recall Section 3.3). The output of **EvenQ** is **True** if the number is even. The program is

```
Do[x=x1; xn={x};
  While[x!=1,If[EvenQ[x],x=x/2,x=3 x+1];
  AppendTo[xn,x]];Print[xn],{x1,{3,19,23}}]
```

Execution of this program gives

```
{3,10,5,16,8,4,2,1}
{19,58,29,88,44,22,11,34,17,52,26,13,40,20,10,5,16,8,4,2,1}
{23,70,35,106,53,160,80,40,20,10,5,16,8,4,2,1}
```

Example 3.10

Usage of Which

We shall obtain the solution to

$$f(x, a, b) = a\cos x + b\sinh x = 0$$

where $0 \le a \le \infty$ and $0 \le b \le \infty$. To improve the accuracy of the solution and to be able to consider the wide range of values that a and b can have, we arbitrarily stipulate that this equation will be used for $a < 500$ and $b < 500$. When either of these values is exceeded, the equation must be divided by that value prior to making the calculation. We shall create a function that computes f for the four possible scenarios by using **Which** as follows.

```
f[x_,a_,b_]:=Which[a<500&&b<500,a Cos[x]+b Sinh[x],
  a<500&&b>=500,a Cos[x]/b+Sinh[x],
  a>=500&&b<500,Cos[x]+b Sinh[x]/a,
  a>=500&&b>=500,Cos[x]/b+Sinh[x]/a]
```

This function would then be used with an appropriate root-finding function, in this case, **FindRoot** (see Section 5.5.)

Example 3.11

Recurrence Relation #3

We shall illustrate the use of **Nest**, **NestList**, and **NestWhile** with the following recurrence formula, which determines the square root of a positive number a starting at any value for x_0. However, a poor choice for x_0 can increase the number of iterations necessary for convergence to within a specified value. The formula is

$$x_{n+1} = \frac{1}{2}\left(x_n + \frac{a}{x_n}\right) \qquad n = 0, 1, 2, \ldots$$

We shall determine the $\sqrt{5}$ using an initial guess of $x_0 = 52.0$. The above expression can be expressed as the pure function **(#+a/#)/2. &**. Then, to compute the first 8 iterations and display the results with 16 digits, we have

```
a=5.;
NumberForm[NestList[(#+a/#)/2. &,52.,8],16]
NumberForm[Nest[(#+a/#)/2. &,52.,8],16]
NumberForm[
  NestWhile[(#+a/#)/2. &,a,Abs[Sqrt[a]-#]>10^(-15) &],16]
NumberForm[Sqrt[a],16]
```

The execution of this procedure displays

```
{52., 26.04807692307692, 13.1200148365846, 6.750555983300693,
    3.745617857904981, 2.540255554540268, 2.254280728154025,
    2.23614154958916, 2.2360679787101}
2.2360679787101
2.23606797749979
2.23606797749979
```

Example 3.12

Recurrence Relation #4

We shall illustrate the use of **NestList** for a mapping of two interrelated variables by examining the recurrence relation

$$\begin{aligned} x_{n+1} &= y_n - \mathrm{sgn}(x_n)\sqrt{|bx_n - c|} \qquad n = 0, 1, 2, \ldots N \\ y_{n+1} &= a - x_n \end{aligned}$$

for $a = 0.4$, $b = c = 1.0$, $x_0 = y_0 = 0$, and $N = 3000$. We shall plot the results using **ListPlot**, which is given in Table 6.1, and label the axes using **AxesLabel**, which is described in Table 6.9. One way to solve this relation is to express the two variables as one variable that is a list with two elements. The first element is identified as **#[[1]],** which corresponds to x and the second element is identified as **#[[2]]** and corresponds to y. Consequently, the output of

NestList is an N×2 list of coordinate pairs, which is in a format that can be directly used by **Listplot**. Then the program is

```
a=0.4;  b=1.;   c=1.;
ListPlot[NestList[{#[[2]]-Sign[#[[1]]]*
  Sqrt[Abs[b #[[1]]-c]],a-#[[1]]}&,{0.,0.},3000],
  AxesLabel->{"x","y"}]
```

The output of this program is shown in Figure 3.1. It is mentioned that this figure changes dramatically as N increases.

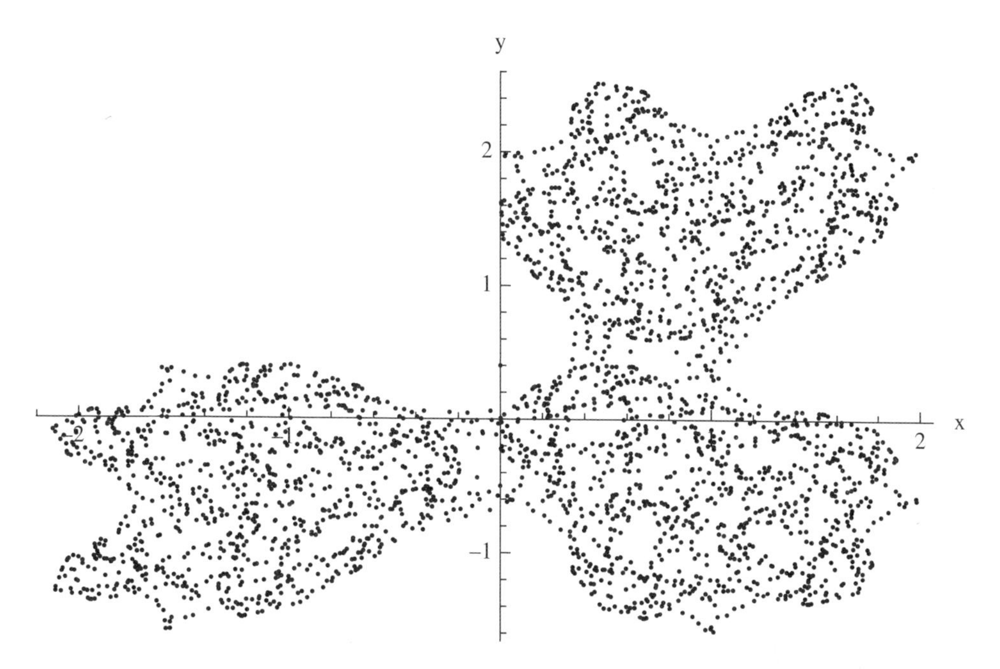

Figure 3.1 Mapping for the recurrence relations of Example 3.12 for $N = 3000$

Example 3.13

Example 3.7 Revisited

In view of the technique used in Example 3.12, we shall repeat Example 3.7 using **NestList** and the fact that we know that the solution converges after three iterations. We shall again display the results with 16 digits. Then,

```
NumberForm[NestList[{2 #[[1]],
  (1-Sqrt[1-#[[2]]^2])/(1+Sqrt[1-#[[2]]^2]),
  #[[3]] (1+(1-Sqrt[1-#[[2]]^2])/(1+Sqrt[1-#[[2]]^2]))^2
    -#[[1]] (1-Sqrt[1-#[[2]]^2])/(1+Sqrt[1-#[[2]]^2])} &,
  {2,1/Sqrt[2.],0.5},5]//Column,16]
```

which displays

```
{2,0.7071067811865475,0.5}
{4,0.1715728752538099,0.3431457505076199}
{8,0.007469666729509568,0.3184125985146642}
{16,0.00001394936942418036,0.3183098869311611}
{32,4.864625369430832×10^-11,0.3183098861837902}
{64,0.,0.3183098861837902}}
```

It is seen that there isn't any difference between y_4 and y_5 and that when $n = 3$, we have the same result that was obtained in Example 3.7.

Another way to implement this recurrence relation is with **NestWhile**. In this case, the above instructions are modified as

```
NumberForm[NestWhile[{2 #[[1]],
  (1-Sqrt[1-#[[2]]^2])/(1+Sqrt[1-#[[2]]^2]),
  #[[3]](1+(1-Sqrt[1-#[[2]]^2])/(1+Sqrt[1-#[[2]]^2]))^2-
   #[[1]](1-Sqrt[1-#[[2]]^2])/(1+Sqrt[1-#[[2]]^2])} &,
  {2,1/Sqrt[2.],0.5},Abs[#[[2]]]>10^(-15) &],16]
```

where the **NestWhile** command is exited when $|x_{n+1}| < 10^{-15}$. The execution of these commands displays

```
{64,0.,0.3183098861837902}
```

Example 3.14

Secant Method

In Example 3.8, the root of a transcendental function was obtained by using interval halving. Another way to find the roots of a function $f(x)$ is to use the secant method, which is determined from the following recurrence relation

$$x_{n+1} = x_n - f(x_n)\frac{x_n - x_{n-1}}{f(x_n) - f(x_{n-1})}$$

To illustrate this, we select

$$f(x) = \cos x \cosh x - 1$$

and $x_0 = 5.0$ and $x_1 = 4.0$. If we specify that a root has been found when $f(x_{root}) < 10^{-6}$, then the root is obtained from

```
fcn[xx_]:=Cos[xx] Cosh[xx]-1
x1=4.;  xo=5.;
sec=First[NestWhile[{#[[2]]-(fcn[#[[2]]](#[[2]]-#[[1]]))/
  (fcn[#[[2]]]-fcn[#[[1]]]),#[[1]]} &,{xo,x1},
  Abs[fcn[#[[2]]]]>10^(-6) &]]
```

which yields

```
4.73004
```

Example 3.15

`Sow[]` and `Reap[]`

`Sow` and **`Reap`** provide a convenient way to create a list of intermediate results that satisfy a specific criterion. It is particularly useful when implementing certain applications of the option **`WhenEvent`** in **`NDSolveValue`**, which is discussed in Section 5.3. A typical form of using this pair of commands is[1]

```
{ans,lst}=Reap[Operation[... Sow[arg] ...]]
```

where **`Operation`** is a series of commands that perform the desired calculations, **`ans`** is the answer resulting from **`Operation`**, and **`lst`** is a list comprised of those values of **`arg`** that satisfy the criterion. The first element in **`lst`** is the result of **`Operation`** and the remaining list of elements are the appropriate values of **`arg`**.

To illustrate these commands, we shall sum n^2 for $n = 1, 2, \ldots, 6$ and form a list of the intermediate values of n^2 that are even numbers. Then,

```
{tot,lst}=Reap[Total[Table[If[EvenQ[n^2],Sow[n^2],n^2],
  {n,1,6}]]]
```

gives

```
{91,{{4,16,36}}}
```

---

[1] An alternate notation that is often used is

```
lst=Reap@Operation[...Sow[...]...]
```

Thus, `tot = 91` and `Flatten[lst] = {4,16,36}`. It is seen that both positions of the `If` command contain n^2. This is because `Total` must sum this value for each value of n. The only difference is that when the square of n is even, it is saved in a separate list, which in this case is `lst`.

3.7 Functions Introduced in Chapter 3

In addition to the functions listed in Tables 3.1 and 3.2, the additional functions introduced in this chapter are listed in Table 3.3.

Table 3.3 Summary of additional commands introduced in Chapter 3

Command	Usage
`Array`	Generates lists and arrays of specified size and with specified elements
`ArrayFlatten`	Flattens arrays according to specified rules
`Do`	Perform a series of operations a specified number of times or using a set of specified values
`EvenQ`	Gives `True` if its argument is an even integer, `False` otherwise
`If`	Evaluates a conditional expression to determine whether it is true or false
`Map`	Applies a function to each element of a list
`Module`	Creates a function in which some or all of the variables can be local
`Negative`	Gives `True` if its argument is a negative number, `False` otherwise
`Nest`	Applies an operation to an expression a specified number of times
`NestList`	Give the intermediate results of `Nest`
`NestWhile`	Performs as `Nest` but continues until a specified condition is satisfied
`OddQ`	Gives `True` if its argument is an odd integer, `False` otherwise
`Positive`	Gives `True` if its argument is a positive number, `False` otherwise
`Reap`	See Example 3.15
`Select`	Identifies all elements of a list that meet a specified criterion
`Sow`	See Example 3.15
`Which`	An efficient way to test numerous conditional expressions
`While`	Repetitively evaluates a conditional expression until it is satisfied
`With`	A means to replace variables with other symbols or values

Exercises

Section 3.2.2

3.1 The elements of an $(N\times N)$ matrix are given by

$$\begin{aligned} h_{nm} &= 0 && n+m-1>N \\ &= n+m-1 && \text{otherwise} \end{aligned}$$

Use `Array` to generate these elements for $N=8$ and display the results in matrix form.

Sections 3.4.1 and 3.5.1

3.2 Consider a function $f(x)$ in the region $a \leq x \leq b$. Sample $f(x)$ at x_n, $n = 1, 2, \ldots, N$, where $x_n - x_{n-1} = (b - a)/(N - 1)$ and $x_1 = a$ and $x_N = b$. Then, for a given $f(x)$, a, b, and N, create a program that determines the first M changes in the sign of $f(x)$ in this region and the values x_k and x_{k-1} that define each region in which this sign change takes place. Recall that a sign change occurs when the product $f(x_k)f(x_{k-1})$ is negative. Recall Example 3.8.

Verify your program by finding the values of x_k and x_{k-1} for the first four sign changes ($M = 4$) of

$$J_1(x) = 0$$

for $1 \leq \Omega \leq 25$ and $N = 41$. The function $J_1(x)$ is the Bessel function of the first kind of order 1.

Section 3.5.1

3.3 Repeat Example 3.2 using nested `Do` commands.

Section 3.5.2

3.4 The complete elliptic integral $K(\alpha)$ can be computed as

$$K(\alpha) = \frac{\pi}{2a_N}$$

where a_N is determined in the following manner. We set $a_0 = 1$, $b_0 = \cos\alpha$ and $c_0 = \sin\alpha$ and employ the recurrence relations

$$\begin{aligned} a_n &= \frac{1}{2}\left(a_{n-1} + b_{n-1}\right) \\ b_n &= \sqrt{a_{n-1}b_{n-1}} \qquad n = 1, 2, \ldots, N \\ c_n &= \frac{1}{2}\left(a_{n-1} - b_{n-1}\right) \end{aligned}$$

such that when $|c_N| < t_o$, where t_o is a specified tolerance, we say that the process has converged to $K(\alpha)$. Setting $t_o = 0.00001$ and $\alpha = \pi/4$, verify that this process converges to the correct value by comparing it to the value obtained from `EllipticK[Sin[α]^2]`.

3.5 Consider the following series

$$\sum_{n=2}^{M} \frac{1}{\left(n^2 - 1\right)^2} = S_M$$

where

$$S_\infty = \frac{\pi^2}{12} - \frac{11}{16}$$

Find the value of M for which

$$\left|\frac{S_M - S_\infty}{S_\infty}\right| < 0.0001$$

Section 3.5.3

3.6 The total interest paid i_T on a loan L over M months at an annual interest rate i_a is obtained from

$$i_T = i_m \sum_{k=1}^{M} b_k$$

where

$$b_k = (1 + i_m)b_{k-1} - p_{mon} \quad k = 1, 2, \ldots, M$$

$$p_{mon} = \frac{i_m L}{1 - (1 + i_m)^{-m}}$$

The quantity p_{mon} is the monthly payment of interest and principal, $i_m = i_a/1200$, and $b_0 = L$. Determine i_T for $i_a = 6.5\%$, $M = 240$ months, and $L = \$150{,}000$.

4

Symbolic Operations

4.1 Introduction

One part of solving engineering problems is to apply one or more solution methods until a form that is amenable to numerical procedures and computer evaluation is obtained. We shall illustrate in this chapter how one can use the symbolic capabilities of Mathematica to meet these goals. One advantage of using symbolic solutions is that once their symbolic solution is known it can be converted to a function and used to perform parametric studies or used in other applications.

We shall illustrate two types of symbolic manipulations. The first type is concerned with the simplification and manipulation of the symbolic expressions to attain a form more suitable for one's end usage. The second type is to perform a mathematical operation on a symbolic expression such as finding its derivative, integrating it, and the like.

Several simplification and manipulation commands that are typically used on symbolic expressions are listed in Table 4.1. Also included in this table are illustrations of the usage of these functions.

The symbolic mathematical operations that will be illustrated are:

Solving equations—**`Solve[]`**
Limits—**`Limit[]`**
Power series—**`Series[]`** and **`Coefficient[]`**
Optimization—**`Maximize[]/Minimize[]`**
Differentiation—**`D[]`**
Integration—**`Integrate[]`**
Solutions to ordinary differential equations—**`DSolve[]`**
Solutions to some partial differential equations—**`DSolve[]`**
Laplace transform—**`LaplaceTransform[]`** and **`InverseLaplaceTransform[]`**

Not all differential equations and integrals have symbolic solutions; in these cases, one must obtain their solutions numerically. Therefore, in the next chapter, we shall re-visit some of these same operations using functions that employ numerical procedures to obtain their solutions.

An Engineer's Guide to Mathematica®, First Edition. Edward B. Magrab.

Companion Website: www.wiley.com/go/magrab

Table 4.1 Manipulation of symbolic expressions

Operation	Mathematica function	Usage	Output
Expand Expressions			
Expand polynomial expressions	**Expand**	`a1=(1+Cos[x])^2`	$(1+\mathrm{Cos}[x])^2$
		`Expand[a1]`	$1+2\ \mathrm{Cos}[x]+\mathrm{Cos}[x]^2$
		`a2=(x+I y)^2`	$(x+i\ y)^2$
		`Expand[a2]`	$x^2+2\ i\ x\ y-y^2$
Expand trigonometric expressions	**TrigExpand**	`TrigExpand[Sin[2 x] Cos[x]^2]`	$\mathrm{Cos}[x]\ \mathrm{Sin}[x]+\mathrm{Cos}[x]^3\ \mathrm{Sin}[x]-\mathrm{Cos}[x]\ \mathrm{Sin}[x]^3$
Expand expressions containing complex variables	**ComplexExpand**	`b2=Re[(x+I y)^2]`	$\mathrm{Re}[(x+i\ y)^2]$
		`ComplexExpand[b2]`	x^2-y^2
		`b3=(1+I)^n`	$(1+i)^n$
		`ComplexExpand[b3]`	$2^{n/2}\ \mathrm{Cos}[n\ \pi/4]+i\ 2^{n/2}\ \mathrm{Sin}[n\ \pi/4]$
Expand certain functions and their arguments	**FunctionExpand**	`b3=Exp[2 ArcTanh[x]]`	$e^{2\ \mathrm{ArcTanh}[x]}$
		`FunctionExpand[b3]`	$-\frac{2x}{-1+x^2}+\frac{-1-x^2}{-1+x^2}$
Expand all products and powers	**PowerExpand**	`e1=Log[a^b]`	$\mathrm{Log}[a^b]$
		`PowerExpand[e1]`	$b\ \mathrm{Log}[a]$
		`e2=x^2 Cos[y]^4 E^(2 x)`	$e^{2x}\ x^2\ \mathrm{Cos}[y]^4$
		`Sqrt[e2]`	$\sqrt{e^{2x}\ x^2\ \mathrm{Cos}[y]^4}$
		`PowerExpand[Sqrt[e2]]`	$e^x\ x\ \mathrm{Cos}[y]^2$
Expand only the numerator	**ExpandNumerator**	`a3=(7-a) (b+3)/(c+d)^2`	$\frac{(7-a)\ (3+b)}{(c+d)^2}$
		`ExpandNumerator[a3]`	$\frac{21-3\ a+7\ b-a\ b}{(c+d)^2}$
Expand only the denominator	**ExpandDenominator**	`a3=(7-a) (b+3)/(c+d)^2`	$\frac{(7-a)\ (3+b)}{(c+d)^2}$
		`ExpandDenominator[a3]`	$\frac{(7-a)\ (3+b)}{c^2+2\ c\ d+d^2}$

Expand ratio of polynomials or trigonometric expressions (partial fractions)	**Apart**	e3=1/((x-b) (x+1))	$\frac{1}{(1+x)(-b+x)}$
		Apart[e3]	$\frac{1}{(-1-b)(1+x)}+\frac{1}{(1+b)(-b+x)}$
		e4=1/(Cos[x]^2-Sin[x]^2)	$\frac{1}{\text{Cos}[x]^2-\text{Sin}[x]^2}$
		Apart[e4,Trig->True]	$1+\frac{\text{Sin}[x]}{\text{Cos}[x]-\text{Sin}[x]}-\frac{\text{Sin}[x]}{\text{Cos}[x]+\text{Sin}[x]}$
Simplify Expressions			
Simplify	Simplify[expr, Assumptions->...] or FullSimplify[expr, Assumptions->...]	Simplify[(2 Sin[2 x]+Sin[4 x])/4]	$2\,\text{Cos}[x]^3\text{Sin}[x]$
		Simplify[1/(a-b)+1/(a+b)]	$\frac{2a}{a^2-b^2}$
		e2=x^2 Cos[y]^4 E^(2 x)	$e^{2x}x^2\,\text{Cos}[y]^4$
		Simplify[Sqrt[e2],Assumptions->{x,y}∈Reals]	$e^x\,\text{Abs}[x]\,\text{Cos}[y]^2$
		Simplify[Sqrt[e2],Assumptions->x>0&&y>0]	$e^x x\,\text{Cos}[y]^2$
		Simplify[Sqrt[e2],Assumptions->x>0]	$e^x x\sqrt{\text{Cos}[y]^4}$
		FullSimplify[ArcCot[Sqrt[2]]-π/2+ArcTan[Sqrt[2]]]	0
		e3=-(1+I)/4 E^(-(1+I) k x)*(-1+I E^(2 I k x)-I E^(2 k x)+E^(2 (1+I) k x))	$-(1+i)\,e^{(-1-i)kx}\times\left(i\,e^{2ikx}-i\,e^{2kx}+e^{(2+2i)kx}-1\right)/4$
		FullSimplify[ComplexExpand[e3]]	Cosh[k x] Sin[k x] - Cos[k x] Sinh[k x]

(continued)

Table 4.1 (*Continued*)

Operation	Mathematica function	Usage	Output
Simplify Expressions			
Combine fractions	`Together`	`b3=1/(a+b)+1/(Cosh[x]+1)`	$\frac{1}{a+b}+\frac{1}{1+Cosh[x]}$
		`Together[b3]`	$\frac{1+a+b+Cosh[x]}{(a+b)(1+Cosh[x])}$
Factor polynomials	`Factor`	`a4=(7 x-b x-7 y+b y)/ (9 c+9 d-6 c z-6 d z+c z^2+d z^2)`	$\frac{7x-bx-7y+by}{9c+9d-6cz-6dz+cz^2+dz^2}$
		`Factor[a4]`	$-\frac{(-7+b)(x-y)}{(c+d)(-3+z)^2}$
Factor trigonometric terms	`TrigFactor`	`a9=1-Cos[2 x]`	`1-Cos[2 x]`
		`TrigFactor[a9]`	`2 Sin[x]`^{2}
		`a5=Cos[x] Sin[x]+Cos[x]^3 Sin[x]+ Cos[x] Sin[x]^3`	`Cos[x] Sin[x]+Cos[x]`^{3} `Sin[x]+ Cos[x] Sin[x]`^{3}
		`TrigFactor[a5]`	`2 Cos[x] Sin[x]`
Factor a specific term	`Collect[expr, term]`	`b4=9 c+9 d-6 c z-6 d z+c z^2+d z^2`	`9 c+9 d-6 c z-6 d z+c z`^{2}`+d z`^{2}
		`Collect[b4,z]`	`9 c+9 d+(-6 c-6 d) z+(c+d) z`^{2}
		`a5=Cos[x] Sin[x]+Cos[x]^3* Sin[x]+Cos[x] Sin[x]^3`	`Cos[x] Sin[x]+Cos[x]`^{3} `Sin[x]+ Cos[x] Sin[x]`^{3}
		`Collect[a5,Sin[x]]`	`(Cos[x]+Cos[x]`^{3}`) Sin[x]+ Cos[x] Sin[x]`^{3}
		`a6=a Exp[-I x]+b I Exp[-I x]+ a Exp[I x]-I b Exp[I x]`	$a\,e^{-ix}+i\,b\,e^{-ix}+a\,e^{ix}-i\,b\,e^{ix}$
		`Collect[a6,Exp[-I x]]`	$(a+i\,b)\,e^{-ix}+(a-i\,b)\,e^{ix}$
		`Collect[a6,{a,b}]`	$b(i\,e^{-ix}-i\,e^{ix})+a(e^{-ix}-e^{ix})$

Simplify Expressions			
Reduce product and powers of trigonometric and hyperbolic functions to trigonometric functions of combined arguments	TrigReduce	e6=Sin[2 x] Cos[x]^2	$Cos[x]^2 Sin[2 x]$
		TrigReduce[e6]	$\frac{1}{4}$ (2 Sin[2 x]+Sin[4 x])
		b6=Sin[x] Cos[a-x]	Cos[a - x] Sin[x]
		TrigReduce[b6]	$\frac{1}{2}$ (Sin[a] - Sin[a - 2 x])
		c6=Sinh[2 x] Cos[x]^2	$Cos[x]^2 Sinh[2 x]$
		TrigReduce[c6]	$-\frac{1}{4}$i (Sin[(2+2i) x]+2 i Sinh[2 x] +i Sinh[(2+2i) x])
Form conversion			
Convert trigonometric functions to exponential functions	TrigToExp	e6=Sin[2 x] Cos[x]^2	$Cos[x]^2 Sin[2 x]$
		TrigToExp[e6]	$\frac{1}{4} i e^{-2ix} - \frac{1}{4} i e^{2ix} + \frac{1}{8} i e^{-4ix} - \frac{1}{8} i e^{4ix}$
		e7=ComplexExpand[(1-I)^n]	$2^{n/2} Cos\left[\frac{\pi n}{4}\right] - i 2^{n/2} Sin\left[\frac{\pi n}{4}\right]$
		TrigToExp[e7]	$2^{n/2} e^{-\frac{1}{4} i \pi n}$
Convert exponential functions to trigonometric functions	ExpToTrig	a8=Exp[a-I b]+Exp[a+I b]	$e^{a-ib}+e^{a+ib}$
		b8=ExpToTrig[a8]	Cosh[a-i b]+Cosh[a+i b]+ Sinh[a-i b]+Sinh[a+i b]
		Simplify[b8]	2 Cos[b] (Cosh[a]+Sinh[a])

(continued)

Table 4.1 (*Continued*)

Operation	Mathematica function	Usage	Output
Form conversion			
		`g=Exp[-2 I x] (Exp[I x]-1)^2` `g1=ExpToTrig[g]`	$e^{-2 i x} (-1+e^{-i x})^2$ `(-1+Cos[x]+I Sin[x])^2* (Cos[2 x]-I Sin[2 x])`
		`Simplify[Expand[g1]]`	`Sin[x/2]^2 (-4 Cos[x]+4 I Sin[x])`
Isolation of numerator and denominator			
Isolation of numerator	**`Numerator`**	`a4=(7 x-b x-7 y+b y)/ (9 c+9 d-6 c z-6 d z+c z^2+d z^2)`	$\frac{7x-bx-7y+by}{9c+9d-6cz-6dz+cz^2+dz^2}$
		`Numerator[a4]`	$7x-bx-7y+by$
Isolation of denominator	**`Denominator`**	`a4=(7 x-b x-7 y+b y)/ (9 c+9 d-6 c z-6 d z+c z^2+d z^2)`	$\frac{7x-bx-7y+by}{9c+9d-6cz-6dz+cz^2+dz^2}$
		`Denominator[a4]`	$9c+9d-6cz-6dz+cz^2+dz^2$

Table 4.2 Common assumptions that can be used in those Mathematica functions that accept them

Assumption*	Mathematica expression for assumption:+ `...,Assumptions->...&&...` or `...,Assumptions->{...,...}`
x is a real number	`Element[x,Reals]` or `x∈Reals`
x is an integer	`Element[x,Integers]` or `x∈Integers`
x is a complex number	`Element[x,Complexes]` or `x∈Complexes`
x is a prime number	`Element[x,Primes]` or `x∈Primes`
$x > n$ or $x \geq n$	`x>n` or `x>=n`
$x < n$ or $x \leq n$	`x<n` or `x<=n`
$m < x < n$ or $m \leq x < n$ or $m < x \leq n$ or $m \leq x \leq n$	`m<x<n` or `m<=x<n` or `m<x<=n` or `m<=x<=n`

*x = variable name or a list of variable names that appear in the expression being operated on by the function; n, m = numerical or symbolic values.
+The symbol ∈ can be found in the *Special Characters* palette by selecting the *Symbols* tab, then the ≠ tab, and then going to the right-most column of the next to last row.

For many uses of the symbolic operations, the most compact final forms of the results are often attained by operating on individual collections of terms and then placing each of them in their most compact form. Thus, the reader is encouraged to interact with the symbolic capabilities of Mathematica one line of code at a time. Several of the examples that follow employ this strategy.

4.2 `Assumption` Options

Many of the functions that we shall discuss in this chapter, including many of those appearing in Table 4.1, allow one to include assumptions concerning one or more of the symbols appearing in the expression being evaluated or manipulated. In those functions that accept **`Assumptions`**, the form can be either

```
...[...,Assumptions->{asum1, asum2, ...}]
```

or

```
...[...,Assumptions->asum1 && asum2 && ...]
```

where **`asumN`** are any of the assumptions appearing in the right-hand column of Table 4.2.

4.3 Solutions of Equations: `Solve[]`

The symbolic solutions to a set of equations, if they exist, are obtained by using either

```
sol=Solve[{eq1L==eq1R,eq2L==eq2R, ...},vars]
```

or

```
sol=Solve[eq1L==eq1R && eq2L==eq2R && ... ,vars]
```

where **eqNR** can be zero and **vars** is a list of the solution variables. The number of solution variables equals the number of equations.

Example 4.1

Solution of Nonlinear Equations

Consider the two equations

$$x^2 + y^2 = 1$$
$$x + y = a$$

Their solution is obtained by using

```
xysol=Solve[x^2+y^2==1 && x+y==a,{x,y}]
```

which yields

$$\left\{\left\{x \to \frac{1}{2}\left(a - \sqrt{2 - a^2}\right), y \to \frac{1}{2}\left(a+\sqrt{2 - a^2}\right)\right\}, \left\{x \to \frac{1}{2}\left(a+\sqrt{2 - a^2}\right), y \to \frac{1}{2}\left(a - \sqrt{2 - a^2}\right)\right\}\right\}$$

It is seen that **xysol** is a 2×2 array.

We shall now detail the procedure that can be used to access the two sets of solutions. Then, a compact syntax will be introduced that we shall use frequently in this chapter and in subsequent chapters. A typical way that these results are accessed is to place them in a list. This is done with

```
z={x,y}/.xysol
```

which displays

$$\left\{\left\{\frac{1}{2}(a - \sqrt{2 - a^2}), \frac{1}{2}(a+\sqrt{2 - a^2})\right\}, \left\{\frac{1}{2}(a+\sqrt{2 - a^2}), \frac{1}{2}(a - \sqrt{2 - a^2})\right\}\right\}$$

Therefore, one solution is (x_1, y_1) = (**z[[1,1]],z[[1,2]]**) and the other solution is (x_2, y_2) = (**z[[2,1]],z[[2,2]]**).

A more compact way to get these results into an accessible list form is by combining the above syntax as follows

```
z={x,y}/.Solve[x^2+y^2==1 && x+y==a,{x,y}]
```

To determine the values of (x_1, y_1) and (x_2, y_2) for, say, $a = 0.5$, we use the substitution rule as follows

```
v=z/.a->0.5
```

which displays

```
{{-0.411438,0.911438},{0.911438,-0.411438}}
```

Thus, for $a = 0.5$, $x_1 =$ `v[[1,1]] = -0.411438`, and $y_1 =$ `v[[1,2]] = 0.911438` and $x_2 =$ `v[[2,1]] = 0.911438`, and $y_2 =$ `v[[2,2]] = -0.411438`.

Example 4.2

Solution of a System of Equations

In Section 2.5, a solution to the following system of equations was obtained by using `LinearSolve`

$$\begin{aligned} 2x_1 + 3x_2 + 5x_3 &= -1 \\ 7x_1 + 11x_2 + 13x_3 &= 7.5 \\ 17x_1 + 19x_2 + 23x_3 &= 31 \end{aligned} \tag{a}$$

We shall change this system of equations slightly by replacing the value on the right-hand side of the second equation to an unknown constant b. Then, Eq. (a) becomes

$$\begin{aligned} 2x_1 + 3x_2 + 5x_3 &= -1 \\ 7x_1 + 11x_2 + 13x_3 &= b \\ 17x_1 + 19x_2 + 23x_3 &= 31 \end{aligned} \tag{b}$$

Using the compact syntax of the previous example, the solution to Eq. (b) is obtained with

```
x123={x1,x2,x3}/.Solve[2 x1+3 x2+5 x3==-1 &&
  7 x1+11 x2+13 x3==b && 17 x1+19 x2+23 x3==31,{x1,x2,x3}]
```

where the appropriate template from the *Basic Math Assistant* palette was used. The execution of this statement yields

$$\left\{\left\{\frac{1}{39}(251 - 13\,b), \frac{b}{2} - \frac{73}{26}, -\frac{b}{6} - \frac{85}{78}\right\}\right\}$$

To place these results in a slightly more convenient form, we use

```
x123=Flatten[x123]
```

which gives

$$\left\{\frac{1}{39}(251-13\,b),\ \frac{b}{2}-\frac{73}{26},\ -\frac{b}{6}-\frac{85}{78}\right\}$$

If an annotated display of the results were desired, the following is used

```
Print["x1=",x123 [[1]]]
Print["x2=",x123 [[2]]]
Print["x3=",x123 [[3]]]
```

which displays

$$x_1 = \frac{1}{39}(251-13\,b)$$

$$x_2 = \frac{b}{2}-\frac{73}{26}$$

$$x_3 = -\frac{b}{6}-\frac{85}{78}$$

To show that these values agree with those obtained in Section 2.5, we set $b = 7.5$. Thus,

```
x123= x123/.b->7.5
```

yields the values obtained in Section 2.5; that is,

```
{3.9359,0.942308,-2.33974}
```

We shall again obtain the solution of Eq. (a), but this time we shall assume that the numbers have units. In particular, it is assumed that the values multiplying x_n have the units of $N{\cdot}m^{-1}$ and those on the right-hand side of the equal sign have the units of N. In this case, we use Section 1.11, the compact syntax introduced in Example 4.1, and the above results to obtain

```
nm="Newtons"/"Meters"; ns="Newtons";
x123=Flatten[{x1,x2,x3}/.Solve[
  {Quantity[2,nm] x1+Quantity[3,nm] x2+
    Quantity[5,nm] x3==Quantity[-1,ns]&&
  Quantity[7,nm] x1+Quantity[11,nm] x2+
    Quantity[13,nm] x3==Quantity[7.5,ns]&&
  Quantity[17,nm] x1+Quantity[19,nm] x2+
    Quantity[23,nm] x3==Quantity[31,ns]},{x1,x2,x3}]]/.b->7.5;
Print["x1=",x123[[1]]]
Print["x2=",x123[[2]]]
Print["x3=",x123[[3]]]
```

which displays

```
x1=3.935 m
x2=0.942308 m
x3=-2.33974 m
```

4.4 Limits: `Limit[]`

The limit of an expression is obtained from

```
Limit[expr,x->xo,Assumptions->...]
```

where **`expr`** is a function of **`x`** and **`xo`** is a symbolic quantity or a numerical quantity, where **`-Infinity < xo < Infinity`**.

For example, to determine

$$\lim_{x\to 0}\frac{1-\cos x}{x}$$

we use

```
Limit[(1-Cos[x])/x,x->0]
```

and obtain 0.

Example 4.3

Heaviside Theta Function

The Heaviside theta function $H(x)$ is defined as follows: $H(x) = 1, x > 0$ and $H(x) = 0, x < 0$. It is undefined for $x = 0$. One mathematical construct that has the properties of the Heaviside theta function is (see also Example 4.20)

$$H(x) = \lim_{t\to 0}\left[\frac{1}{2}+\frac{1}{\pi}\tan^{-1}\left(\frac{x}{t}\right)\right]$$

To verify this, we determine the limits for $\pm x, x \neq 0$. Thus,

```
hxp=Limit[1/2+1/π ArcTan[x/t],t->0]
hxm=Limit[1/2+1/π ArcTan[-x/t],t->0]
```

which gives

$$\frac{x+\sqrt{x^2}}{2x}$$

$$\frac{x-\sqrt{x^2}}{2x}$$

To simplify these expressions, we use **PowerExpand**. Thus,

```
hxp=PowerExpand[hxp]
hxm=PowerExpand[hxm]
```

which gives that **hxp = 1** and **hxm = 0**.

Example 4.4

Limit Using Assumptions

As indicated above, the **Limit** function also accepts **Assumptions**. To see what difference this can make on the final form of the limit, consider the limit

$$\lim_{x \to \pi} \left[\frac{\sin nx}{\sin 2nx} \right]$$

If we make no assumptions about n, then the limit is obtained from

```
Limit[Sin[n x]/Sin[2 n x],x->π]
```

which displays

$$\frac{1}{2}\text{Sec}[n\pi]$$

On the other hand, if n is an integer we can rewrite the limit statement as

```
Limit[Sin[n x]/Sin[2 n x],x->π,Assumptions->n ∈ Integers]
```

which gives

$$\frac{(-1)^n}{2}$$

This is the same result that one obtains from the previous result, except an additional step is necessary.

Example 4.5

Natural Frequencies of Beams

The limit process can also be used to obtain special cases of very general solutions to engineering applications. Consider the following characteristic equation in terms of a frequency parameter Ω for an Euler beam that is supported at each end by translational springs expressed by the nondimensional constants $0 \le K_L \le \infty$ and $0 \le K_R \le \infty$ and torsional springs expressed by the nondimensional constants $0 \le T_L \le \infty$ and $0 \le T_R \le \infty$

$$\begin{aligned} &z_1\left[\cos\Omega_n \sinh\Omega_n + \sin\Omega_n \cosh\Omega_n\right] + z_2\left[\cos\Omega_n \sinh\Omega_n - \sin\Omega_n \cosh\Omega_n\right] \\ &\quad -2z_3 \sin\Omega_n \sinh\Omega_n + z_4(\cos\Omega_n \cosh\Omega_n - 1) \\ &\quad +z_5(\cos\Omega_n \cosh\Omega_n + 1) + 2z_6 \cos\Omega_n \cosh\Omega_n = 0 \end{aligned}$$

where

$$\begin{aligned} z_1 &= \left[b_{1n}b_{2n}(a_{1n}+a_{2n}) + (b_{1n}-b_{2n})\right] \\ z_2 &= \left[a_{1n}a_{2n}(b_{1n}-b_{2n}) - (a_{1n}+a_{2n})\right] \\ z_3 &= (a_{1n}a_{2n} + b_{1n}b_{2n}) \\ z_4 &= (1 - a_{1n}a_{2n}b_{1n}b_{2n}) \\ z_5 &= (a_{2n}b_{2n} - a_{1n}b_{1n}) \\ z_6 &= (a_{1n}b_{2n} - a_{2n}b_{1n}) \end{aligned}$$

and

$$a_{1n} = \frac{K_L}{\Omega_n^3} \quad a_{2n} = \frac{K_R}{\Omega_n^3}$$

$$b_{1n} = \frac{T_L}{\Omega_n} \quad b_{2n} = -\frac{T_R}{\Omega_n}$$

Special cases of this general result can be obtained by selecting the appropriate limiting values for the spring constants. Thus, for the case of a beam clamped at both ends, $K_L \to \infty$, $K_R \to \infty$, $T_L \to \infty$, and $T_R \to \infty$. This is equivalent to letting $a_{1n} \to \infty$, $a_{2n} \to \infty$, $b_{1n} \to \infty$, and $b_{2n} \to \infty$. Thus, we divide the characteristic equation sequentially by these quantities and after each division we take the limit to arrive at the characteristic equation for a beam clamped at both ends.

The program is

```
z1=b1 b2 (a1+a2)+b1-b2;
z2=a1 a2 (b1-b2)-a1-a2;
z3=a1 a2+b1 b2;
z4=1-a1 a2 b1 b2;
z5=a2 b2-a1 b1;
z6=a1 b2-a2 b1;
ce=z1 (Cos[Ω] Sinh[Ω]+Sin[Ω] Cosh[Ω])+2 z6 Cos[Ω] Cosh[Ω]+
  z2 (Cos[Ω] Sinh[Ω]-Sin[Ω] Cosh[Ω])-2 z3 Sin[Ω] Sinh[Ω]+
  z4 (Cos[Ω] Cosh[Ω]-1)+z5 (Cos[Ω] Cosh[Ω]+1);
p=Limit[Limit[Limit[Limit[ce/a1,a1->∞]/a2,a2->∞]/b1,b1->∞]/
  b2,b2->∞]
```

which, upon execution, gives

```
1-Cos[Ω] Cosh[Ω]
```

Example 4.6

Determination of the Poles of an Expression

In using the inversion theorem for Laplace transforms, it is necessary to know whether a pole (zero of a denominator) is a simple pole; that is, the expression for the pole of the denominator appears as $(z - z_o)$, where z_o is the pole. If the denominator appears as $(z - z_o)^m$ it is called an mth-order pole.

Consider the expression

$$f(z) = \frac{\eta z \cosh z - \sinh \eta z}{z^3 \cosh z} e^{zt}$$

To determine if $z = 0$ is a pole, we take the limit of $f(z)$ as $z \to 0$. Thus,

```
Limit[Exp[z t] (η z Cosh[z]-Sinh[η z])/(z^3 Cosh[z]),z->0]
```

gives

$$-\frac{1}{6}\eta(-3+\eta^2)$$

Thus, $z = 0$ is not a pole; if it were a pole, the limit would have been ∞.

4.5 Power Series: Series[], Coefficient[], and CoefficientList[]

One can obtain an nth-order power series approximation of an expression about the point $x = x_o$ with

```
Series[f,{x,xo,n}]
```

where **f** is the expression for which the series is to be obtained.

To illustrate the use of **Series**, we shall get a fourth-order series of the cosine of x about the point a. The statement to do this is

```
p=Series[Cos[x],{x,a,4}]
```

which gives

```
Cos[a] - Sin[a] (x - a) - 1/2 Cos[a] (x - a)^2 + 1/6 Sin[a] (x - a)^3
  + 1/24 Cos[a] (x - a)^4 + O[x - a]^5
```

The expression `O[x-a]`^{5} represents a term on the order of $(x - a)^5$. When a series is given in this form, one cannot use it for further numerical evaluation; that is, for example, **`p/.x->2`** results in an error message. To obtain an expression without this term and in a form that permits further numerical evaluation, the function **`Normal`** is used as follows

```
p=Normal[Series[Cos[x],{x,a,4}]]
```

Execution of this command results in the previous result, but without the last term `O[x-a]`^{5}.

To be able to compare this result to those that follow, we shall expand **`p`** as follows

```
Expand[p]
```

which results in

```
1/24 a^4 Cos[a] - 1/6 a^3 x Cos[a] - 1/6 a^3 Sin[a] + 1/4 a^2 x^2 Cos[a] +
  1/2 a^2 x Sin[a] - 1/2 a^2 Cos[a] + 1/24 x^4 Cos[a] + 1/6 x^3 Sin[a] -
  1/6 a x^3 Cos[a] - 1/2 a x^2 Sin[a] - 1/2 x^2 Cos[a] - x Sin[a] +
  a x Cos[a] + a Sin[a] + Cos[a]
```

The coefficients of the series can be accessed two ways. If **`expans`** is a polynomial in the powers of $(x - a)$ up to the nth power, then the first way that the coefficients can be accessed is by using

```
Coefficient[expans, (x-a),m]]
```

which retrieves the coefficient associated with $(x - a)^m$, $n \geq m \geq 1$, or with

```
CoefficientList[expans,x]
```

which gives a list of all the coefficients up to and including x^{n-1} $(n \geq 1)$.

For example, continuing with the expansion of cosine of x above,

```
Coefficient[p, (x-a),2]]
```

gives

```
- Cos[a]
  ------
    2
```

whereas

```
Coefficient[p,x,2]]
```

gives

$$\frac{1}{4}a^2 \text{Cos}[a] - \frac{1}{2}a\,\text{Sin}[a] - \frac{\text{Cos}[a]}{2}$$

On the other hand,

```
CoefficientList[p,x]
```

displays

$$\left\{\frac{1}{24}a^4 \text{Cos}[a] - \frac{1}{6}a^3 \text{Sin}[a] - \frac{1}{2}a^2 \text{Cos}[a] + a\,\text{Sin}[a] + \text{Cos}[a],\right.$$
$$-\frac{1}{6}a^3 \text{Cos}[a] + \frac{1}{2}a^2 \text{Sin}[a] - \text{Sin}[a] + a\,\text{Cos}[a],$$
$$\left.\frac{1}{4}a^2 \text{Cos}[a] - \frac{1}{2}a\,\text{Sin}[a] - \frac{\text{Cos}[a]}{2}, \frac{\text{Sin}[a]}{6} - \frac{1}{6}a\,\text{Cos}[a], \frac{\text{Cos}[a]}{24}\right\}$$

This list contains the coefficients of the expansion shown in Eq. (4.1), where the first element in the list contains the terms multiplying x^0, the second element the terms multiplying x^1, and so on.

Example 4.7

Perturbation Solution #1

The solution to

$$\frac{d^2y}{dt^2} + \varepsilon\frac{dy}{dt} = -1$$

with the initial conditions of $y(0) = 0$ and $dy(0)/dt = 1$ is

$$y = \frac{1+\varepsilon}{\varepsilon^2}\left(1 - e^{-\varepsilon t}\right) - \frac{t}{\varepsilon}$$

It is assumed that $\varepsilon \ll 1$ and the expansion of this solution up to terms including ε^2 is sought. The statement to do this is

```
Normal[Series[(1+ε) (1-Exp[-ε t])/ε^2-t/ε,{ε,0,2}]]
```

which displays

$$t - \frac{t^2}{2} + \frac{1}{6}\left(-3\,t^2 + t^3\right)\varepsilon + \left(\frac{t^3}{6} - \frac{t^4}{24}\right)\varepsilon^2$$

Example 4.8

Perturbation Solution #2

Perturbation theory can be used to obtain the approximate roots of transcendental equations. For example, consider the expression

$$x^2 - 1 = \mu e^x \tag{a}$$

where μ is a small quantity. To determine the value of x that satisfies this equation using the perturbation technique, one assumes a solution of the form

$$x = \sum_{n=0}^{N} X_n \mu^n \tag{b}$$

where X_n is to be determined.

If we assume that $N = 3$ and substitute Eq. (b) into Eq. (a), we have

$$\left(X_0 + X_1\mu + X_2\mu^2 + X_3\mu^3\right)^2 - \mu e^{X_0+X_1\mu+X_2\mu^2+X_3\mu^3} - 1 = 0$$

To obtain the values of X_n, the above expression is expanded by expressing the exponential function as a series and then by collecting the terms multiplying like powers of μ. These terms are then set equal to zero and a system of equations is obtained from which X_n can then be determined.

We shall introduce the solution process in stages. First, we determine the coefficients of the powers of μ using

```
trans=Normal[(X0+μ X1+μ^2 X2+μ^3 X3)^2-1
  -μ Series[Exp[X0+μ X1+μ^2 X2+μ^3 X3],{μ,0,2}]]
```

which gives

$$-1+X_0^2+\mu\left(2\,X_0\,X_1 - e^{X_0}\right)+\mu^2\left(X_1^2 - e^{X_0}\,X_1+2\,X_0\,X_2\right)+$$
$$\mu^3\left(2\,X_1\,X_2 - \frac{1}{2}\,e^{X_0}\left(X_1^2+2\,X_2\right)+2\,X_0\,X_3\right)$$

Next, we create a list of equations in the format required by **`Solve`**. First, we create a list of the coefficients of μ as follows

```
zz=CoefficientList[trans,μ]
```

This gives

$$\left\{X_0^2 - 1,\, 2\,X_0\,X_1 - e^{X_0},\, X_1^2 - e^{X_0}\,X_1+2\,X_0\,X_2,\, 2\,X_1\,X_2 - \frac{1}{2}e^{X_0}\left(X_1^2+2\,X_2\right)+2\,X_0\,X_3\right\}$$

We now convert the elements of this list into equations equaling zero by appending "==0" to each of these coefficients. Thus,

```
q=Map[#==0 &,zz]
```

which produces

$$\left\{X_0^2 - 1 == 0,\ 2\,X_0\,X_1 - e^{X_0} == 0,\ X_1^2 - e^{X_0}\,X_1 + 2\,X_0\,X_2 == 0,\ 2\,X_1\,X_2 - \frac{1}{2}e^{X_0}\left(X_1^2 + 2\,X_2\right) + 2\,X_0\,X_3 == 0\right\}$$

Lastly, we solve these four equations appearing in **q** for X_0 to X_3 using **Solve** as follows

```
Solve[q,{X0,X1,X2,X3}]
```

which gives

$$\left\{\left\{X_0 \to -1,\ X_1 \to -\frac{1}{2e},\ X_2 \to \frac{3}{8e^2},\ X_3 \to -\frac{7}{16e^3}\right\},\ \left\{X_0 \to 1,\ X_1 \to \frac{e}{2},\ X_2 \to \frac{e^2}{8},\ X_3 \to \frac{e^3}{16}\right\}\right\}$$

Then using the above result and Eq. (b), the two roots are approximated by

$$x^+ \cong 1 + \frac{e}{2}\mu + \frac{e^2}{8}\mu^2 + \frac{e^3}{16}\mu^3$$

$$x^- \cong -1 - \frac{1}{2e}\mu + \frac{3}{8e^2}\mu^2 - \frac{7}{16e^3}\mu^3$$

4.6 Optimization: Maximize[]/Minimize[]

For a function $f(x)$, the value of $x = x_m$ at which this function is a maximum/minimum and the magnitude of $f(x_m)$ is obtained from

```
Maximize[{f,con},x]
Minimize[{f,con},x]
```

where **f** is the function to be maximized or minimized, **con** are any constraints that **f** is subject to, and **x** is the independent variable. The output of these functions is often fairly complex so that several other functions are frequently used in conjunction with them.

To isolate the expression corresponding to the value of $f(x_m)$, **Maximize** and **Minimize**, respectively, are replaced with

```
MaxValue[{f,con},x]
MinValue[{f,con},x]
```

and to isolate the expression corresponding to x_m, **Maximize** and **Minimize**, respectively, are replaced with

```
ArgMax[{f,con},x]
ArgMin[{f,con},x]
```

One additional function that is used to further reduce the results to a specific region is

```
Refine[exp,assum]
```

where **exp** is a symbolic expression, in this case the output from one of the above functions, and **assum** are the assumptions that specify the region of interest.

The use of these functions is illustrated in the following example.

Example 4.9

Peak Amplitude Response of a Single Degree-of-Freedom System

The amplitude response function for a single degree-of-freedom system is given by

$$H(\Omega) = \frac{1}{\sqrt{\left(1-\Omega^2\right)^2 + (2\zeta\Omega)^2}}$$

where Ω is a frequency ratio and $0 < \zeta < 1$. Expressions for the maximum value and the value of Ω at which this maximum occurs are obtained from

```
homega[Ω_,ζ_]:=1/Sqrt[(1-Ω^2)^2+(2 ζ Ω)^2]
hmax=Maximize[{homega[Ω,ζ],0<ζ<1,Ω>0},Ω]//Simplify
```

which displays

$$\left\{\begin{cases} 1 & \frac{1}{\sqrt{2}} < \zeta < 1 \\ \frac{1}{2}\sqrt{\frac{1}{\zeta^2-\zeta^4}} & 0 < \zeta < \frac{1}{\sqrt{2}} \\ -\infty & \text{True} \end{cases},\ \left\{\Omega \to \begin{cases} \text{Root}\left[4+\frac{1}{\zeta^2(\zeta^2-1)}+\left(\frac{4}{\zeta^2-1}-\frac{2}{\zeta^2(\zeta^2-1)}\right)\#1^2+\frac{\#1^4}{\zeta^2(\zeta^2-1)}\&,3\right] & 0<\zeta<\frac{1}{\sqrt{2}} \\ \text{Indeterminate} & \text{True} \end{cases}\right\}\right\}$$

It is seen that the solution of interest is that given for the region $0 < \zeta < 1/\sqrt{2}$. Hence, the elements of this solution can be accessed in the following manner. To isolate the expression for $H(\Omega_m)$, we instead obtain the solution with

```
hmax=MaxValue[{homega[Ω,ζ],0<ζ<1,Ω>0},Ω]//Simplify;
sh=Refine[hmax,0<ζ<1/Sqrt[2]]
```

which displays

$$\frac{1}{2}\sqrt{\frac{1}{\zeta^2 - \zeta^4}}$$

Further simplification is done by using

```
hmax=PowerExpand[Factor[sh],Assumptions->0<ζ<1/Sqrt[2]]
```

which displays

$$\frac{1}{2\,\zeta\,\sqrt{1-\zeta^2}}$$

To obtain Ω_m, we re-solve the expression using

```
omegmax=ArgMax[{homega[Ω,ζ],0<ζ<1,Ω>0},Ω]//Simplify;
so=Refine[omegmax,0<ζ<1/Sqrt[2]][[1]]
Solve[so[Ω]==0,Ω]
```

which results in

$$\left\{\left\{\Omega \to -\sqrt{1-2\,\zeta^2}\right\},\left\{\Omega \to -\sqrt{1-2\,\zeta^2}\right\},\left\{\Omega \to \sqrt{1-2\,\zeta^2}\right\},\left\{\Omega \to \sqrt{1-2\,\zeta^2}\right\}\right\}$$

Thus,

$$\Omega_m = \sqrt{1-2\,\zeta^2}$$

and, therefore,

$$H_m(\Omega_m) = H\left(\sqrt{1-2\zeta^2}\right) = \frac{1}{2\zeta\sqrt{1-\zeta^2}} \qquad 0<\zeta<\frac{1}{\sqrt{2}}$$

4.7 Differentiation: D[]

Differentiation with respect to one or more variables is performed using

```
D[f,{x1,n1},{x2,n2}, ...]
```

where `f` is an expression in one or more of the variables `{x1,x2, ...}` and `n1,n2, ...` are the orders of the derivatives in each variable. Thus, `D` also obtains the partial derivative. There are several equivalent syntaxes for `D` that, along with examples of their usage, are shown in Table 4.3. Also presented in this table are examples of `D` applied to several classes of functions.

Table 4.3 Examples of the symbolic derivative of different types of functions

Type	Mathematical form	Mathematica expression‡	Output
Arbitrary function of one variable	$\frac{d^4 f}{dx^4}$	`D[f[x],{x,4}]` or `D[f[x],x,x,x,x]` or `Derivative[4][f][x]` or `f''''[x]` or $\partial_{x,x,x,x}$ `f[x]`	$f^{(4)}[x]$
Arbitrary function of one variable evaluated at a point b	$\left.\frac{df(x)}{dx}\right\rvert_{x=b}$	`D[f[x],x]/.x->b` or `Derivative[1][f][b]` or `f'[x]/.x->b` or ∂_x `f[x] /.x- > b`	`f'[b]`
Arbitrary function of two variables	$\frac{\partial^3 f(x,y)}{\partial x^2 \partial y}$	`D[f[x,y],{x,2},{y,1}]` or `D[f[x,y],x,x,y]` or `D[f[x,y],y,x,x]` or `Derivative[2,1][f][x,y]` or $\partial_{x,x,y}$ `f[x, y]`	$f^{(2,1)}[x,y]$
Arbitrary function of two variables evaluated at a point b	$\left.\frac{\partial f(x,y)}{\partial x}\right\rvert_{x=b}$	`D[f[x,y],x]/.x->b` or `Derivative[1,0][f][b,y]` or ∂_x `f[x] /.x- > b`	$f^{(1,0)}[b,y]$
Differentiate a function within an integral#	$\frac{d}{dy}\int_{c(y)}^{d(y)} f(y,x)\,dx$	`D[Integrate[f[y,x],` `{x,c[y],d[y]}],y]`	$\int_{c[y]}^{d[y]} f^{(1,0)}(y,x)\,dx - c'[y]\,f[y,c[y]]$ $+d'[y]\,f[y,d[y]]$
	$\left[\frac{d}{dy}\int_{c(y)}^{d(y)} f(y,x)\,dx\right]_{y=r}$	`D[Integrate[f[y,x],` `{x,c[y],d[y]}],y]/.y->r`	$\int_{c[r]}^{d[r]} f^{(1,0)}(r,x)\,dx - c'[r]\,f[r,c[r]]$ $+d'[r]\,f[r,d[y]]$

(*continued*)

Table 4.3 (*Continued*)

Type	Mathematical form	Mathematica expression[‡]	Output
Power	x^n	D[x^n,x] or ∂_xx^n	n x^{-1+n}
	$[f(ax)]^n$	D[f[ax]^n,x] or ∂_xf[ax]^n	a n f[ax]$^{-1+n}$ f′[ax]
	a^{nx}	D[a^(nx),x] or ∂_xa^{nx}	a^{nx} n Log[a]
Trigonometric	$\sin ax$	D[Sin[ax],x] or ∂_xSin[ax]	a Cos[ax]
	$\sin^{-1} ax$	D[ArcSin[ax],x] or ∂_xArcSin[ax]	$\frac{a}{\sqrt{1 - a^2 x^2}}$
Hyperbolic	$\cosh ax$	D[Cosh[ax],x] or ∂_xCosh[ax]	a Sinh[ax]
	$\cosh^{-1} ax$	D[ArcCosh[ax],x] or ∂_xArcCosh[ax]	$\frac{a}{\sqrt{-1+ax}\sqrt{1+ax}}$
Exponential and logarithmic	e^{ax}	D[E^(ax),x] or ∂_xE^{ax}	a e^{ax}
	$\ln ax$	D[Log[ax],x] or ∂_xLog[ax]	$\frac{1}{x}$
	$\log_{10} ax$	D[Log10[ax],x] or ∂_xLog10[ax]	$\frac{1}{x \text{ Log}[10]}$
Special–Bessel	$J_n(ax)$	D[BesselJ[n,ax],x] or ∂_xBesselJ[n,ax]	$\frac{1}{2}$ a (BesselJ[-1+n, a x] -BesselJ[1+n, a x])
Special–Legendre	$P_n(ax)$	D[LegendreP[n,ax],x] or ∂_xLegendreP[n,ax]	$\frac{a}{-1+a^2 x^2}$(a (-1 - n) x LegendreP[n, a x] +(1+n) LegendreP[1+n, a x])

[‡]As indicated, the differentiation also can be accomplished with ∂_x`f`, where `f` is the expression appearing in the column labeled *Mathematical form* and $\partial_\square\square$ is obtained from a template in the *Basic Math Assistant*.
[#]See Section 4.8 for the use of `Integrate[]`.

For an example, consider the expression

$$z = x^3 y^4$$

To determine $\partial^5 z / \partial x^2 \partial y^3$, we proceed as follows

```
D[x^3 y^4,{x,2},{y,3}]  (* or D[x^3 y^4,x,x,y,y,y] *)
```

which gives

```
144 x y
```

When f is a function of one variable, `D` can be replaced by a prime (′) provided that f is explicitly represented as a function of the independent variable. Consider the following expression

$$y(x) = x^3 \cos ax$$

Then the second derivative of y can be determined using primes as follows

```
y[x_]=x^3 Cos[a x];
y''[x]   (* or D[y[x],x,x] *)
```

The execution of these statements displays

$$6\ x\ \mathrm{Cos}[a\ x] - a^2\ x^3\ \mathrm{Cos}[a\ x] - 6\ a\ x^2\ \mathrm{Sin}[a\ x]$$

The command `D` can also be used to perform differentiation using functional representation and a change of variables. Consider the function $f(u)$, where $u = z(x)$. Then df/dx is obtained from

```
u=z[x];
D[f[u],x]   (* or in a compact form as D[f[z[x]],x] *)
```

which displays

```
f'[z[x]] z'[x]
```

where the prime denotes the derivative with respect to its argument; that is,

$$\texttt{f'[z[x]]z'[x]} \rightarrow \frac{df}{dz}\frac{dz}{dx}$$

For example, if $z = bx^g$, then we can use the previous command and the transformation rule to determine df/dx as follows

```
D[f[z[x]],x]/.D[z[x],x]->D[b x^g,x]
```

which gives

```
b g x^(-1+g) f'[z[x]]
```

An expression for d^2f/dx^2 is obtained from

```
D[f[z[x]],x,x]
```

which displays

```
z'[x]^2 f''[z[x]]+f'[z[x]] z''[x]
```

where the prime denotes the derivative with respect to its argument; that is,

$$\text{z'[x]}^2\ \text{f''[z[x]]+f'[z[x]]}\ \ \text{z''[x]} \rightarrow \left(\frac{dz}{dx}\right)^2 \frac{d^2f}{dz^2}+\frac{df}{dz}\frac{d^2z}{dx^2}$$

Again letting $z = bx^g$, we can use the previous command and the transformation rule to determine d^2f/dx^2 as follows

```
Simplify[D[f[z[x]],x,x]/.{D[z[x],x]->D[b x^g,x],
  D[z[x],x,x]->D[b x^g,x,x]}]
```

which yields

```
b g x^(-2+g)((-1+g) f'[z[x]]+b g x^g f''[z[x]])
```

We shall now illustrate the use of `D` with several additional examples.

Example 4.10

Radius of Curvature

The curvature of a plane curve in terms of the parametric quantities $x(t)$ and $y(t)$ is given by

$$\kappa = \frac{x'y'' - y'x''}{\left(x'^2 + y'^2\right)^{3/2}}$$

where the prime indicates the derivative with respect to t. We shall determine κ when

$$x = a(t - \tanh t)$$
$$y = a \operatorname{sech} t$$

where a is a constant.

The program to determine κ is

```
xp=D[a (t-Tanh[t]),t];
yp=D[a Sech[t],t];
kap=Simplify[(xp D[yp,t]-yp D[xp,t])/(xp^2+yp^2)^(3/2)]
```

Execution of these instructions gives

$$\frac{\texttt{Sech[t]}}{\sqrt{\texttt{a}^2\ \texttt{Tanh[t]}^2}}$$

This result can be further simplified by using **PowerExpand**. Thus, we add the statement

```
kap=PowerExpand[kap]
```

which, upon execution, gives the final result

$$\frac{\texttt{Csch[t]}}{\texttt{a}}$$

Example 4.11

Euler–Lagrange Equation

The governing differential for the dynamic response of a thin beam resting on an elastic foundation can be obtained from the Euler–Lagrange equation

$$F_w + \frac{\partial^2 F_{w_{xx}}}{\partial x^2} - \frac{\partial F_{\dot{w}}}{\partial t} = 0$$

where F is an expression relating the kinetic energy, the potential energy, and the external work of the system and

$$F_\alpha = \frac{\partial F}{\partial \alpha}, \quad w_x = \frac{\partial w}{\partial x}, \quad w_{xx} = \frac{\partial^2 w}{\partial x^2}, \quad \dot{w} = \frac{\partial w}{\partial t}$$

The quantity F for this beam system is given by

$$F = \frac{1}{2}c_1\left(\frac{\partial w}{\partial t}\right)^2 - \frac{1}{2}c_2\left(\frac{\partial^2 w}{\partial x^2}\right)^2 - \frac{1}{2}c_3 w^2 + g(x,t)w$$

where c_k are constants.

We shall now use these equations to derive the governing equation as follows

```
fF=1/2 c1 D[w[x,t],t]^2-1/2 c2 D[w[x,t],{x,2}]^2
  -1/2 c3 w[x,t]^2+g[x,t] w[x,t];
eqn=D[fF,w[x,t]]+D[D[fF,D[w[x,t],x,x]],x,x]
  -D[D[fF,D[w[x,t],t]],t]
```

Upon execution, we obtain

```
g[x, t] - c3 w[x, t] - c1 w(0,2) [x, t] - c2 w(4,0) [x, t]
```

which is the governing equation of the beam. The superscripts in this expression are interpreted as follows. The first value in the superscript indicates the order of the derivative of the first variable in **w[x,t]**, x in this case, and the second superscript indicates the order of the derivative of the second variable in **w[x,t]**, t in this case. Thus, in standard form, this equation is, after rearrangement,

$$c_2 \frac{\partial^4 w}{\partial x^4} + c_3 w + c_1 \frac{\partial^2 w}{\partial t^2} = g(x, t)$$

Variational Methods Package: The results in this example could also have been obtained with the *Variational Methods Package*. First, the package is loaded into the Mathematica session with

```
Needs["VariationalMethods`"]
```

Then, the governing equation is obtained with

```
EulerEquations[fF,w[x,t],{x,t}]
```

which yields the previous result.

4.8 Integration: `Integrate[]`

The indefinite integral with respect to one or more variables is given by

```
Integrate[expr,x,y, ... ,Assumptions-> ... ]
```

where **expr** is, in general, a function of **x, y**, ...

The definite integral with respect to one or more variables is obtained using

```
Integrate[expr,{x,xlow,xup},{y,ylow,yup}, ... ,
  Assumptions-> ... ]]
```

where **expr** is, in general, a function of **x, y**, ... , **xlow** is the lower **x** limit and **xup** is the upper **x** limit and similarly for the **y** variable **ylow** is the lower **y** limit and **yup** is the upper **y** limit. Either or both of the limits can be infinity; that is, **Infinity** (or ∞ from the appropriate template.)

When the integral cannot be found in terms of known functions, the system returns the **Integrate** statement as entered.

Before proceeding with several examples of the usage of **Integrate**, a command that has utility in evaluating integrals and in finding Laplace transforms, as well as other applications, is **Piecewise**, which is defined as

```
Piecewise[{{expr1,cond1},{expr2,cond2}, ...}]
```

where **exprN** are expressions and **condN** are most typically the ranges over which each **exprN** is to apply and is most often of the form $a_N \le x \le b_N$. An illustration of its usage is shown in Example 4.17.

Integration can also be performed using the appropriate template from the *Advanced* portion of the *Calculator* section of the *Basic Math Assistant*. Its usage is illustrated in Example 4.12.

Example 4.12

Integration #1

We shall evaluate the definite integral

$$y(t) = \frac{e^{-\zeta t}}{\sqrt{1-\zeta^2}} \int_0^t e^{\zeta\eta} \sin\left(\sqrt{1-\zeta^2}(t-\eta)\right) d\eta$$

where $0 \le \zeta < 1$. Then,

```
Simplify[Exp[-ζ t]/Sqrt[1-ζ^2]*
  Integrate[Exp[ζ η] Sin[Sqrt[1-ζ^2] (t-η)],{η,0,t}]]
```

yields

$$1 - e^{-t\zeta} \text{Cos}\left[t\sqrt{1-\zeta^2}\right] - \frac{e^{-t\zeta}\zeta \,\text{Sin}\left[t\sqrt{1-\zeta^2}\right]}{\sqrt{1-\zeta^2}}$$

Another way to program this function is to use the appropriate templates from the *Basic Math Assistant*. In this case, the above program becomes

$$\text{Simplify}\left[e^{-\eta t}/\sqrt{1-\zeta^2} \int_0^t e^{\eta\zeta} \,\text{Sin}\left[(t-\eta)\sqrt{1-\zeta^2}\right] d\eta\right]$$

which yields the previously obtained result.

Example 4.13

Fourier Coefficients

We shall determine the value of

$$c_n = \sqrt{a_n^2 + b_n^2}$$

where

$$a_n = \frac{1}{\pi} \int_0^{2\pi k/m} \sin m\tau \cos n\tau d\tau$$

$$b_n = \frac{1}{\pi} \int_0^{2\pi k/m} \sin m\tau \sin n\tau d\tau$$

and k, n, and m are integers.

We start by computing c_n for the case where $n \neq m$. Then,

```
an=Simplify[Integrate[Sin[m t] Cos[n t]/π,{t,0,2 π k/m}],
  Assumptions->{k,n}∈Integers];
bn=Simplify[Integrate[Sin[m t] Sin[n t]/π,{t,0,2 π k/m}],
  Assumptions->{k,n}∈Integers];
cn=PowerExpand[Simplify[Sqrt[an^2+bn^2]]]
```

which yields

$$\frac{2\,\mathrm{m}\,\mathrm{Sin}\left[\frac{\mathrm{k\,n}\,\pi}{\mathrm{m}}\right]}{(\mathrm{m}^2 - \mathrm{n}^2)\,\pi}$$

For the case of $n = m$, we have

```
an=Simplify[Integrate[Sin[n t] Cos[n t]/π,{t,0,2 π k/n}],
  Assumptions->{k,n}∈Integers];
bn=Simplify[Integrate[Sin[n t]^2/π,{t,0,2 π k/n}],
  Assumptions->{k,n}∈Integers];
cn=PowerExpand[Simplify[Sqrt[an^2+bn^2]]]
```

and obtain

$$\frac{\mathrm{k}}{\mathrm{n}}$$

Thus,

$$c_n = \frac{2m\sin(kn\pi/m)}{\pi\left(m^2 - n^2\right)} \quad m \neq n$$
$$= \frac{k}{n} \quad m = n$$

Example 4.14

Integration #2

Consider the following integral, which has infinity as one of its limits

$$h = \int_0^\infty e^{-at^2} dt \quad a > 0$$

Its evaluation is obtained with

```
Integrate[Exp[-a t^2],{t,0,∞},Assumptions->a>0]
```

which produces

$$\frac{\sqrt{\pi}}{2\sqrt{a}}$$

Example 4.15

Integration #3

Integrals of special functions can also be determined. Consider the following integral that involves the Bessel function of the first kind of order m, $m \geq 0$,

$$\int_0^a J_m^2(r/a) r dr$$

Its evaluation is obtained with

```
Integrate[r BesselJ[m,r/a]^2,{r,0,a},Assumptions->m>=0]
```

which gives

$$\frac{1}{2}a^2\left(\text{BesselJ[m, 1]}^2 - 2\,\text{m BesselJ[m, 1] BesselJ[1+m, 1]} + \text{BesselJ[1+m, 1]}^2\right)$$

If the traditional form of the result is desired, then

```
TraditionalForm[Integrate[r BesselJ[m,r/a]^2,{r,0,a},
  Assumptions->m>=0]
```

is used and the following is displayed

$$\frac{1}{2}a^2\left(J_{m+1}(1)^2 - 2mJ_m(1)J_{m+1}(1)+J_m(1)^2\right)$$

Example 4.16

Integration #4

As part of performing an inversion of a Laplace transform, the convolution integral leads to the following integral

$$f(\tau) = \frac{1}{\pi}\int_0^\tau \frac{\cos\left(\eta^2/(4\mu)\right)}{\sqrt{\mu}\sqrt{(\tau-\mu)}}d\mu \quad \eta > 0$$

The evaluation of this integral is obtained with

```
Integrate[Cos[η^2/(4 μ)]/(Sqrt[μ] Sqrt[(τ-μ)]),{μ,0,τ},
  Assumptions->η>0]/π
```

The execution of this statement displays

```
ConditionalExpression[1 - FresnelC[η/(√(2π)√τ)] - FresnelS[η/(√(2π)√τ)],
  Re[t] > 0&&Im[τ] ==0]
```

where **FresnelC** and **FresnelS** are the Fresnel integrals given by

$$C(x) = \int_0^x \cos\left(\pi t^2/2\right)dt$$

$$S(x) = \int_0^x \sin\left(\pi t^2/2\right)dt$$

Thus,

$$f(\tau) = 1 - C\left(\eta\big/\sqrt{2\pi t}\right) - S\left(\eta\big/\sqrt{2\pi t}\right) \quad t > 0 \quad \eta > 0$$

Example 4.17

Integration #5

Consider the integral

$$m = \int_0^c f(x)dx$$

where

$$\begin{aligned} f(x) &= a_1 && 0 \le x < b \\ &= 0 && b \le x < d \\ &= -a_2 && d \le x < c \end{aligned}$$

The evaluation of this integral is obtained with the aid of **Piecewise** as follows

```
f=Piecewise[{{a1,0<=x<b},{0,b<=x<d},{-a2,d<=x<=c}}];
Simplify[Integrate[f,{x,0,c},Assumptions->0<b<d<c]]
```

which gives

```
a1 b+a2 (-c+d)
```

Example 4.18

Cauchy Integral Formula

The Cauchy integral formula is

$$\int_C g(z)dz = 2\pi j \sum_{k=1}^{K} \text{Res}[g(z); z_k]$$

where $g(z)$ has z_k, $k = 1, 2, \ldots, K$, singular points all contained within C and Res stands for the residue of $g(z)$ evaluated at each z_k. The singular points are those values of z_k for which $g(z) \to \infty$. The residues are determined by using for each z_k

```
Residue[gz,{z,zk}]
```

where **gz** $= g(z)$ and **zk** $= z_k$.

As an example, let

$$g(z) = \frac{\cos(z)}{z\left(z^2 + 9\right)}$$

The values of z_k are determined from

```
zk=z/.Solve[z (z^2+9)==0,z]
```

which yields

```
{0,-3 I,3 I}
```

Then, the value of the integral is obtained from

```
2 π I Total[Table[Residue[Cos[z]/(z (z^2+9)),{z,zk[[n]]}],
  {n,1,3}]]//Simplify
```

which displays

$$-\frac{2}{9}\, i\, \pi\, (\text{Cosh}[3] - 1)$$

4.9 Solutions of Ordinary Differential Equations: `DSolve[]`

The symbolic solutions to both ordinary and partial differential equations are obtained with

```
DSolve[{eqn1,eqn2, ...},{y1,y2, ...},{x1,x2, ...}]
```

where `{eqn1,eqn2, ...}` is a list of coupled differential equations and their boundary conditions, `{y1,y2, ...}` is a list of the dependent variables appearing in `{eqn1,eqn2, ...}`, and `{x1,x2, ...}` is a list of the independent variables. When the system of equations is composed of ordinary differential equations, there is only one independent variable. When the boundary conditions are omitted, the system obtains a solution in terms of unknown constants `C[n]`, where `n` goes from 1 to the order of the differential equation. In the list `{eqn1,eqn2, ...}`, the representation of each dependent variable has to be given explicitly as `y1[x1,x2, ...], y2[x1,x1, ...]`, ... Not all equations will have a symbolic solution. However, the solutions to many classical equations and some variations thereof are available and these solutions are presented in terms of special functions where appropriate. Several examples are shown in Table 4.4.

We shall illustrate the use of `DSolve` with the following examples. In several of these cases, the results will be plotted using the basic form of `Plot`, which is given in Table 6.1.

Example 4.19

Solution to an Inhomogeneous Ordinary Differential Equation

Consider the following equation

$$\frac{d^2y}{dt^2} + y = ae^{-\gamma t}$$

where $\gamma > 0$ and the initial conditions are $y(0) = y'(0) = 0$.

Table 4.4 Examples of the symbolic solution to well-known ordinary differential equations: $n = 0, 1, 2, \ldots$

Type	Mathematical form: equation and solution	Mathematica expression	Output for `expr`, which appears in `{{y->expr}}`
Bessel	$x^2y'' + xy' + (x^2 - n^2)y = 0$ $y = C_1 J_n(x) + C_2 Y_n(x)$	`DSolve[x^2 y''[x]+x y'[x] +(x^2-n^2) y[x]==0,y[x],x]`	`BesselJ[n,x] C[1]+BesselY[n,x] C[2]`

Bessel (Modified)	$x^2y'' + xy' - (x^2 + n^2)y = 0$
	$y = C_1 I_n(x) + C_2 K_n(x)$, where
	$I_n(x) = (-j)^{-n} J_n(-jx)$
	$K_n(x) = -\frac{\pi}{2} e^{-jn\pi/2}\left(jJ_n(-jx) + Y_n(-jx)\right)$
		`DSolve[x^2 y''[x]+x y'[x] -(x^2+n^2) y[x]==0,y[x],x]`	`BesselJ[n,-i x] C[1]+BesselY[n,-i x] C[2]`
			Numerical equivalent:
			`BesselI[n,x]=(-I)^(n) BesselJ[n,-x I]`
			`BesselK[n,x]=π/2 Exp[I n π/2]* (-I BesselJ[n,-x I]-BesselY[n,-x I])`

Bessel (Kelvin)	$x^2y'' + xy' - (jx^2 + n^2)y = 0$
	$y = C_1\left(\text{ber}_n x + j\text{bei}_n x\right) + C_2\left(\text{ker}_n x + j\text{kei}_n x\right)$, where
	$\text{ber}_n x + j\text{bei}_n x = J_n(xe^{3\pi j/4})$
	$\text{ker}_n x + j\text{kei}_n x = j^{-n} K_n(xe^{\pi j/4}) = -\frac{\pi}{2} e^{-jn\pi}\left(jJ_n(-(-1)^{3/4}x) + J_n(-(-1)^{3/4}x)\right)$
		`DSolve[x^2 y''[x]+x y'[x] -(I x^2+n^2) y[x]==0, y[x],x]`	`BesselJ[n,-(-1)^(3/4) x] C[1] +BesselY[n,-(-1)^(3/4) x] C[2]`
			Numerical equivalent:
			`KelvinBer[n,x]+I KelvinBei[n,x]= BesselJ[n,-(-1)^(3/4) x]`
			`KelvinKer[n,x]+I KelvinKei[n,x]=-π/2.* Exp[-I n π] (I BesselJ[n,-(-1)^(3/4) x] +BesselY[n,-(-1)^(3/4) x])`

Legendre	$(1 - x^2)y'' - 2xy' + n(n+1)y = 0$ $y = C_1 P_n(x) + C_2 Q_n(x)$	`DSolve[(1-x^2) y''[x] -2 x y'[x]+n (n+1) y[x] ==0,y[x],x]`	`C[1] LegendreP[n,x]+C[2] LegendreQ[n,x]`
Mathieu	$y'' + (a - 2q\cos 2x)y = 0$ $y = C_1 C(a,q,x) + C_2 S(a,q,x)$	`DSolve[y''[x] +(a-2 q Cos[2 x]) y[x] ==0,y[x],x]`	`C[1] MathieuC[a,q,x]+ C[2] MathieuS[a,q,x]`

The solution is obtained as follows

```
v=DSolve[{y''[t]+y[t]==a Exp[-γ t],y[0]==0,y'[0]==0},y,t]
```

which yields

$$\left\{\left\{y \to \text{Function}\left[\{t\}, \frac{a\, e^{-\gamma t}}{1+\gamma^2}\left(-e^{\gamma t}\,\text{Cos}[t] + \text{Cos}[t]^2 + e^{\gamma t}\gamma\,\text{Sin}[t] + \text{Sin}[t]^2\right)\right]\right\}\right\}$$

To access the solution and simplify it, we use

```
p=FullSimplify[y[t]/.v[[1]]]
```

which gives

$$\frac{a}{1+\gamma^2}\left(e^{-\gamma t} - \text{Cos}[t] + \gamma\,\text{Sin}[t]\right)$$

This result can also be obtained by using the more compact syntax

```
p=FullSimplify[y[t]/.DSolve[{y''[t]+y[t]==a Exp[-t γ],
  y[0]==0,y'[0]==0},y,t][[1]]]
```

Additionally, if both $y(t)$ and $y'(t)$ were of interest, then the previous statement would be written as

```
p=FullSimplify[{y[t],y'[t]}/.DSolve[{y''[t]+y[t]==
  a Exp[-t γ],y[0]==0,y'[0]==0},y,t][[1]]]
```

which results in the two-element array

$$\left\{\frac{a}{1+\gamma^2}\left(e^{-\gamma t} - \text{Cos}[t] + \gamma\,\text{Sin}[t]\right), \frac{a}{1+\gamma^2}\left(\gamma e^{-\gamma t} + \gamma\,\text{Cos}[t] + \text{Sin}[t]\right)\right\}$$

The first element `p[[1]]` $= y(t)$ contains the result previously obtained and the second element `p[[2]]` $= y'(t)$ contains the expression for the first derivative.

These results can be evaluated for plotting or evaluated at specific values of `a`, γ, and `t`. For example, if we are interested in $y(t)$ evaluated at `a = 1, t = 0.4`, and γ `= 1, 3, 5, 7`, then because `a`, γ, and `t` are global variables

```
TableForm[Table[{γ,p[[1]]/.{a->1.,t->0.4}},{γ,1,7,2}],
  TableHeadings->{None,{"γ","y(0.4)"}}]
```

yields

γ	y(0.4)
1	0.0693387
3	0.0548388
5	0.0446679
7	0.0373135

The symbolic solution of the differential equation for, say, $y(t)$ can be made into a function for subsequent use as follows

```
uu[t_,γ_,a_]=p[[1]];
```

Then,

```
uu[0.4,7,1]
```

displays

```
0.0373135
```

which agrees with one of the values previously obtained. Notice that in the definition of **uu** there is no colon before the equal sign.

The expressions for $y(t)$ and $y'(t)$ can be plotted for $a = 1.0$ and $\gamma = 1.0$ as follows

```
a=1.; γ=1.;
Plot[p,{t,0,20},AxesLabel->{"t","y(t)"}]
```

which produces Figure 4.1. In this figure, the curve with the largest initial magnitude is $y(t)$.

It is mentioned that **p** is only available during this notebook session. Therefore, if one wanted to use this result in another application where a and γ varied over a range of values and it was not computational efficient to solve the differential equation symbolically each time, then one can copy the symbolic result into another procedure and go from there. For example, using the previous result for the solution to the differential equation, one could create the following function in the current notebook or in another notebook

```
yag[t_, a_, γ_] := (a e^(-γ t))/(1+γ^2) (-e^(γ t) Cos[t] + Cos[t]^2 + e^(γ t) γ Sin[t] + Sin[t]^2)
```

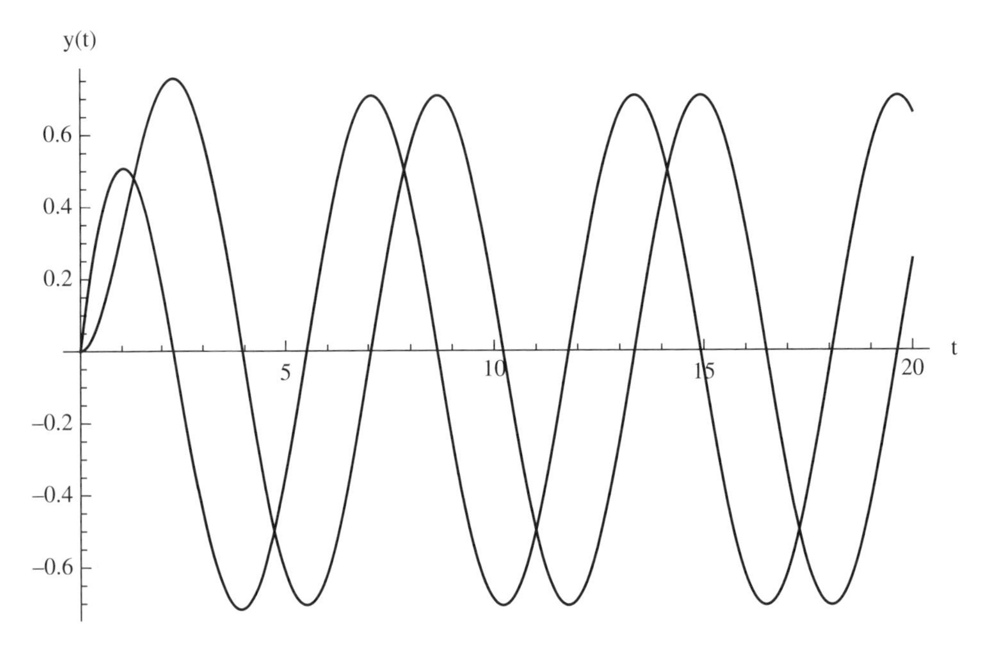

Figure 4.1 Graph of the numerical evaluation of the symbolic results of Example 4.19

Example 4.20

Analysis of Beams

The nondimensional equation of a thin beam subjected to a concentrated load of unit magnitude at x_o is given by

$$\frac{d^4y}{dx^4} = \delta(x - x_o) \quad 0 \le x \le 1$$

where $\delta(x)$ is the delta function. If it is assumed that the beam is hinged at each end, then the boundary conditions are

$$y(0) = y(1) = 0$$
$$y''(0) = y''(1) = 0$$

Noting that the delta function is implemented by using **`DiracDelta`**, the symbolic solution is obtained from

```
hhbeam=Simplify[y[x]/.DSolve[{y''''[x]==DiracDelta[x-xo],
  y[0]==0,y''[0]==0,y''[1]==0,y[1]==0},y,x][[1]],
  Assumptions->1>xo>0]
```

The execution of this statement gives

```
1
─ (x(-1+xo) (x²+(-2+xo)xo)+(x - xo)³ HeavisideTheta[x - xo]
6
```

where **HeavisideTheta[x]** is defined as 0 for **x** < 0 and 1 for **x** > 0. It is undefined at **x** = 0. This function differs from **UnitStep[x]**, which is defined as 0 for **x** < 0 and 1 for **x** ≥ 0. In addition, the derivative of **HeavisideTheta[x]**; that is, **D[HeavisideTheta[x],x]**, gives **DiracDelta**. The derivative of **UnitStep[x]** does not reduce to **DiracDelta**.

We shall obtain a graph of y, y', y'', and y''' when **xo** = 0.7. Since we are interested in y and its derivatives, the above statement is rewritten as follows (recall Example 4.19)

```
hhbeam=Simplify[{y[x],y'[x],y''[x],y'''[x]}/.DSolve[
  {y''''[x]==DiracDelta[x-xo],y[0]==0,y''[0]==0,y''[1]==0,
  y[1]==0},y,x][[1]],Assumptions->1>xo>0]
```

Hence, $y(x)$ = **hhbeam[[1]]**, $y'(x)$ = **hhbeam[[2]]**, and so on. We use **GraphicsGrid** to display the results on a 2×2 grid. See Table 6.17 for the use of **GraphicsGrid**. Then

```
xo=0.7;
plt[n_]:=Plot[hhbeam[[n]],{x,0,1}]
GraphicsGrid[{{plt[1],plt[2]},{plt[3],plt[4]}}]
```

gives the results shown in Figure 4.2. The function **plt** was created to improve readability of the program.

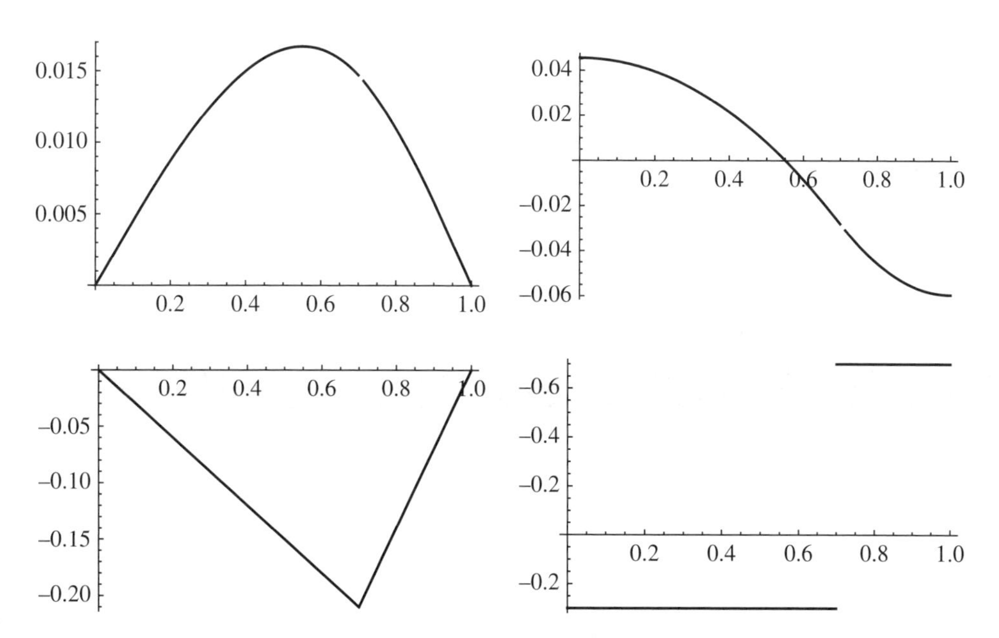

Figure 4.2 Graph of the numerical evaluation of the symbolic results of Example 4.20

Example 4.21

Deformation of a Timoshenko Beam

Consider the following coupled second-order equations in terms of nondimensional quantities

$$\frac{d^2W}{dx^2} - \frac{d\Psi}{dx} = q(x)$$

$$\alpha\frac{d^2\Psi}{dx^2} - \Psi + \frac{dW}{dx} = 0$$

which are valid over the region $0 \le x \le 1$. These equations describe the static deformation of a Timoshenko beam. We shall assume that the boundary conditions are $W(0) = W(1) = 0$ and $\Psi(0) = \Psi(1) = 0$. Then, if we let $q(x) = 1$, the solutions are obtained from

```
sol=Flatten[Simplify[{w[x],s[x]}/.DSolve[{w''[x]-s'[x]==1,
  α s''[x]-s[x]+w'[x]==0,w[0]==0,w[1]==0,s[0]==0,s[1]==0},
  {w[x],s[x]},x]]]
```

where we have employed the compact notation of Examples 4.19 and 4.20. The execution of this instruction gives

$$\left\{\frac{(-1+x)\,x\,(12\,\alpha+x-x^2)}{24\,\alpha},\ -\frac{x\,(1-3\,x+2\,x^2)}{12\,\alpha}\right\}$$

Therefore, the individual solutions are accessed with $W(x)$ = `sol[[1]]` and $\Psi(x)$ = `sol[[2]]`. If the value of $W(0.3)$ when $\alpha = 2$ is desired, then its value is obtained with

```
sol[[1]]/.{α->2.,x->0.3}
```

which yields `-0.105919`.

If a plot of the $W(x)$ and $\Psi(x)$ is desired for, say, $\alpha = 0.1$, then the following is used

```
α=0.1;
Plot[sol,{x,0,1}]
```

Example 4.22

Logistic Equation

Consider the following nonlinear equation, called the logistic equation,

$$\frac{dx}{dt} = x(1-x)$$

which has the initial condition $x(0) = h$. The solution is obtained from

```
ss=x[t]/.DSolve[{x'[t]==x[t] (1-x[t]),x[0]==h},x[t],t][[1]]
```

which yields

$$\frac{e^t h}{1 - h + e^t h}$$

To plot these results for $h = 0.01$, we use

```
h=0.01;
Plot[ss,{t,0,10},AxesLabel->{"t","x"}]
```

which generates Figure 4.3.

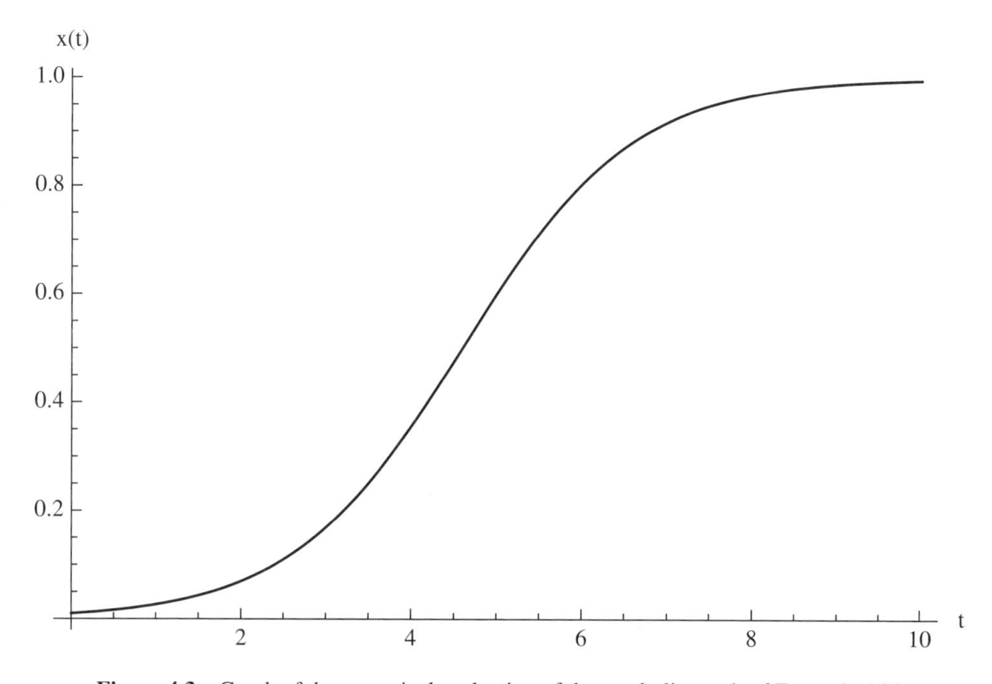

Figure 4.3 Graph of the numerical evaluation of the symbolic result of Example 4.22

Example 4.23

System of First-Order Equations and the Matrix Exponential

Consider the following system of first-order equations with constants coefficients

$$\frac{dx_1}{dt} = 6x_1 + 3x_2 - 2x_3$$

$$\frac{dx_2}{dt} = -4x_1 - x_2 + 2x_3$$

$$\frac{dx_3}{dt} = 13x_1 + 9x_2 - 3x_3$$

subject to the following initial conditions: $x_1(0) = -2$, $x_2(0) = 1$, and $x_3(0) = 4$. The solution is obtained with **DSolve** as follows.

```
Expand[Flatten[{x1[t],x2[t],x3[t]}/.DSolve[
  {x1'[t]==6 x1[t]+3 x2[t]-2 x3[t],
  x2'[t]==-4 x1[t]- x2[t]+2 x3[t] ,
  x3'[t]==13 x1[t]+9 x2[t]-3 x3[t],
  x1[0]==-2,x2[0]==1,x3[0]==4},{x1,x2,x3},t]]]
```

which yields

$$\left\{8\,e^{-t}-11\,e^{t}+e^{2\,t},\,-8\,e^{-t}+11\,e^{t}-2\,e^{2\,t},\,16\,e^{-t}-11\,e^{t}-e^{2\,t}\right\}$$

Another way to solve this type of system of equations is to note that the equations and the initial conditions can be written in matrix form as

$$\{\dot{x}\} = [A]\{x\}, \quad \{x(0)\} = \{a\}$$

where

$$\{\dot{x}\} = \begin{Bmatrix} \dot{x}_1 \\ \dot{x}_2 \\ \dot{x}_3 \end{Bmatrix}, \quad [A] = \begin{bmatrix} 6 & 3 & -2 \\ -4 & -2 & 2 \\ 13 & 9 & -3 \end{bmatrix}, \quad \{x\} = \begin{Bmatrix} x_1 \\ x_2 \\ x_3 \end{Bmatrix}$$

$$\{x(0)\} = \begin{Bmatrix} x_1(0) \\ x_2(0) \\ x_3(0) \end{Bmatrix}, \quad \{a\} = = \begin{Bmatrix} -2 \\ 1 \\ 4 \end{Bmatrix}$$

and the over dot indicates the derivative with respect to t. The solution is given by

$$\{x\} = [p]e^{t[d]}[p]^{-1}\{x(0)\}$$

where the square matrices $[p]$ and $[d]$ are solutions to $[A] = [p][d][p]^{-1}$. The matrices $[p]$ and $[d]$ are obtained from

```
{p,d}=JordanDecomposition[A]
```

Then, the solution to the matrix form of the system of first-order equations with constant coefficients is obtained with the following program.

```
A={{6,3,-2},{-4,-1,2},{13,9,-3}};
{p,d}=JordanDecomposition[A];
ee=p.MatrixExp[d t].Inverse[p];
x=Flatten[ee.Transpose[{{-2,1,4}}]]//Simplify
```

The output of this program is identical to that obtained from **DSolve**, as given above.

Example 4.24

Rearrangement of a Symbolic Solution

The solution to the following fourth-order differential equation

$$\frac{d^4y}{d\eta^4}+4q^4y=0$$

is obtained from

```
z=y[x]/.DSolve[y''''[x]+4 q^4 y[x]==0,y[x],x][[1]]
```

which displays

$$\text{C[1]}\,e^{(-1)^{3/4}\sqrt{2}qx}+\text{C[2]}\,e^{-\sqrt[4]{-1}\sqrt{2}qx}+\text{C[3]}\,e^{-(-1)^{3/4}\sqrt{2}qx}+\text{C[4]}\,e^{\sqrt[4]{-1}\sqrt{2}qx}$$

We shall transform this solution into one that is in terms of trigonometric and hyperbolic functions and real quantities. The interactive procedure is as follows. To obtain **z** in terms of trigonometric and hyperbolic functions, we use

```
z1=ComplexExpand[ExpToTrig[z]]//Simplify
```

which gives

```
C[1] Cos[(1+I) q x]+C[3] Cos[(1+I) q x]+
  (C[2]+C[4]) Cosh[(1+I) q x]+I C[1] Sin[(1+I) q x]-
  I C[3] Sin[(1+I) q x]+(-C[2]+C[4]) Sinh[(1+I) q x]
```

We again employ **ComplexExpand** to have the arguments of the trigonometric and hyperbolic function in terms of real quantities only. Then,

```
z2=ComplexExpand[z1]
```

results in

```
C[1] Cos[q x] Cosh[q x]+C[2] Cos[q x] Cosh[q x]+
  C[3] Cos[q x] Cosh[q x]+C[4] Cos[q x] Cosh[q x]-
  C[1] Cos[q x] Sinh[q x]-C[2] Cos[q x] Sinh[q x]+
  C[3] Cos[q x] Sinh[q x]+C[4] Cos[q x] Sinh[q x]+
  I (C[1] Cosh[q x] Sin[q x]-C[2] Cosh[q x] Sin[q x]-
    C[3] Cosh[q x] Sin[q x]+C[4] Cosh[q x] Sin[q x]-
    C[1] Sin[q x] Sinh[q x]+C[2] Sin[q x] Sinh[q x]-
    C[3] Sin[q x] Sinh[q x]+C[4] Sin[q x] Sinh[q x])
```

To simplify this result further, we collect similar terms using

```
z3=Collect[z2,{Cos[q x] Cosh[q x],Cosh[q x] Sin[q x],
  Cos[q x] Sinh[q x],Sin[q x] Sinh[q x]}]
```

which displays

```
(C[1]+C[2]+C[3]+C[4]) Cos[q x] Cosh[q x]+
  (I C[1]-I C[2]-I C[3]+I C[4]) Cosh[q x] Sin[q x]+
  (-C[1]-C[2]+C[3]+C[4]) Cos[q x] Sinh[q x]+
  (-I C[1]+I C[2]-I C[3]+I C[4]) Sin[q x] Sinh[q x]
```

The final step is to redefine the collection of constants. Hence

```
z4=z3/.{(C[1]+C[2]+C[3]+C[4])->B1,
  (I C[1]-I C[2]-I C[3]+I C[4])->B2,
  (-C[1]-C[2]+C[3]+C[4])->B3,
  (-I C[1]+I C[2]-I C[3]+I C[4])->B4}
```

and the final result is

```
Cos[q x] Cosh[q x] B1+(Cosh[q x] Sin[q x] B2+Cos[q x] Sinh[q x] B3+
  Sin[q x] Sinh[q x] B4
```

4.10 Solutions of Partial Differential Equations: DSolve[]

The symbolic solutions to a limited set of partial differential equations also can be obtained by using **DSolve**. We shall illustrate this capability with several examples. Before proceeding, it is noted that an important difference between the solutions obtained for ordinary differential equations and those obtained for partial differential equations is that the solution constants for ordinary differential equations are constants whereas those for partial differential equations may be functions of one or more of the independent variables.

Example 4.25

Solution to Partial Differential Equation #1

Consider the partial differential equation

$$x\frac{\partial u}{\partial x} - y\frac{\partial u}{\partial y} + y^2 u = y^2$$

Its solution is obtained with the following statement

```
sol=u[x,y]/.DSolve[x D[u[x,y],x]-y D[u[x,y],y]+
  y^2 u[x,y]==y^2,u[x,y],{x,y}][[1]]
```

which gives

$$1+e^{\frac{y^2}{2}}\,C[1][x\,y]$$

It is seen that the `C[1]` is a function of the product of x and y.

To obtain the value of **sol** when $x = 2$ and $y = 3$, we use

```
sol/.{x->2.,y->3.}
```

to obtain

```
1+90.0171 C[1][6.]
```

Example 4.26

Solution to Partial Differential Equation #2

Consider the partial differential equation

$$\frac{\partial v}{\partial x} + \frac{\partial v}{\partial t} + \sigma v = 0$$

with the boundary condition $v(x,0) = \sin x$. Its solution is obtained with

```
sol2=v[x,t]/.Simplify[DSolve[{D[v[x,t],x]+D[v[x,t],t]+
  σ v[x,t]==0,v[x,0]==Sin[x]},v[x,t],{x,t}]][[1]]
```

which yields

```
-e^(-σt) Sin[t - x]
```

Example 4.27

Solution to Partial Differential Equation #3

Consider the partial differential equation

$$\frac{\partial^2 u}{\partial x^2} + 5\frac{\partial^2 u}{\partial x \partial y} + 6\frac{\partial^2 u}{\partial y^2} = 0$$

Its solution is obtained with

```
sol3= w[x,y]/.DSolve[D[w[x,y],x,x]+5 D[w[x,y],x,y]+
  6 D[w[x,y],y,y]==0,w[x,y],{x,y}][[1]]
```

which gives

```
C[1][-2 x+y]+C[2][-3 x+y]
```

4.11 Laplace Transform: LaplaceTransform[] and InverseLaplaceTransform[]

The Laplace transform is obtained from

```
LaplaceTransform[expr,t,s,Assumptions->{ ... }]
```

where **s** is the Laplace transform parameter, **t** is the independent variable that is being transformed, **expr** is an expression that is a function of **t** or is a constant, and **Assumptions** is used to place restrictions on any parameters appearing in **expr**. The initial conditions are dealt with by using **Simplify** and its **Assumptions** option.

Before proceeding, it is again noted that the definition of the unit step function $u(x)$ as given by **UnitStep[x]** is used in Mathematica to represent a piecewise function such that $u(x) = 0$ for $x < 0$ and $u(x) = 1$ for $x \geq 0$. Note that this definition includes 0. This is different from the Heaviside Theta function, which is given by **HeavisideTheta**. As mentioned in Example 4.20, this function is defined as 0 when its argument is < 0 and equal to 1 when its argument is > 0. It is not defined for an argument equal to 0.

We now consider the following example.

Example 4.28

Laplace Transform of a Half Sine Wave

We shall determine the Laplace transform of the half sine wave

$$f(t) = (1 - u(t - \pi/a)) \sin at \quad t \geq 0, \quad a > 0$$

which is obtained from

```
lt=Simplify[LaplaceTransform[(1-UnitStep[t-π/a]) Sin[a t],
  t,s,Assumptions->a>0]]
```

Its execution gives

$$\frac{a(1+e^{-\frac{\pi s}{a}})}{a^2+s^2}$$

The inverse Laplace transform is obtained from

```
ilt=InverseLaplaceTransform[lt,s,t]
```

which displays

$$-(-1+\text{HeavisideTheta}[-\frac{\pi}{a}+t])\,\text{Sin}[at]$$

To verify that **ilt** is correct, we use **Simplify** with two different assumptions as follows

```
Simplify[ilt,Assumptions->t<π/a]
Simplify[ilt,Assumptions->t>π/a]
```

which gives, respectively,

```
Sin[a t]
0
```

Example 4.29

Laplace Transform Solution of an Inhomogeneous Differential Equation #1

We continue with Example 4.28 in the following manner. Consider the ordinary differential equation

$$\frac{d^2y}{dt^2} + y = (1 - u(t - \pi/a)) \sin at \quad t \geq 0, \quad a > 0 \tag{a}$$

with $y(0) = y'(0) = 0$. Because of the way the general solution from **LaplaceTransform** is presented, we shall obtain the Laplace transform of this equation and its inverse one expression at a time. The Laplace transform of Eq. (a) is obtained from

```
ltde=Simplify[LaplaceTransform[y''[t]+y[t]==
  (1-UnitStep[t-π/a]) Sin[a t],t,s,Assumptions->a>0],
  Assumptions->{a>0,y[0]==0,y'[0]==0}]
```

which displays

```
(1+s^2) LaplaceTransform[y[t], t, s, Assumptions → a > 0] ==
```

$$\frac{a\,(1+e^{-\frac{\pi s}{a}})}{a^2+s^2}$$

The unknown quantity is **LaplaceTransform[y[t],t,s,Assumptions->a>0]**. This is a little awkward to use in this form. Therefore, to improve readability, we shall use the transformation rule and redefine it (arbitrarily) as **laps**. Then, we start over and use the modified instruction

```
ltde=Simplify[LaplaceTransform[y''[t]+y[t]==
  (1-UnitStep[t-π/a]) Sin[a t],t,s,Assumptions->a>0]/.
  LaplaceTransform[y[t],t,s,Assumptions->a>0]->laps,
  Assumptions->{a>0,y[0]==0,y'[0]==0}]
```

which displays

$$\text{laps}\ (1+s^2) == \frac{a\ (1+e^{-\frac{\pi s}{a}})}{a^2+s^2}$$

Next, we solve for the Laplace transform of $y(t)$, which is now represented by the quantity `laps`, by using `Solve` as follows

```
sols=laps/.Simplify[Solve[ltde,laps]][[1]]
```

which results in

$$\frac{a\ (1+e^{-\frac{\pi s}{a}})}{(1+s^2)\ (a^2+s^2)}$$

The inverse Laplace transform is determined from

```
ytt=InverseLaplaceTransform[sols,s,t]
```

which yields

$$\frac{1}{-1+a^2}\ (\text{a Sin[t] - Sin[a t]} + \text{HeavisideTheta}\left[-\frac{\pi}{a}+t\right]\left(-\text{a Sin}\left[\frac{\pi}{a} - t\right] + \text{Sin[a t]}\right))$$

It is noted that when $a \rightarrow 1$

```
Limit[ytt,a->1]
```

we obtain

$$\left\{\frac{1}{2}\ (\text{-t Cos[t]+Sin[t] - HeavisideTheta[-}\pi\text{+t]((}\pi\text{ - t) Cos[t]+Sin[t]))}\right\}$$

To plot `ytt` with, say, $a = 2$, we use

```
a=2.;
Plot[ytt,{t,0,30},AxesLabel->{"t","y(t)"}]
```

which results in Figure 4.4.

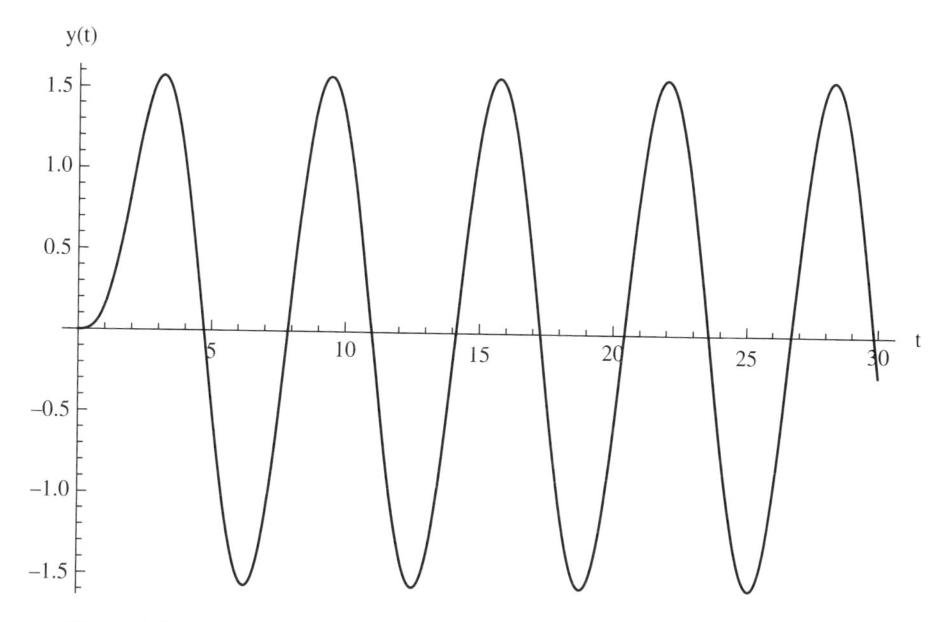

Figure 4.4 Graph of the numerical evaluation of the symbolic results of Example 4.29

Example 4.30

Response of a Two Degrees-of-Freedom System

The nondimensional governing equations of motion of a two degrees-of-freedom system are [1]

$$\frac{d^2x_1}{d\tau^2} + \left(2\zeta_1 + 2\zeta_2 m_r\omega_r\right)\frac{dx_1}{d\tau} + \left(1 + m_r\omega_r^2\right)x_1 - 2\zeta_2 m_r\omega_r\frac{dx_2}{d\tau} - m_r\omega_r^2 x_2 = \frac{f_1(\tau)}{k_1}$$

$$\frac{d^2x_2}{d\tau^2} + 2\zeta_2\omega_r\frac{dx_2}{d\tau} + \omega_r^2 x_2 - 2\zeta_2\omega_r\frac{dx_1}{d\tau} - \omega_r^2 x_1 = \frac{f_2(\tau)}{k_1 m_r}$$

where x_j is the displacement of mass m_j as a function of the nondimensional time τ. The quantities ω_r, m_r, and ζ_j are nondimensional parameters, k_1 is the spring stiffness, and f_j is the force applied to mass m_j.

We shall solve this system of ordinary differential equations using the Laplace transform. It is assumed that the initial conditions are zero; that is, $x_1(0) = dx_1(0)/d\tau = x_2(0) = dx_2(0)/d\tau = 0$, and that $f_1(t) = 0$. In addition, to improve readability of the results we shall replace the system's output variable indicating the Laplace transform of the function by a shorter designation using the transformation rule. In particular, we shall use the following transformations

```
LaplaceTransform[x1[t],t,s]->x1s
LaplaceTransform[x2[t],t,s]->x2s
LaplaceTransform[f1[t],t,s]->f1s
LaplaceTransform[f2[t],t,s]->f2s
```

Then the Laplace transform of the governing equations are obtained with

```
lt1 = Simplify[LaplaceTransform[
  x1''[t]+(2 ζ1+2 ζ2 mr ωr) x1'[t]+(1+mr ωr^2) x1 [t]
  -2 ζ2 mr ωr x2'[t]-mr ωr^2 x2[t]==f1[t]/k1,t,s]/.
  {LaplaceTransform[x1[t],t,s]->sx1,
  LaplaceTransform[x2[t],t,s]->sx2,
  LaplaceTransform[f1[t],t,s]->sf1},
  Assumptions->{x1'[0]==0,x1[0]==0,x2[0]==0,sf1==0}]
lt2=Simplify[LaplaceTransform[
  x2''[t]+2 ζ2 ωr x2'[t]+ωr^2 x2[t]-2 ζ2 ωr x1'[t]
  -ωr^2 x1[t]==f2[t]/(k1 mr),t,s]/.
  {LaplaceTransform[x1[t],t,s]->sx1,
  LaplaceTransform[x2[t],t,s]->sx2,
  LaplaceTransform[f2[t],t,s]->sf2},
  Assumptions->{x2'[0]==0,x2[0]==0,x1[0]==0}]
```

which gives

$$sx_2\, m_r\, \omega_r \left(\omega_r+2\,\zeta_2\, s\right) = sx_1 \left(m_r\, \omega_r \left(\omega_r+2\,\zeta_2\, s\right)+s^2+2\,\zeta_1\, s+1\right)$$

$$sx_2 \left(2\,\zeta_2\, s\, \omega_r+\omega_r^2+s^2\right) = \frac{sf_2}{k_1 m_r}+sx_1\, \omega_r \left(\omega_r+2\,\zeta_2\, s\right)$$

We now solve for sx_1 and sx_2 using

```
sols2=Flatten[{sx1,sx2}/.Simplify[Solve[lt1&&lt2,
  {x1s,x2s}]]]
```

which results in the two-element list

$$\left\{\frac{sf_2\, \omega_r \left(\omega_r+2\,\zeta_2\, s\right)}{s^4+s^3 \left(2\,\zeta_1+2\,\zeta_2\, m_r\, \omega_r+2\,\zeta_2\, \omega_r\right)+s^2 \left(m_r\, \omega_r^2+4\,\zeta_1\,\zeta_2\, \omega_r+\omega_r^2+1\right)+s \left(2\,\zeta_1\, \omega_r^2+2\,\zeta_2\, \omega_r\right)+\omega_r^2},\right.$$
$$\left.\frac{sf_2 \left(m_r\, \omega_r \left(\omega_r+2\,\zeta_2\, s\right)+s^2+2\,\zeta_1\, s+1\right)}{s^4+s^3 \left(2\,\zeta_1+2\,\zeta_2\, m_r\, \omega_r+2\,\zeta_2\, \omega_r\right)+s^2 \left(m_r\, \omega_r^2+4\,\zeta_1\,\zeta_2\, \omega_r+\omega_r^2+1\right)+s \left(2\,\zeta_1\, \omega_r^2+2\,\zeta_2\, \omega_r\right)+\omega_r^2}\right\}$$

The symbolic inverse Laplace transform of these results is not very useful. Therefore, we shall assign numerical values to the constants and assume that the force on m_2 is an impulse force; that is, $f_2(t) = \delta(t)$. The parameters are: $\omega_r = 0.85$, $m_r = 0.45$, $k_1 = 1$, and $\zeta_1 = \zeta_2 = 0.15$. Then the plotted inverse Laplace transforms are obtained with

```
f2s=LaplaceTransform[DiracDelta[t],t,s];
mr=0.45; ωr=0.85; ζ1=0.15; ζ2=0.15; k1=1;
ilt1=InverseLaplaceTransform[sols2[[1]],s,t];
ilt2=InverseLaplaceTransform[sols2[[2]],s,t];
Plot[{Chop[ilt1],Chop[ilt2]},{t,0,50},PlotRange->All,
  AxesLabel->{"τ","x1(τ), x2(τ)"}]
```

which results in Figure 4.5.

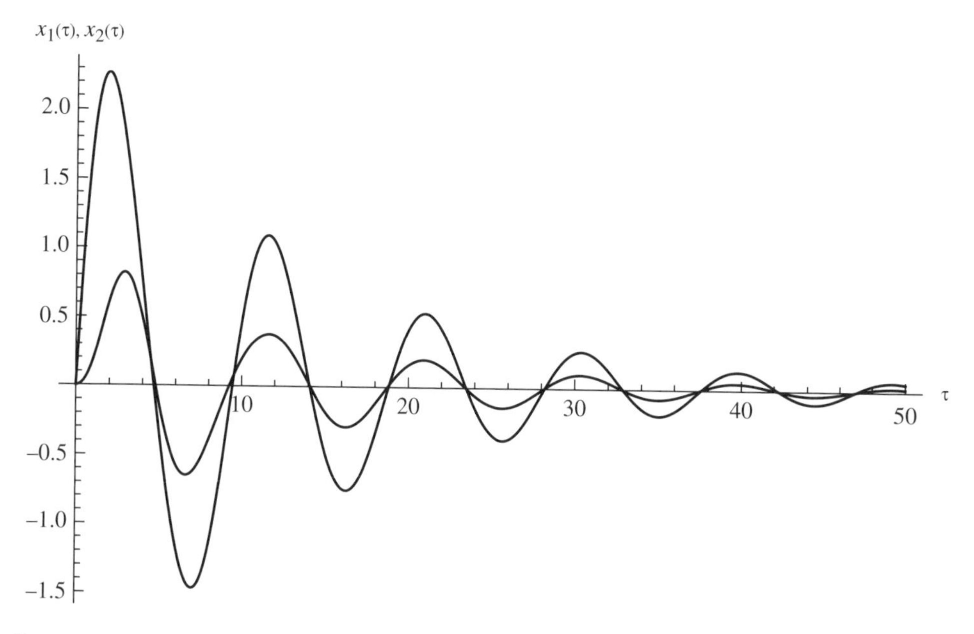

Figure 4.5 Graph of the numerical evaluation of the inverse Laplace transform of Example 4.30. The curve with the larger magnitude is x_2.

Example 4.31

Laplace Transform Solution of an Inhomogeneous Differential Equation #2

Consider the following differential equation

$$\frac{d^2r}{dt^2} + 2p\frac{dr}{dt} + r = g(t)$$

where

$$\begin{aligned} g(t) &= 1 \qquad 0 \le t < t_1 \\ &= -1 \quad\; t_1 \le t < 2t_1 \\ &= 0 \qquad 2t_1 < t \end{aligned}$$

We shall solve this ordinary differential equation using the Laplace transform. It is assumed that the initial conditions are zero; that is, $r(0) = dr(0)/dt = 0$. In addition, to improve readability of the results we shall replace the system's output variable indicating the Laplace transform of the function by a shorter designation using the transformation rule. Then the statements to obtain the Laplace transform of the governing equations are

```
rs=Simplify[LaplaceTransform[r''[t]+2 p r'[t]+r[t]==
  Piecewise[{{1,0<=t<t1},{-1,t1<=t<2 t1}}],t,s]/.
  LaplaceTransform[r[t],t,s]->lts,
  Assumptions->t1>0&&r[0]==0&&r'[0]==0];
rts=lts/.Solve[rs,lts][[1]]
```

The output of this program is

$$\frac{e^{-2\,s\,t1}\left(-1+e^{s\,t1}\right)^2}{s\,\left(1+2\,p\,s+s^2\right)}$$

To determine the solution in the time domain, we take the inverse Laplace transform using

```
rt=Simplify[InverseLaplaceTransform[rts,s,t]];
```

The result is rather lengthy and has been omitted. However, it can be plotted. Therefore, we shall assume that $t_1 = 0.5$ and $p = 0.15$ and the plotted results, which are shown in Figure 4.6, are obtained with

```
t1=0.5; p=0.15;
Plot[Chop[rt],{t,0,25},AxesLabel->{"t","r(t)"}]
```

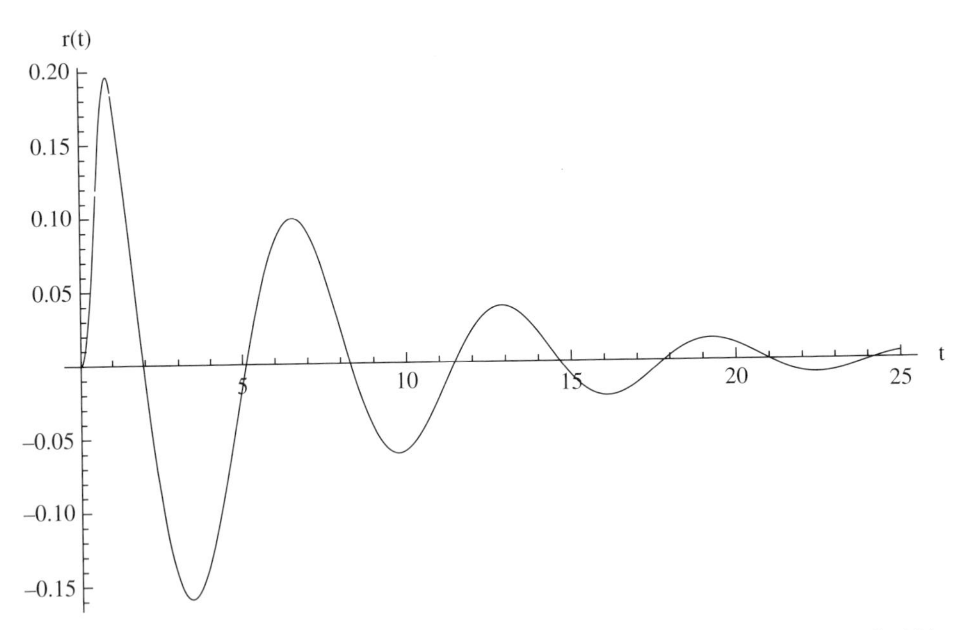

Figure 4.6 Graph of the numerical evaluation of the inverse Laplace transform of Example 4.31

Table 4.5 Summary of additional commands introduced in Chapter 4

Command	Usage
`Coefficient`	Gives the coefficient of a polynomial or power series
`CoefficientList`	Gives a list of the coefficients of a polynomial or a power series
`D`	Determines the ordinary or partial derivative or multiple derivative of a symbolic function
`DiracDelta`	Represents the delta function
`Dsolve`	Attempts to obtain a symbolic solution to an ordinary or partial differential equation
`EulerEquations`	Obtains the Euler–Lagrange equations [Requires loading of the *Variational Methods Package*]
`HeavisideTheta`	Represents the Heaviside theta function of argument x: when $x < 0$, output is 0; when $x > 0$, output is 1; undefined at $x = 0$
`Integrate`	Attempts to obtain a symbolic solution of an indefinite or definite integral
`InverseLaplaceTransform`	Attempts to determine the inverse Laplace transform
`JordanDecomposition`	Obtains $[p]$ and $[d]$ as the solution to $[A] = [p][d][p]^{-1}$ for square matrices $[A]$, $[p]$, and $[d]$ when $[A]$ is given
`LaplaceTransform`	Attempts to determine the Laplace transform of an expression or an ordinary differential equation
`Limit`	Finds the limiting value of an expression
`Normal`	Converts some special forms of results to a normal form
`Piecewise`	Represents a piecewise function over the regions specified in its argument
`Residue`	Finds the residue of an expression at one of its singularities
`Series`	Gives the power series expansion of an expression or of a common function
`Solve`	Attempts to find a symbolic solution to an equation or system of equations
`UnitStep`	Represents the unit step function of argument x: when $x < 0$, output is 0; when $x \geq 0$, output is 1

4.12 Functions Introduced in Chapter 4

Listed in Table 4.5 are the functions introduced in this chapter that do not appear in Table 4.1.

References

[1] B. Balachandran and E. B. Magrab, *Vibrations*, 2nd edn, Cengage Learning, Toronto, Ontario, 2009, p. 472.

[2] S. Timoshenko and S. Woinowsky-Krieger, *Theory of Plates and Shells*, 2nd edn, McGraw-Hill, New York, 1959, pp. 298–300.

Exercises

Table 4.1

4.1 The command `TrigExpand` and `TrigReduce` also work with hyperbolic functions. Consider the expression

$$\tanh(a+b)+\tanh(a-b)$$

Show that this can be reduced to

$$\frac{2\sinh 2a}{\cosh 2a+\cosh 2b}$$

4.2 Find an expression for the real part of w in terms of trigonometric and hyperbolic functions when

$$w = aj\left[1-\frac{\cosh[b(1+j)]}{\cosh[c(1+j)]}\right]e^{jd}$$

Section 4.3

4.3 Solve for x when

$$\eta = \sqrt{2}\left[\tanh^{-1}\left(\sqrt{\frac{2}{3}}\right)-\tanh^{-1}\left(\frac{\sqrt{x+2}}{\sqrt{3}}\right)\right]$$

4.4 Given the function

$$f(x) = e^{\sin(x+c)}$$

The value of $x = x_{max}$ is that value at which the maximum value of $f(x)$ occurs. It can be determined from the solution to $df(x)/dx = 0$ provided that $d^2f(x_{max})/dx^2 < 0$. Determine x_{max}.

4.5 Given the following system of equations

$$\begin{aligned} -6Ri_1+4R\left(i_2-i_1\right) &= -V_1 \\ 2R\left(i_3-i_2\right)-3Ri_2-4R\left(i_2-i_1\right) &= -V_2 \\ -Ri_3-2R\left(i_3-i_2\right) &= V_3 \end{aligned}$$

Determine expressions for i_k.

4.6 Given the following set of equations

$$\begin{aligned} x_2-x_1\left(1+k_c\right) &= 0 \\ \left(x_1-x_2\right)k_1-\left(x_2-x_3\right)h &= 0 \\ \left(x_2-x_3\right)k_2-\left(1+x_3\right)h &= 0 \end{aligned}$$

Solve for x_j and then use the results in the following equation to obtain a cubic equation in h from which h can be determined

$$h - y - x_2 - x_3 = 0$$

Do not solve for h.

4.7 If

$$f_m(\alpha) = m + \frac{\alpha^2}{m} \qquad m = 1, 2, ...$$

then determine an expression for α, $\alpha > 0$, that satisfies $f_m(\alpha) = f_{m+1}(\alpha)$

4.8 The nondimensional buckling load α of bar under a distributed axial load and resting on an elastic foundation can be determined from the determinant of the coefficients a_j of

$$\left[\left(n^4 + \gamma\right)\pi^2 - 2\alpha\left(\frac{n^2\pi^2}{3} - 1\right)\right] a_n + 16\alpha \sum_{\substack{m=1,3,5 \\ m \neq n}}^{K} a_m \frac{mn\left(m^2 + n^2\right)}{\left(m^2 - n^2\right)^2} = 0 \qquad n = 1, 3, 5 ... K$$

where γ is a nondimensional quantity that is proportional to the spring modulus of the elastic foundation.

(a) Find the symbolic solution for α when $K = 1$ and when $K = 3$.

(b) Find the numerical value of α when $K = 1$, 3, and 5 and $\gamma = 0$ and 0.4. Present the results in a table with the appropriate labels.

4.9 Given the following determinant

$$\begin{vmatrix} -\Delta + k_a^2 + \frac{1}{2}(1-\nu)n^2 & \frac{1}{2}(1+\nu)nk_a & \nu k_a \\ \frac{1}{2}(1+\nu)nk_a & -\Delta + \frac{1}{2}(1-\nu)k_a^2 + n^2 & n \\ \nu k_a & n & -\Delta + 1 + \beta^2\left(k_a^2 + n^2\right)^2 \end{vmatrix} = 0$$

where $\Delta = \Omega^2$ is a nondimensional frequency coefficient. Determine the smallest positive value of Ω when $k_a = \pi/4$, $\nu = 0.3$, $\beta = 0.05/\sqrt{12}$, and $n = 0, 1, \ldots, 10$. Display the results in tabular form.

Section 4.4

4.10 Find the following limits

$$\lim_{\varepsilon \to 0} \frac{x^\varepsilon - 1}{\varepsilon} \qquad \lim_{x \to 0} (1 - \sin(2x))^{1/x}$$

$$\lim_{t \to 0} \left(\frac{e^t - 1}{t}\right)^{-a} \qquad \lim_{x \to 1} \frac{\ln x^n}{1 - x^2}$$

4.11 Find the limit of the following expression when n is an integer

$$\lim_{\Omega \to n} \frac{\sin \pi\Omega}{n^2 - \Omega^2}$$

Section 4.5

4.12 Obtain the coefficients of a five-term expansion of the following expression about $\varepsilon = 0$.

$$f = \frac{\pi\varepsilon\sqrt{\pi^2\left(1-\varepsilon^2\right)+16\varepsilon^2}}{\left(1-\varepsilon^2\right)^2}$$

4.13 Obtain the coefficients of the polynomial in γ that result from expanding the determinant

$$\begin{vmatrix} 1+k^2\left(\gamma^4-\omega^2\right) & j\gamma\left(1+k^2\gamma^2\right) \\ j\gamma\left(1+k^2\gamma^2\right) & -\gamma^2-k^2\left(\gamma^2-\omega^2\right) \end{vmatrix} = 0$$

4.14 Find the sum of the squares of the coefficients of the first five terms of the series expansion around $x = 0$

$$\frac{1-4x^2-\sqrt{1-4x^2}}{2\left(2x^3-x^2\right)}$$

Section 4.7

4.15 Using the expression for the curvature of a parametric curve given in Example 4.10, show that when

$$x = a\sin t$$

$$y = \frac{a(2+\cos t)\cos^2 t}{3+\sin^2 t}$$

the curvature is given by

$$\kappa = \frac{6\sqrt{2}(\cos t-2)^3(3\cos t-2)\sec t}{a[73-80\cos t+9\cos(2t)]^{3/2}}$$

Section 4.8

4.16 Evaluate the following integral when $b > 0$ and b is real.

$$f(b) = \int_0^b \frac{2x+5}{x^2+4x+5}dx$$

4.17 Evaluate the following integral, which is proportional to the total radiation by a black body at all frequencies.

$$\int_0^\infty \frac{\eta^3}{e^\eta - 1} d\eta$$

Section 4.9

4.18 The nondimensional transverse displacement y of a uniformly loaded annular circular plate with Poisson's ratio ν is given by

$$\frac{d^4y}{d\xi^4} + \frac{2}{\xi}\frac{d^3y}{d\xi^3} - \frac{1}{\xi^2}\frac{d^2y}{d\xi^2} + \frac{1}{\xi^3}\frac{dy}{d\xi} = 1$$

Determine an expression for y when the boundary conditions at the inner boundary $\xi = \alpha$, $0 < \alpha < 1$, are

$$y(\alpha) = \frac{dy(\alpha)}{d\xi} = 0$$

and those at the outer boundary $\xi = 1$ are

$$\left(\frac{d^2y}{d\xi^2} + \frac{\nu}{\xi}\frac{dy}{d\xi}\right)_{\xi=1} = 0$$

$$\left(\frac{d^3y}{d\xi^3} + \frac{1}{\xi}\frac{d^2y}{d\xi^2} - \frac{1}{\xi^2}\frac{dy}{d\xi}\right)_{\xi=1} = 0$$

Plot the results for $y(\xi)$, $y'(\xi)$, $y''(\xi)$, and $y'''(\xi)$ when $\alpha = 0.2$ and $\nu = 0.3$ using the grid plot commands of Example 4.20.

4.19 Determine the general solution to the following equation

$$\left(\frac{d^2}{dr^2} + \frac{d}{rdr} - \frac{m^2}{r^2}\right)\left(\frac{d^2w}{dr^2} + \frac{dw}{rdr} - \frac{m^2}{r^2}w\right) = q_o$$

where $w = w(r)$, q_o is a constant, and $m = 0, 1, \ldots, 5$.

4.20 Consider a solid circular plate whose thickness varies as

$$y = e^{-\beta\eta^2/6}$$

where β is a constant. The governing equation for this type of plate when it is subjected to a nondimensional uniform load of magnitude p over the entire plate is given by [2, p. 298–300]

$$\frac{d^2\varphi}{d\eta^2} + \left(\frac{1}{\eta} - \beta\eta\right)\frac{d\varphi}{d\eta} - \left(\frac{1}{\eta^2} + \beta\nu\right)\varphi = -p\eta e^{\beta\eta^2/2}$$

and w is the transverse displacement.

(a) Verify that a particular solution to this equation is

$$\varphi_p = -\frac{p}{(3-\nu)\beta}\eta e^{\beta\eta^2/2}$$

(b) Determine the solution to the homogeneous equation; that is, when $p = 0$.

(c) If the solution is to remain finite at $\eta = 0$, then take the solution obtained in (b) and determine which part of the solution remains finite at $\eta = 0$.

4.21 In terms of nondimensional quantities, the governing equation of an annular circular plate whose thickness varies linearly and is subjected to a uniform load Q_o distributed over the entire plate and a line load along the inner boundary of magnitude P_o is given by [2, p. 303–305]

$$\eta^2\frac{d^2\varphi}{d\eta^2} + 4\eta\frac{d\varphi}{d\eta} + (3\nu - 1)\,\varphi = -Q_o\left(1 - \frac{\alpha^2}{\eta^2}\right) - \frac{P_o}{\eta^2}$$

where ν is the Poisson's ratio. Determine the solution to this equation when $\nu \neq 1/3$ and when $\nu = 1/3$.

4.22 Given the following differential equation

$$\frac{d^2y}{dt^2} + \varepsilon\frac{dy}{dt} + k^2y = \cos t$$

where $\varepsilon \ll 1$. Assume that an approximate solution to this equation is

$$y(t) = x_0(t) + \varepsilon x_1(t) + \varepsilon^2 x_2(t)$$

(a) Substitute this solution into the differential equation and generate a list of three equations of the form

```
{-Cos(t)+q² x0(t)+x0''(t) ==0, q² x1(t)+x0'(t)+x1''(t) ==0,
  q² x2(t)+x1'(t)+x2''(t) ==0}
```

(b) Obtain a solution to these three, coupled equations.

Section 4.11

4.23 Find the inverse Laplace transform of

$$X(s) = \frac{0.1s^3 + 0.0282s^2 - 0.0427s + 0.0076}{s^4 + 0.282s^3 + 4.573s^2 + 0.4792s + 2.889}$$

and plot the result.

5

Numerical Evaluations of Equations

5.1 Introduction

In this chapter, we shall introduce several Mathematica functions that have a wide range of uses in obtaining numerical solutions to engineering applications. These functions are:

Integration—**NIntegrate[]**
Solutions of ordinary and partial differential equations—**NDSolveValue[]** and
  ParametricNDSolveValue[]
Solutions of equations and polynomials—**NSolve[]**
Numerical roots of transcendental equations—**FindRoot[]**
Optimization: Find maximum or minimum of a function—**FindMaximum[]** and
  FindMinimum[]
Fit data to a specified function—**FindFit[]**
Piece-wise fit to data—**Interpolation[]**
Discrete Fourier transform and correlation—**Fourier[]**, **InverseFourier**, and
  ListCorrelate[]

Since the outputs of many of the functions to be introduced in this chapter are plotted, we shall be using the basic forms of **Plot** and **ParametricPlot** as given in Table 6.1. In addition, two options will be used: **PlotRange->All**, which ensures that all points are plotted; and **PlotStyle->{styl1,styl2,...}**, where each **stylN** corresponds to the line style of each curve being plotted. The line styles that we shall use are solid and dashed.

5.2 Numerical Integration: NIntegrate[]

Numerical integration of a function $f(x)$ over the region $x_1 \le x \le x_2$ is performed with

```
NIntegrate[f,{x,x1,x2},opts]
```

An Engineer's Guide to Mathematica®, First Edition. Edward B. Magrab.

Companion Website: www.wiley.com/go/magrab

and the numerical integration of a function $g(x,y)$ over the region $x_1 \le x \le x_2$ and $y_1 \le y \le y_2$ is performed with

```
NIntegrate[g,{x,x1,x2},{y,y1,y2},opts]
```

where the first variable x corresponds to the inner integral; that is, integration with respect to this variable will be performed first. The quantity **opts** represents options that can be used, one of which is discussed below.

Important option: In implementing **NIntegrate**, Mathematica attempts to evaluate the integral of the argument symbolically, **g** or **f** in this case. Occasionally it is necessary to suppress this symbolic attempt, which is done with the following option

```
NIntegrate[f,{x,x1,x2},Method->{Automatic,
 "SymbolicProcessing"->False}]
```

For examples of the use of this option, see Sections 8.4.2 and 11.6.2.

Example 5.1

Numerical Integration #1

We shall determine numerically the value of the integral

$$f(t) = \frac{e^{-\zeta t}}{\sqrt{1-\zeta^2}} \int_0^t e^{\zeta\tau} \sin\left((t-\tau)\sqrt{1-\zeta^2}\right) d\tau$$

for $\zeta = 0.15$ and $t = 20$. Then,

```
ζ=0.15; t = 20.;
f=Exp[-ζ t]/Sqrt[1-ζ^2] NIntegrate[
 Exp[ζ τ] Sin[(t-τ) Sqrt[1-ζ^2]],{ τ,0,t}]
```

yields **0.963975**.

If we would like to graph $f(t)$ over a range $0 \le t \le t_e$ for $t_e = 20$ and for several values of ζ, say $\zeta = 0.05$, 0.25, and 0.7, then we make **f** into a function and rewrite the above program as

```
fp[ζ_,t_]:=Exp[-ζ t]/Sqrt[1-ζ^2] NIntegrate[
 Exp[ζ τ] Sin[(t-τ) Sqrt[1-ζ^2]],{τ,0,t}]
Plot[{fp[0.05,t],f[0.25,t],f[0.7,t]},{t,0,20},
 AxesLabel->{"t","f(t)"}]
```

which yields Figure 5.1.

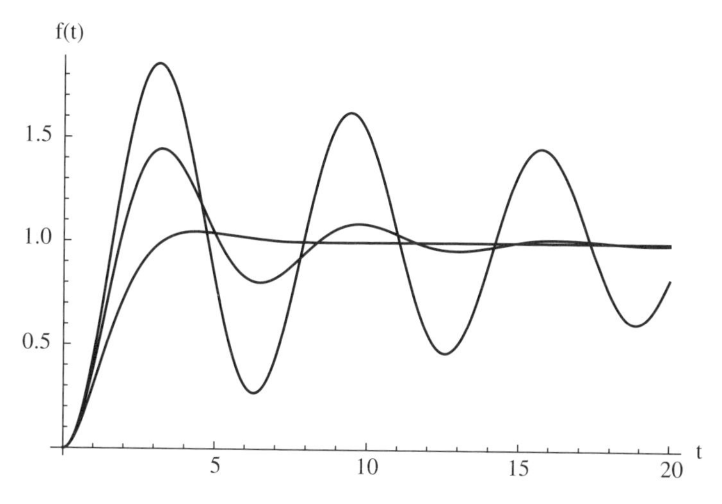

Figure 5.1 Results for $f(t)$ of Example 5.1 for several values of ζ

Example 5.2

Numerical Integration #2

We shall determine the value of the integral

$$p = \frac{1}{2\pi\sqrt{1-r^2}} \int_{-2}^{2} \int_{-3}^{3} e^{-(x^2-2rxy+y^2)} dxdy$$

for $r = 0.5$. Then,

```
r=0.5;
p=NIntegrate[Exp[-(x^2-2 r x y+y^2)]/(2 π Sqrt[1-r^2]),
  {x,-3,3},{y,-2,2}]
```

yields the value `0.657016`.

Example 5.3

Numerical Integration #3

We shall determine the value of the integral

$$C(\omega_0) = 2\sqrt{\pi}e^{-\omega_0^2} \int_{-\infty}^{\infty} \frac{1}{|\omega|} e^{-\omega^2} \left(e^{\omega\omega_0} - 1\right)^2 d\omega$$

and plot the result as a function of ω_0 for $-10 \leq \omega_0 \leq 10$. Note that the integrand is finite at $\omega = 0$. Then,

```
cwo[ω0_]:=2 Sqrt[π] Exp[-ω0^2] NIntegrate[
  Exp[-ω^2] (Exp[ω ω0]-1)^2/Abs[ω],{ω,-∞,∞}]
Plot[cwo[ω0],{ω0,-10,10},AxesLabel->{"ω0","C(ω0)"}]
```

The results are shown in Figure 5.2.

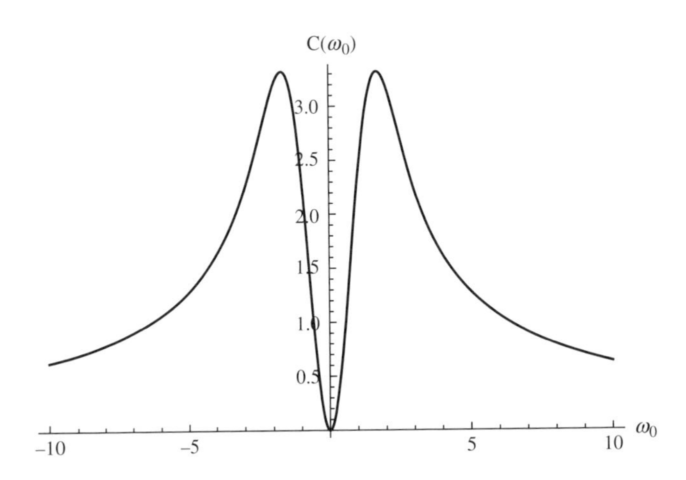

Figure 5.2 Results for the integral of Example 5.3 as a function of ω_0

5.3 Numerical Solutions of Differential Equations: NDSolveValue[] and ParametricNDSolveValue[]

The numerical solution of a single ordinary differential equation, denoted **eqn**, that is a function of the dependent variable $u(x)$ over the range $x_s \leq x \leq x_e$ is obtained with

```
sol=NDSolveValue[{eqn,bc,WhenEvent[ev,act]},u,{x,xs,xe},
  opts]
```

where **bc** are the boundary conditions or initial conditions. In the option **WhenEvent**, **ev** is the event initiated by the dependent variable **u**, which, when **True**, triggers an action **act**. The action can involve altering **u** in some manner or recording **u** and/or **x** for subsequent use.[1] When **WhenEvent** is recording values, it is typically used with **Reap** and **Sow**. **WhenEvent** can appear more than once and **ev** can contain a relational statement such as > and **If**. The number of boundary or initial conditions equals the order of the differential equation. The quantity **opts** are options that one uses to change the default settings in special cases, two of which are discussed subsequently in the paragraph labeled **Important Options**.

[1] See **WhenEvent** in the *Documentation Center*.

The output `sol` is

```
InterpolatingFunction[{{xs,xe}},<>]
```

which is an approximate function where its values are found by interpolation. One can take the derivative of this function by using either `sol'[x]`, `Derivative[1][sol][x]`, or `D[sol[x],x]` as indicated in Table 4.3.

The numerical solution of a set of M coupled ordinary differential equation, denoted `eqns`, that are a function of the dependent variables $u_k(x)$, $k = 1, 2, \dots, M$, over the range $x_s \le x \le x_e$ is obtained with

```
{so1,so2, ... ,soM}=NDSolveValue[{eqns,bck,WhenEvent[ev,act]},
  {u1,u2, ... ,uM},{x,xs,xe}]
```

where `bck` are the boundary conditions or the initial conditions for each $u_k(x)$. The number of boundary/initial conditions equals the sum of the order of each dependent variable.

The output `{so1,so2, ... ,soM}` is the list

```
{InterpolatingFunction[{{xs,xe}},<>],
  InterpolatingFunction[{{xs,xe}},<>], ...}
```

such that `so1` is the first interpolation function in the list, `so2` is the second, and so on. For each interpolation function, one can take the derivative in the manner described above.

The numerical solution of a partial differential equation, denoted `eqnp`, that is a function of the dependent variable $u(x,t)$ over the range $x_s \le x \le x_e$ and $t_s \le t \le t_e$ is obtained with

```
solp=NDSolveValue[{eqnp,bcp,WhenEvent[ev,act]},u,{x,xs,xe},
  {t,ts,te}}]
```

where `bcp` are the boundary conditions and/or the initial conditions. The number of boundary conditions and initial conditions equals the sum of the order of the differential equation in each of the dependent variables.

The output `solp` is

```
InterpolatingFunction[{{xs,xe},{ts,te}},<>]
```

One can take the partial derivative of this function as indicated in Table 4.3. For example, the first derivative with respect to t is obtained with `D[solp[x,t],t]`.

The numerical solution of a set of M coupled partial differential equations, denoted `eqnps`, that is a function of the dependent variables $u_k(x,t)$, $k = 1, \dots, M$, over the range $x_s \le x \le x_e$ and $t_s \le t \le t_e$ is obtained with

```
{sop1,sop2, ... ,sopM}=NDSolveValue[
  {eqnps,bcps,WhenEvent[ev,act]},{u1,u2, ... ,uM},{x,xs,xe},
  {t,ts,te}]
```

where `bcps` are the boundary conditions and initial conditions for each $u_k(x,t)$.

The output {sop1,sop2, ... ,sopM} is the list

```
{InterpolatingFunction[{{xs,xe}},{ts,te}},<>],
 InterpolatingFunction[{{xs,xe}},{ts,te}},<>], ...}
```

For each interpolation function, one can take the partial derivative in the manner described previously.

When a single ordinary differential equation and/or its boundary conditions contain unspecified parameters **p1**, **p2**, … , the numerical solution in terms of these parameters is obtained with

```
solpa=ParametricNDSolveValue[{eqn,bc,WhenEvent[ev,act]},u,
 {x,xs,xe},{p1,p2, ...}]
```

The definitions of the terms in these commands are the same as those corresponding to **NDSolveValue**.

The output **solpa** is

```
ParametricFunction[<>]
```

One can take the derivative of this function as indicated in Table 4.3. For example, the first derivative of **solpa** is obtained by using **D[solpa[pr,p2, ...][x],x]**.

We shall now illustrate **NDSolveValue** and **ParametricNDSolveValue** with several examples, a few of which will illustrate the use of **WhenEvent**.

Important Options: (1) **MaxSteps->Infinity** is often used when **xe** is a very large number, as shown in Example 5.11. (2) **InterpolationOrder->All** is used to create an interpolation function that produces curves that are of the same degree polynomial as that used by **NDSolveValue** to fit values between data points. This can be necessary when higher-order derivatives are taken. Its usage is shown in Examples 5.4 to 5.6.

Example 5.4

Beam with a Concentrated Load

We shall repeat Example 4.20, in which a thin beam was subjected to a concentrated load at an interior location x_o, $0 < x_o < 1$. The concentrated load for the numerical implementation will be approximated by

$$f(x) = \frac{e^{-(x-x_o)^2/a^2}}{a\sqrt{\pi}}$$

for $a = 0.002$. It is mentioned that for this value of a, the integral of $f(x)$ over the region is 1. This function is plotted in Figure 5.3a for $x_o = 0.7$. From this figure, it is seen that the "width" of the approximation is on the order of 0.01.

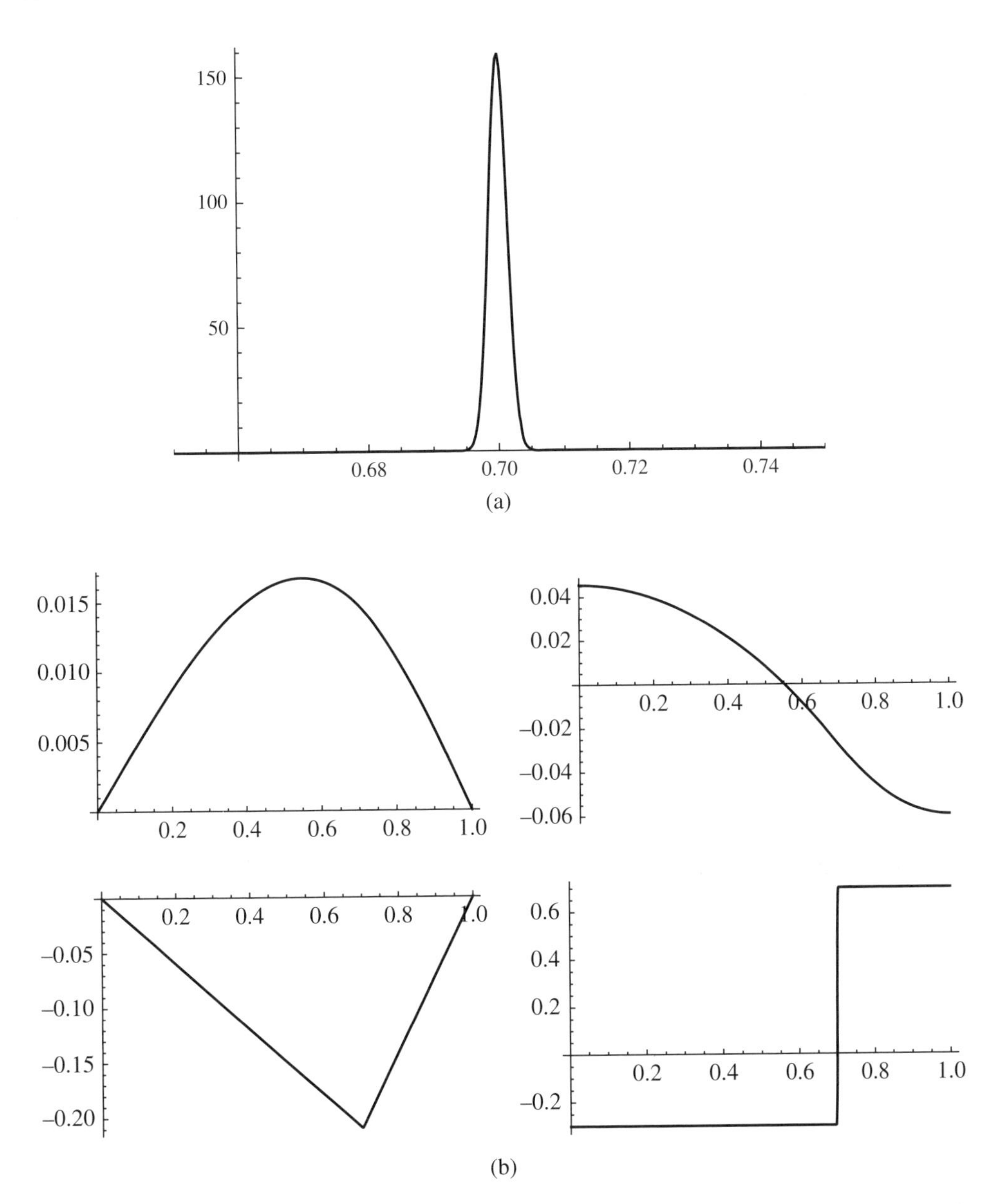

Figure 5.3 (a) Graph of an approximation to a concentrated load centered at 0.7 with $a = 0.002$; (b) graph of y (upper left), $dy/d\eta$ (upper right), $d^2y/d\eta^2$ (lower left), and $d^3y/d\eta^3$ (lower right)

The nondimensional equation of a thin beam subjected to this approximation of a concentrated load is

$$\frac{d^4y}{dx^4} = \frac{e^{-(x-x_o)^2/a^2}}{a\sqrt{\pi}} \quad 0 \leq x \leq 1 \tag{a}$$

If it is assumed that the beam is hinged at each end, then the boundary conditions are

$$\begin{aligned} y(0) &= y''(0) = 0 \\ y(1) &= y''(1) = 0 \end{aligned} \tag{b}$$

We shall obtain a solution to Eq. (a) subject to the boundary conditions given by Eq. (b) for $x_o = 0.7$ and plot the results for y, $dy/d\eta$, $d^2y/d\eta^2$, and $d^3y/d\eta^3$. The four solutions are obtained from

```
a=0.002; xo=0.7;
hhbeam=NDSolveValue[{y''''[x]==Exp[-(x-xo)^2/a^2]/
  (a Sqrt[π]),y[0]==0,y''[0]==0,y[1]==0,y''[1]==0},
  y,{x,0,1},InterpolationOrder->All,MaxStepSize->1/100];
plp[m_]:=Plot[D[hhbeam[v],{v,m}]/.v->x,{x,0,1},
  PlotRange->All];
GraphicsGrid[{{plp[0],plp[1]},{plp[2],plp[3]}}]
```

The results are shown in Figure 5.3b and they are virtually identical to those shown in Figure 4.2. It is mentioned that two options had to be used to obtain the results shown. The first was to decrease the maximum step size to match approximately the "width" of the concentrated load. This was done setting the option **MaxStepSize** to 0.01. The second option that had to be changed was the interpolation order used to produce the curves so that the output interpolation function was the same as that used by **NDSolveValue**. This was accomplished by selecting the option for **InterpolationOrder** as **All**. **GraphicsGrid** is illustrated in Table 6.17.

Example 5.5

Beam with an Overhang

Consider a uniformly loaded beam with an overhang as shown in Figure 5.4. The left end of the beam is hinged, the right end is free, and at the interior support the displacement is zero but the beam is free to rotate. The governing equation for each portion of the beam in terms of nondimensional quantities is

$$\frac{d^4w}{d\eta^4} = 1 \quad 0 \le \eta \le 1$$

$$\frac{d^4g}{d\xi^4} = 1 \quad 0 \le \xi \le b$$

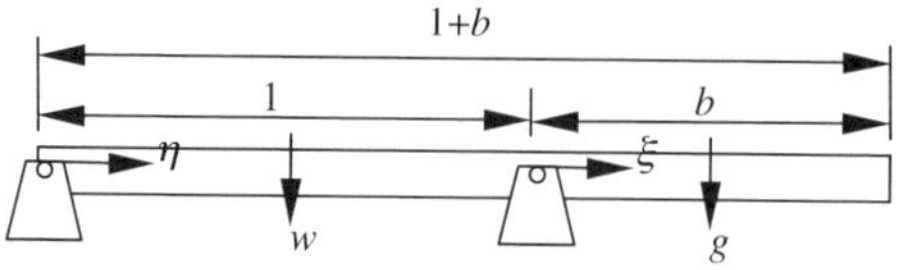

Figure 5.4 Nomenclature for a uniformly loaded simply supported beam with an overhang

where w and g are the displacements of each portion of the beam. The boundary conditions are

$$w(0) = \frac{d^2w(0)}{d\eta^2} = 0$$

$$\frac{d^2g(b)}{d\xi^2} = \frac{d^3g(b)}{d\xi^3} = 0$$

and the continuity conditions at the common boundary are that the displacements are zero, the slope of each portion of the beam is equal, and the moment of each portion of the beam is equal. That is,

$$w(1) = 0 \quad g(0) = 0$$

$$\frac{dw(1)}{d\eta} = \frac{dg(0)}{d\xi}$$

$$\frac{d^2w(1)}{d\eta^2} = \frac{d^2g(0)}{d\xi^2}$$

In order for **`NDSolveValue`** to be able to solve these equations, the range of each solution has to be over the same nondimensional length; $0 \le \eta \le 1$ in this case. To meet this objective, we introduce the coordinate transformation $z = (\xi - 1)/(b - 1)$, $1 \le \xi \le b$, into the second equation, the boundary conditions, and the continuity conditions. Then the second governing equation becomes

$$\frac{d^4g}{dz^4} = (b-1)^4 \quad 0 \le z \le 1$$

The boundary conditions become

$$w(0) = \frac{d^2w(0)}{d\eta^2} = 0$$

$$\frac{d^2g(1)}{dz^2} = \frac{d^3g(1)}{dz^3} = 0$$

and the continuity conditions become

$$w(1) = 0 \quad g(0) = 0$$

$$\frac{dw(1)}{d\eta} = \frac{1}{(b-1)}\frac{dg(0)}{dz}$$

$$\frac{d^2w(1)}{d\eta^2} = \frac{1}{(b-1)^2}\frac{d^2g(0)}{dz^2}$$

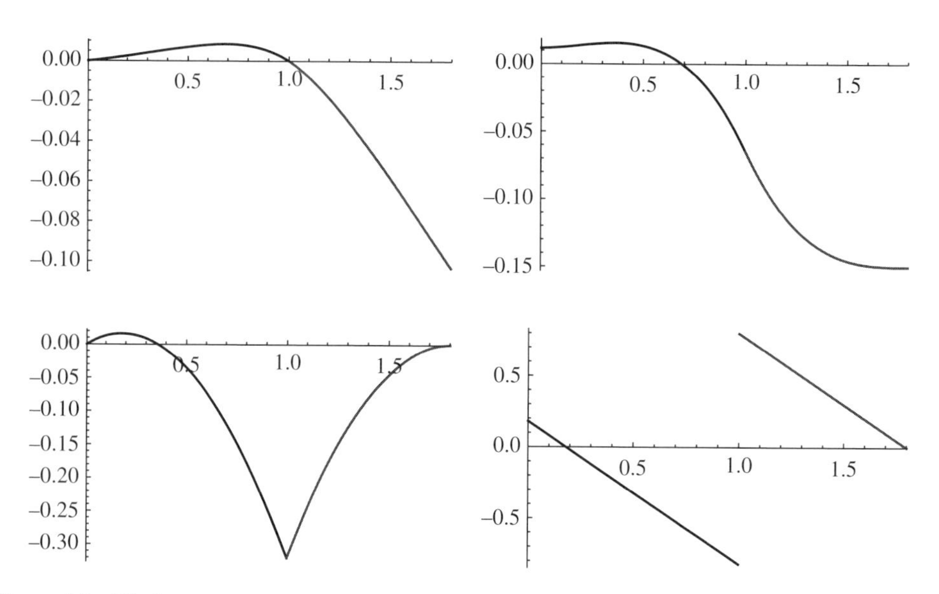

Figure 5.5 Displacement, rotation of the cross section, moment, and shear force for a uniformly loaded simply supported beam with an overhang

The program to solve this system of equations, boundary conditions, and continuity conditions is given below. In order for the displacement to plot in the downward direction, we plot the negative of the results; hence, the introduction of the minus signs in the differential equations. Whereas the solutions that we obtain from **NDSolveValue** are valid over the range $0 \le \eta \le 1$ and $0 \le z \le 1$, the range for g is from $0 \le \xi \le b$. Therefore, when the results are plotted, w is plotted from $0 \le \eta \le 1$ and g is plotted from $1 \le \xi \le 1 + b$. This requires a coordinate shift of the horizontal coordinates prior to plotting by using $\xi = 1 + (b - 1)z$. This translation of the results is obtained using **Table**, as shown in the program. The results of **Table** are plotted using **ListLinePlot**, which is described in Table 6.1 and **GraphicsGrid**, which is illustrated in Table 6.17. As discussed in Section 6.2.5, the **Show** command permits one to combine several graphs obtained from different plotting commands. The results are given in Figure 5.5 for the case where $b = 1.8$.

```
b=1.8; d=b-1;
{ww,gg}=NDSolveValue[{w''''[x]==-1,g''''[x]==-d^4,w[0]==0,
  w''[0]==0,g''[1]==0,g'''[1]==0,w[1]==0,g[0]==0,
  w''[1] d^2==g''[0],w'[1] d==g'[0]},{w,g},{x,0,1},
  InterpolationOrder->All];
p[n_]:=Show[Plot[D[ww[v],{v,n}]/.v->x,{x,0,1},
  PlotRange->{{0,b},All}],ListLinePlot[Table[{1+d x,
  (D[gg[v],{v,n}]/.v->x)/d^(n)},{x,0,1,0.05}]]]
GraphicsGrid[{{p[0],p[1]},{p[2],p[3]}}]
```

As was done in Example 5.4, the option for **InterpolationOrder** was set to **All**.

Example 5.6

Beam with Abrupt Change in Properties

Consider a uniformly loaded beam that is hinged at both ends. The beam has an abrupt change in its geometric and physical properties over a portion of the beam. In terms of nondimensional quantities, the governing equation for each section of the beam is given by

$$\frac{d^4y_1}{d\eta^4} = 1 \quad 0 \le \eta \le \beta$$

$$\frac{d^4y_2}{d\xi^4} = \alpha \quad \beta \le \xi \le 1$$

where α is a parameter relating the properties of each section of the beam and β is the fraction representing the segment of the beam with properties different from the remainder of the beam.

The boundary conditions are

$$y_1(0) = \frac{d^2y_1(0)}{d\eta^2} = 0$$

$$y_2(1) = \frac{d^2y_2(1)}{d\xi^2} = 0$$

The continuity conditions are that the displacements, slopes, moments, and shear forces are equal at $\xi = \eta = \beta$; that is,

$$y_1(\beta) = y_2(\beta)$$

$$\frac{dy_1(\beta)}{d\eta} = \frac{dy_2(\beta)}{d\xi}$$

$$\frac{d^2y_1(\beta)}{d\eta^2} = \frac{d^2y_2(\beta)}{d\xi^2}$$

$$\frac{d^3y_1(\beta)}{d\eta^3} = \frac{d^3y_2(\beta)}{d\xi^3}$$

As noted in Example 5.5, in order for **NDSolveValue** to be able to solve these equations, the range of each solution has to be over the same nondimensional length; $0 \le \eta \le \beta$ in this case. To meet this objective, we introduce the coordinate transformation $z = \beta(\xi - \beta)/(1 - \beta)$, $\beta \le \xi \le 1$, into the second governing equation, the boundary conditions, and the continuity conditions. Then, the second governing equation becomes

$$\frac{d^4y_2}{dz^4} = \frac{(1-\beta)^4}{\beta^4}\alpha \quad 0 \le z \le \beta$$

The boundary conditions become

$$y_1(0) = \frac{d^2 y_1(0)}{d\eta^2} = 0$$

$$y_2(\beta) = \frac{d^2 y_2(\beta)}{dz^2} = 0$$

and the continuity conditions become

$$y_1(\beta) = y_2(0)$$

$$\frac{dy_1(\beta)}{d\eta} = \frac{\beta}{(1-\beta)}\frac{dy_2(0)}{dz}$$

$$\frac{d^2 y_1(\beta)}{d\eta^2} = \frac{\beta^2}{(1-\beta)^2}\frac{d^2 y_2(0)}{dz^2}$$

$$\frac{d^3 y_1(\beta)}{d\eta^3} = \frac{\beta^3}{(1-\beta)^3}\frac{d^3 y_2(0)}{dz^3}$$

The program to solve this system of equations, boundary conditions, and continuity conditions is given below. In order for the displacement to plot in the downward direction, we have to plot the negative of the results; hence, the introduction of the minus signs in the differential equations. The solutions that we obtain from **NDSolveValue** are valid over the range $0 \le \eta \le \beta$ and $0 \le z \le \beta$, whereas the range for y_2 is from $\beta \le \xi \le 1$. Therefore, when the results are plotted, y_1 is plotted from $0 \le \eta \le \beta$ and y_2 from $\beta \le \xi \le 1$. This requires a coordinate shift of the horizontal coordinates prior to plotting by using $\xi = \beta + (1-\beta)z/\beta$. This translation of the results is obtained by using **Table** as shown in the program. The results of **Table** are plotted using **ListLinePlot** as described in Table 6.1. As discussed in Section 6.2.5, the **Show** command permits one to combine several graphs obtained from different plotting commands. We have displayed the results in Figure 5.6 for the case where $\beta = 0.35$ and $\alpha = 6$. Lastly, in the program, we have used $w = y_1$ and $g = y_2$.

```
β=0.35; α=6.; dd=(1-β)/β;
{ww,gg}=NDSolveValue[{w''''[x]==-1,g''''[x]==-α dd^4,
  w[0]==0,w''[0]==0,g[β]==0,g''[β]==0,w[β]==g[0],
  w'[β] dd==g'[0],w''[β] dd^2==g''[0],
  w'''[β] dd^3==g'''[0]},{w,g},{x,0,β},
  InterpolationOrder->All];
pp[n_]:=Show[Plot[D[ww[v],{v,n}]/.v->x,{x,0,β},
  PlotRange->{{0,1},All}],ListLinePlot[ Table[{β+dd x,
  (D[gg[v],{v,n}]/.v->x)/dd^(n)},{x,0,β,0.05}]]]
GraphicsGrid[{{pp[0],pp[1]},{pp[2],pp[3]}}]
```

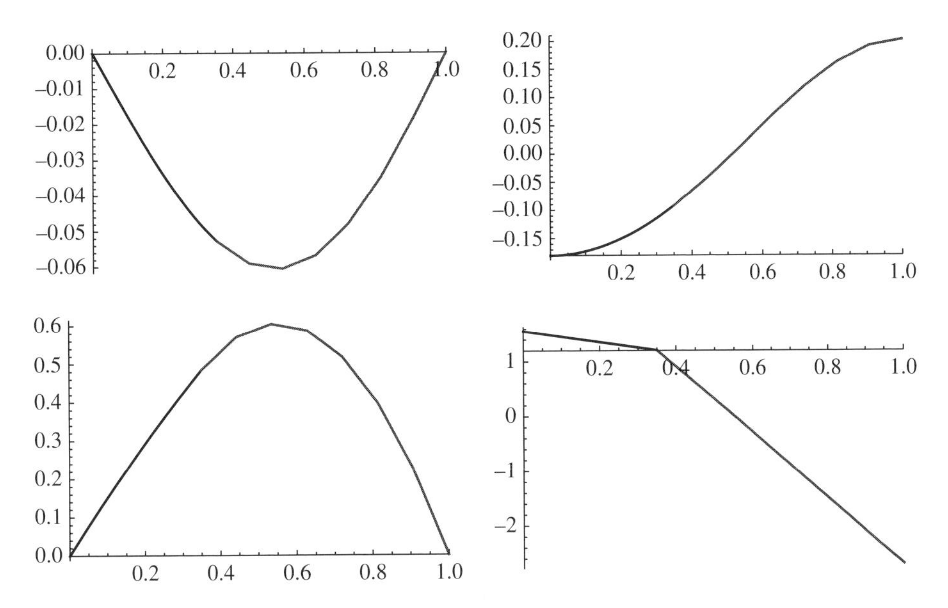

Figure 5.6 Displacement, rotation of the cross section, moment, and shear force for a uniformly loaded simply supported beam with an abrupt change in cross section properties

---

Example 5.7

Two Degrees-of-Freedom System Revisited

We shall again obtain a solution to the two degrees-of-freedom system described in Example 4.30, whose governing equations of motions in terms of nondimensional quantities are

$$\frac{d^2x_1}{d\tau^2} + \left(2\zeta_1 + 2\zeta_2 m_r\omega_r\right)\frac{dx_1}{d\tau} + \left(1 + m_r\omega_r^2\right)x_1 - 2\zeta_2 m_r\omega_r\frac{dx_2}{d\tau} - m_r\omega_r^2 x_2 = \frac{f_1(\tau)}{k_1}$$

$$\frac{d^2x_2}{d\tau^2} + 2\zeta_2\omega_r\frac{dx_2}{d\tau} + \omega_r^2 x_2 - 2\zeta_2\omega_r\frac{dx_1}{d\tau} - \omega_r^2 x_1 = \frac{f_2(\tau)}{k_1 m_r}$$

It is assumed that the initial conditions are zero; that is, $x_1(0) = dx_1(0)/d\tau = x_2(0) = dx_2(0)/d\tau = 0$, and that $f_1(t) = 0$ and $f_2(t) = \delta(t)$, where $\delta(t)$ is the delta function. In addition, the following values are assigned to the parameters: $\omega_r = 0.85$, $m_r = 0.45$, $k_1 = 1$, and $\zeta_1 = \zeta_2 = 0.15$. In this example, we approximate the delta function by a rectangular pulse of duration a and magnitude $1/a$, where we choose $a = 0.001$. This rectangular pulse is represented by the difference of two unit step functions. In addition, as was done in Example 5.4, **MaxStepSize** is set to 0.01. The program is

```
a=0.001; mr=0.45; ωr=0.85; ζ1=0.15; ζ2=0.15; k1=1;
{x11,x22}=NDSolveValue[{x1''[t]+(2 ζ1+2 ζ2 mr ωr) x1'[t]+
  (1+mr ωr^2) x1[t]-2 ζ2 mr ωr x2'[t]-mr ωr^2 x2[t]==0,
```

```
  x2''[t]+2 ζ2 ωr x2'[t]+ωr^2 x2[t]-2 ζ2 ωr x1'[t]-
  ωr^2 x1[t]==(1-UnitStep[t-a])/(a k1 mr),x1[0]==0,
  x1'[0]==0,x2[0]==0,x2'[0]==0},{x1,x2},{t,0,50},
  MaxStepSize->1/100];
Plot[{x11[t],x22[t]},{t,0,50},PlotRange->All]
```

The plot of x_1 and x_2 is identical to that given in Figure 4.5.

Example 5.8

Particle Impact Damper

Consider a single degree-of-freedom system that undergoes a nondimensional displacement y_1. The mass is hollowed out so that another mass, which has a nondimensional displacement y_2, can slide within it without friction a specified maximum distance before impacting either opposing wall of the hollow mass. The walls of the cavity are assumed to have stiffness and damping. Such a system forms a single particle impact damper. The governing equations for this system in terms of nondimensional quantities when the hollow mass is subjected to a rectangular pulse of magnitude f_o and duration τ_d are [1]

$$\frac{d^2y_1}{d\tau^2} + 2\zeta\frac{dy_1}{d\tau} + y_1 + k_{21}h(y_1, y_2) + 2\zeta c_{21}g(\dot{y}_1, \dot{y}_2) = f_o\left[u(\tau) - u(\tau - \tau_d)\right]$$

$$\frac{d^2y_2}{d\tau^2} - \frac{k_{21}}{m_{21}}h(y_1, y_2) - 2\zeta\frac{c_{21}}{m_{21}}g(\dot{y}_1, \dot{y}_2) = 0$$

where $u(\tau)$ is the unit step function and

$$h(y_1, y_2) = \left(y_1 - y_2 - 1\right)u\left(y_1 - y_2 - 1\right) + \left(y_1 - y_2 + 1\right)u\left(-y_1 + y_2 - 1\right)$$

$$g(\dot{y}_1, \dot{y}_2) = \left(\dot{y}_1 - \dot{y}_2\right)u\left(y_1 - y_2 - 1\right) + \left(\dot{y}_1 - \dot{y}_2\right)u\left(-y_1 + y_2 - 1\right)$$

The program that solves this system of equations for $\tau_d = 0.025$, $\zeta = 0.005$, $m_{21} = 0.04$, $k_{21} = 100$, and $f_o = 50$ is

```
h12=(y1[t]-y2[t]-1) UnitStep[y1[t]-y2[t]-1]+
  (y1[t]-y2[t]+1) UnitStep[-y1[t]+y2[t]-1];
g12=(y1'[t]-y2'[t]) UnitStep[y1[t]-y2[t]-1]+
  (y1'[t]-y2'[t]) UnitStep[-y1[t]+y2[t]-1];
ζ=0.005; m21=0.04; k21=100.; c21=15.; fo=50.; tend=300;
{y11,y22}=NDSolveValue[{y1''[t]+2 ζ y1'[t]+
  y1[t]+k21 h12+2 ζ c21 g12==fo (1-UnitStep[t-0.0250]),
  y2''[t]-k21/m21 h12-2 ζ c21/m21 g12==0,y1[0]==0,y1'[0]==0,
  y2[0]==0,y2'[0]==0},{y1,y2},{t,0,tend}];
Plot[y11[t],{t,0,tend},PlotRange->All,
  AxesLabel->{"τ","y1(τ)"}]
```

```
Plot[y22'[t],{t,0,tend},PlotRange->All,
 AxesLabel->{"τ","dy2(τ)/dτ"}]
```

The displacement y_1 is given in Figure 5.7a and the velocity $dy_2/d\tau$ is given in Figure 5.7b.

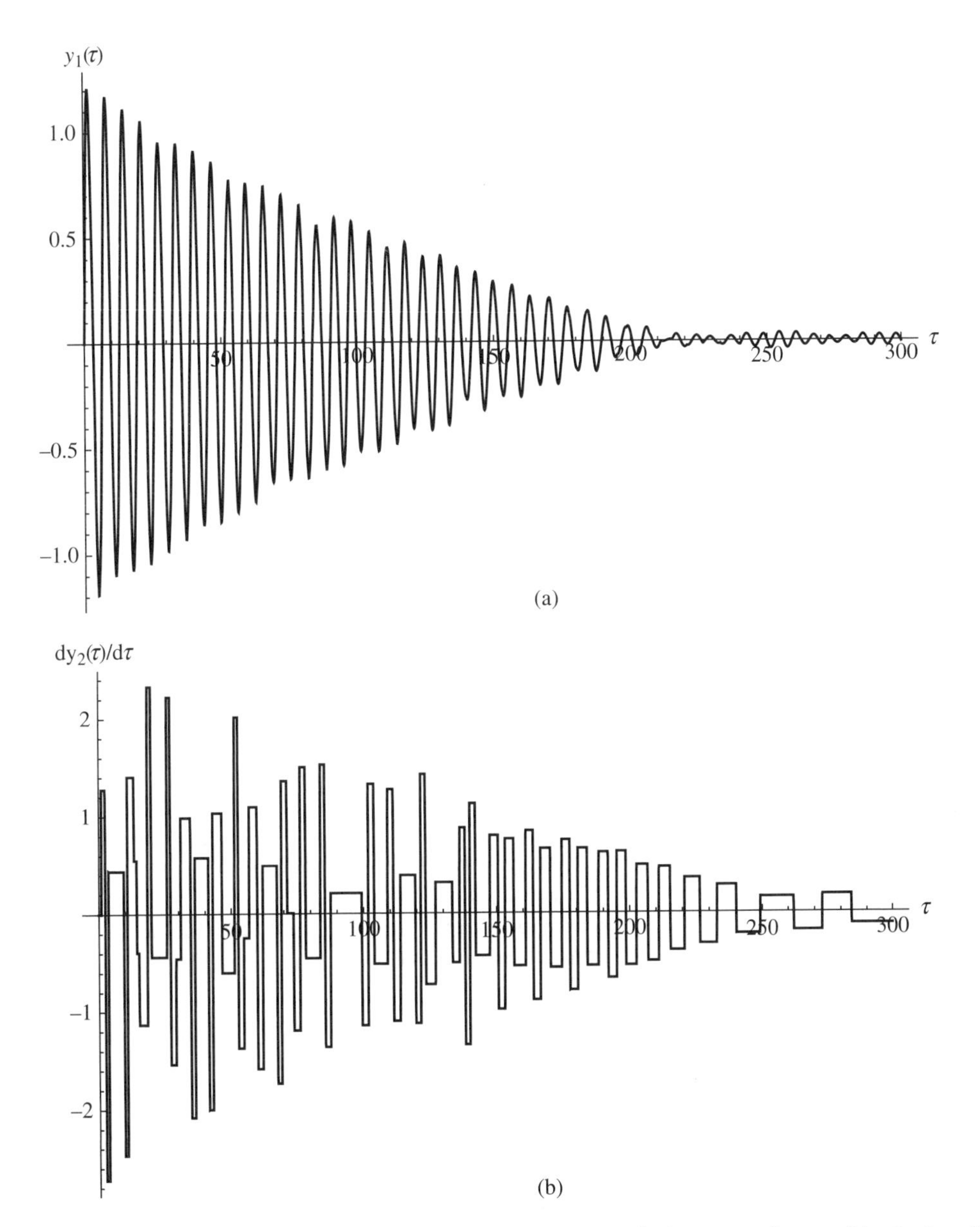

Figure 5.7 Response of a particle impact damper to a pulse (a) displacement of mass; (b) velocity of particle

Example 5.9

Change in Period of a Nonlinear System

We shall illustrate the use of **WhenEvent** by determining the times at which the response of a single degree-of-freedom subjected to an initial velocity equals zero. The spring for this system is a piece-wise linear spring; that is, after the mass traverses a certain distance it encounters another spring whose stiffness is proportional to μ, $\mu \geq 0$. The governing equation in terms of nondimensional parameters for this system is

$$\frac{d^2y}{d\tau^2} + 2\zeta\frac{dy}{d\tau} + y + \mu h(y) = 0$$

where ζ is a parameter and

$$\begin{aligned} h(y) &= 0 & |y| \leq 1 \\ &= y - \text{sgn}(y) & |y| > 1 \end{aligned}$$

We shall determine the values of the response of the system to an initial velocity whose nondimensional magnitude is 10 when $\mu = 10$, $\zeta = 0.15$, and $0 \leq \tau \leq 30$. From this response, the times at which the response has zero amplitude will be determined and saved. From these values, we shall compute their differences and show that as time progresses these differences become a constant value; that is, the second spring is no longer encountered. We shall capture the values at which the displacement equals zero with **Reap** and **Sow**. The program is

```
ζ=0.15; μ=10.; ten=30;
{oden,times}=Reap[NDSolveValue[{y''[t]+2 ζ y'[t]+y[t]+
  μ If[Abs[y[t]]<=1,0,y[t]-Sign[y[t]]]==0,y[0]==0,
  y'[0]==10.,WhenEvent[y[t]==0,Sow[t],
  "DetectionMethod"->"Interpolation"]},y,{t,0,ten}]];
Plot[oden[t],{t,0,ten},PlotRange->All,
  AxesLabel->{"τ","y(τ)"}]
Differences[Flatten[times]]
```

which produces Figure 5.8 and displays the following list of time differences

```
{1.19679,1.25915,1.34535,1.47039,1.66457,1.99869,
  2.68289,3.17754,3.17754,3.17754,3.17754,3.17754}
```

In obtaining these results, we selected the option for the detection method as that shown to ensure that no zero crossings were missed.

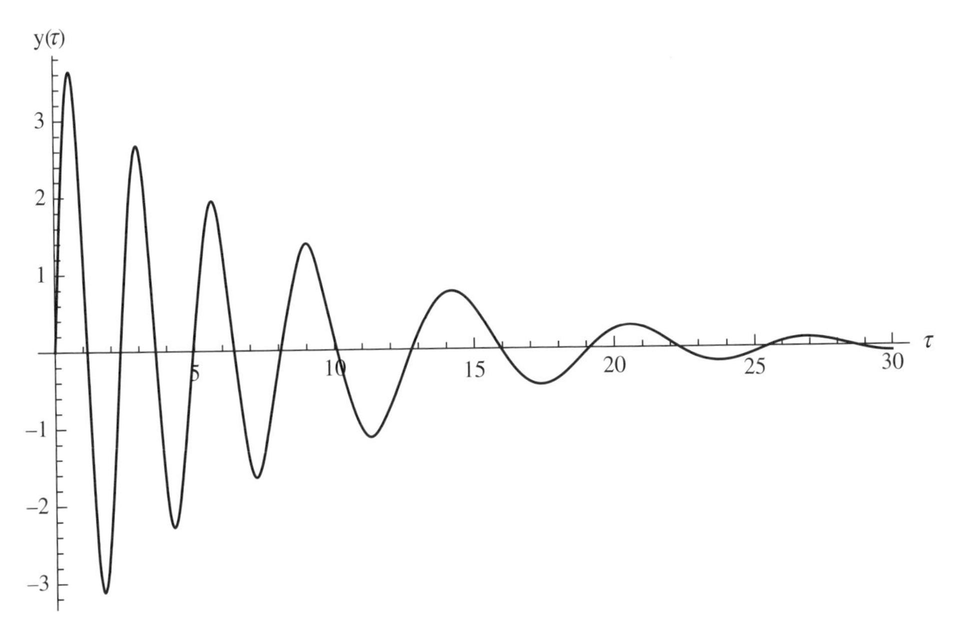

Figure 5.8 Displacement response of a single degree-of-freedom system with a piece-wise linear spring

Example 5.10

Single Degree-of-Freedom System

We shall determine the times at which the amplitude of the displacement response of a single degree-of-freedom system equals 0.3 when it is subjected to an initial displacement of 1 and zero initial velocity. The governing equation is

$$\frac{d^2y}{d\tau^2} + 0.15\frac{dy}{d\tau} + y = 0$$

The program is

```
{uu,ptz}=Reap[NDSolveValue[{u''[t]+0.15 u'[t]+u[t]==0,
  u[0]==1,u'[0]==0,WhenEvent[u[t]==0.3,Sow[{t,u[t]}]]},
  u,{t,0,25}]];
Show[Plot[uu[t],{t,0,25},AxesLabel->{"τ","y(τ)"}],
  ListPlot[Tooltip[ptz],PlotMarkers->Automatic]]
```

The results are plotted in Figure 5.9. The times at which this amplitude occurs can be identified by passing the cursor over the indicated points. This feature is obtained by using **`Tooltip`**, which is described in Table 6.18.

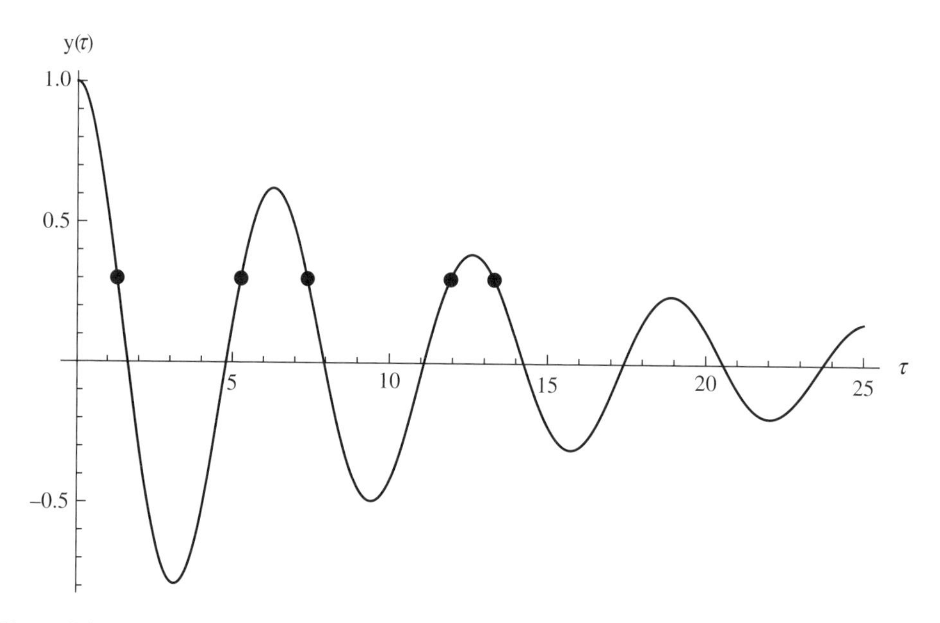

Figure 5.9 Times of occurrences when the response of a single degree-of-freedom system equals 0.3

Example 5.11

Poincare Plot

Consider the following nonlinear differential equation [2]

$$\frac{d^2x}{d\tau^2} + \beta\frac{dx}{d\tau} - \frac{1}{2}x + \frac{1}{2}x^3 = \varepsilon\cos\Omega\tau$$

with zero initial conditions. For $\beta = 0.1$, $\Omega = 1.2$, and $\varepsilon = 0.38$, we shall obtain a Poincare plot, which is a plot of the values of x versus $dx/d\tau$ when $\tau = \tau_n = 2n\pi/\Omega$, $n = 0, 1, \ldots$. The multiples of $2\pi/\Omega$ are determined by using **Mod**. The initial conditions are zero. The program that performs these operations and then produces Figure 5.10 is

```
Ω=1.2; β=0.1; ε=0.38; tend=50000.;
{xx,pst}=Reap[NDSolveValue[
  {x''[t]+β x'[t]-0.5 x[t]+0.5 x[t]^3==ε Cos[Ω t],
  x[0]==0,x'[0]==0,WhenEvent[Mod[t,2 π/Ω]==0,
  Sow[{x[t],x'[t]}]]},x,{t,0,tend},MaxSteps->Infinity]];
ListPlot[pst,PlotStyle->PointSize[0.0025],PlotRange->All,
  AxesLabel->{"x(τn)","dx(τn)/dτ"}]
```

The list `pst` is a list of pairs of points in a format required by `ListPlot`.

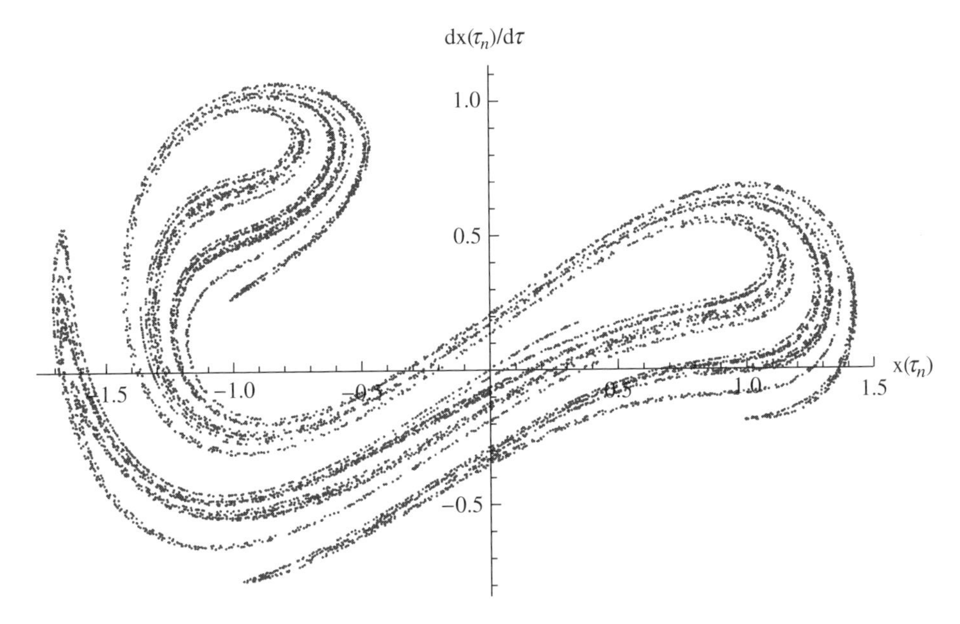

Figure 5.10 Poincare plot of the equation in Example 5.11

Example 5.12

Nonlinear Ordinary Differential Equation

Consider the following nondimensional nonlinear equation

$$\frac{d^2\theta}{d\tau^2} + \alpha\frac{d\theta}{d\tau} - \sin\theta + \beta\left(1 - \frac{1}{\sqrt{5 - 4\cos\theta}}\right)\sin\theta = 0$$

We shall determine the solution to this equation for $0 \leq \tau \leq 50$ when $\alpha = 0.1$, $\beta = 10$, and for the initial conditions $\theta(0) = \pi/4$ and $d\theta(0)/d\tau = 0$. The solutions will be plotted in two ways. The first figure will plot $\theta(\tau)$ and $d\theta(\tau)/d\tau$ as a function of τ. The second figure will be a parametric plot of $\theta(\tau)$ versus $d\theta(\tau)/d\tau$. The program is

```
β=10; α=0.1;
sol=NDSolveValue[{θ''[t]+α θ'[t]-Sin[θ[t]]+
  β (1-1/Sqrt[5-4 Cos[θ[t]]]) Sin[θ[t]]==0,θ[0]==π/4,
  θ'[0]==0},θ,{t,0,50}];
Plot[{sol[t],sol'[t]},{t,0,50},
  PlotStyle->{Dashing[{}],Dashing[Medium]},
  AxesLabel->{"τ","θ(τ), dθ(τ)/dτ"}]
ParametricPlot[{sol[t],sol'[t]},{t,0,50},
  AxesLabel->{"θ(τ)","dθ(τ)/dτ"}]
```

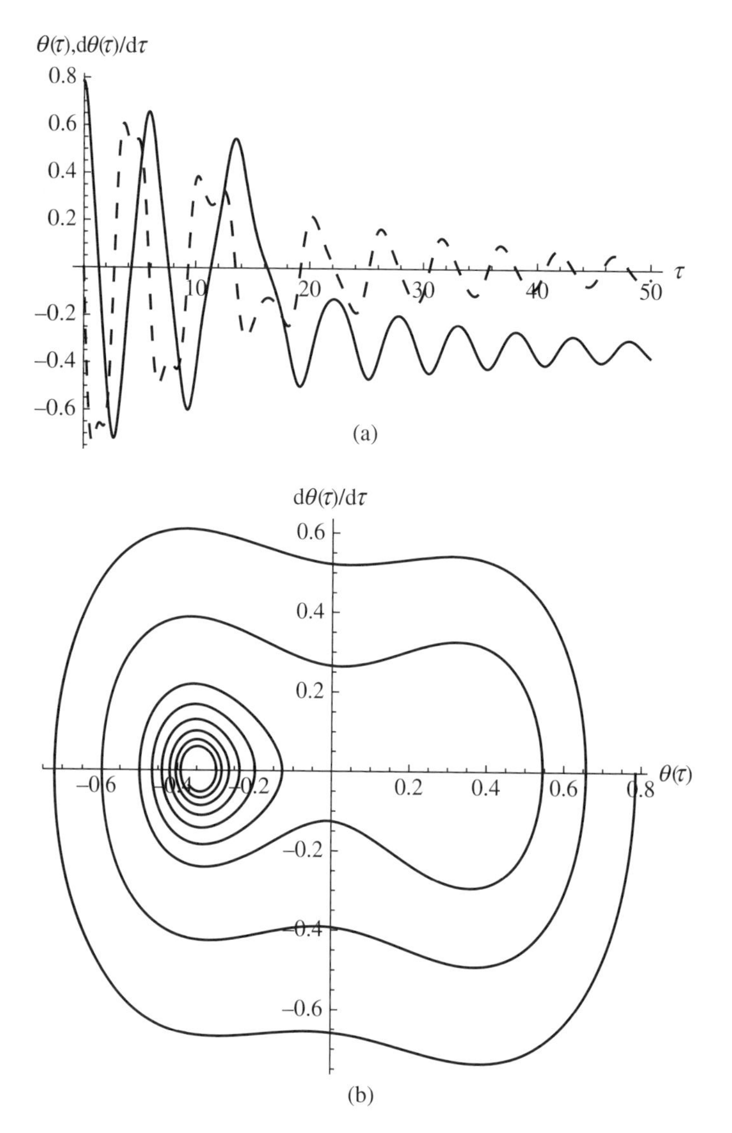

Figure 5.11 (a) Graph of θ (solid) and $d\theta/d\tau$ (dashed); (b) parametric plot of θ versus $d\theta/d\tau$

The results are shown in Figure 5.11.

To explore the solution as a function of α and/or β, we use **`ParametricNDSolveValue`**. For example, to obtain a parametric plot of θ versus $d\theta/d\tau$ for the pairs $\alpha = 0.1$ and $\beta = 10$, $\alpha = 0.1$ and $\beta = 2$, and $\alpha = 0.05$ and $\beta = 6$, we proceed as follows.

```
solp=ParametricNDSolveValue[{θ''[t]+α θ'[t]-Sin[θ[t]]+
 β (1-1/Sqrt[5-4 Cos[θ[t]]]) Sin[θ[t]]==0,θ[0]==π/4,
 θ'[0]==0},θ,{t,0,40},{α,β}];
```

```
sx[t_,α_,β_]:={solp[α,β][t],D[solp[α,β][z],z]/.z->t}
ParametricPlot[{sx[v,0.1,10],sx[v,0.1,2],sx[v,0.05,6]},
 {v,0,40},PlotStyle->{{Black,Dashing[{}]},
 {Black,Dashing[Small]},{Black,Dashing[Large]}},
 AxesLabel->{"θ(τ)","dθ(τ)/dτ"}]
```

The results are shown in Figure 5.12. For clarity, we have used dashed and solid lines to delineate each case. The function **sx** was introduced to improve readability.

With the use of **ParametricNDSolveValue**, one can examine the variation of the results as a function of one of the parameters. For example, assume that $\alpha = 0.1$ and that β is to be varied. The results can be examined two ways. For the first way, $\theta(\tau)$ is plotted for $\beta = 2$, 5, and 10. For the second way, the quantity $\theta(7)$ is plotted as a function of β over the range $2 \le \beta \le 10$. The program that creates these two sets of results is

```
α=0.1;
solp2=ParametricNDSolveValue[{θ''[t]+α θ'[t]-Sin[θ[t]]+
 β (1-1/Sqrt[5-4 Cos[θ[t]]]) Sin[θ[t]]==0,θ[0]==π/4,
 θ'[0]==0},θ,{t,0,50},{β}];
Plot[Table[solp2[β][t],{β,{2,5,10}}],{t,0,50},
 AxesLabel->{"τ","θ(τ)"}]
Plot[solp2[β][7.],{β,2,10},PlotRange->All,
 AxesLabel->{"β","θ(7)"}]
```

Execution of these commands results in Figure 5.13.

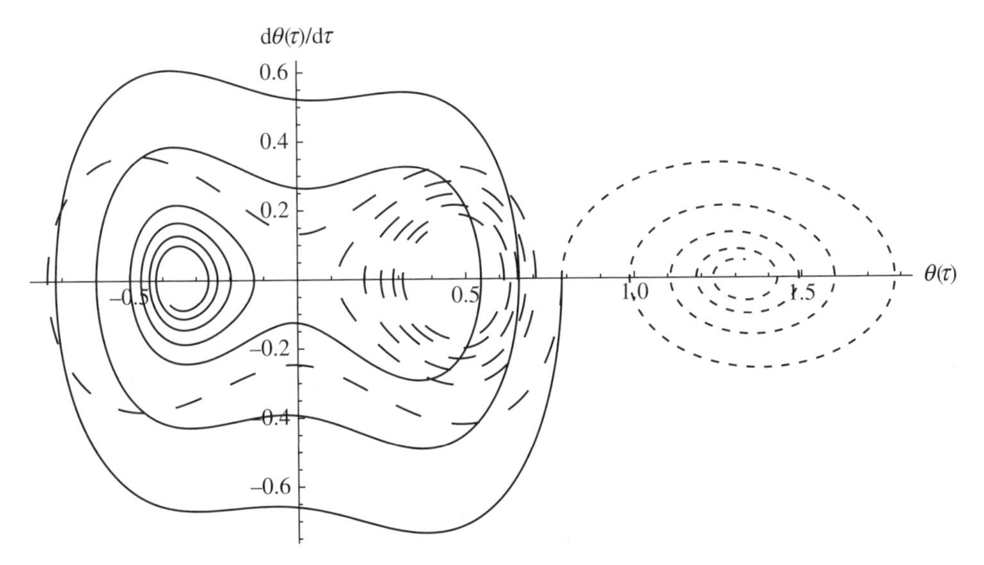

Figure 5.12 Parametric plot of θ versus $d\theta/d\tau$ for $\alpha = 0.1$ and $\beta = 10$ (solid); $\alpha = 0.1$ and $\beta = 2$ (small dashes); and $\alpha = 0.05$ and $\beta = 6$ (large dashes)

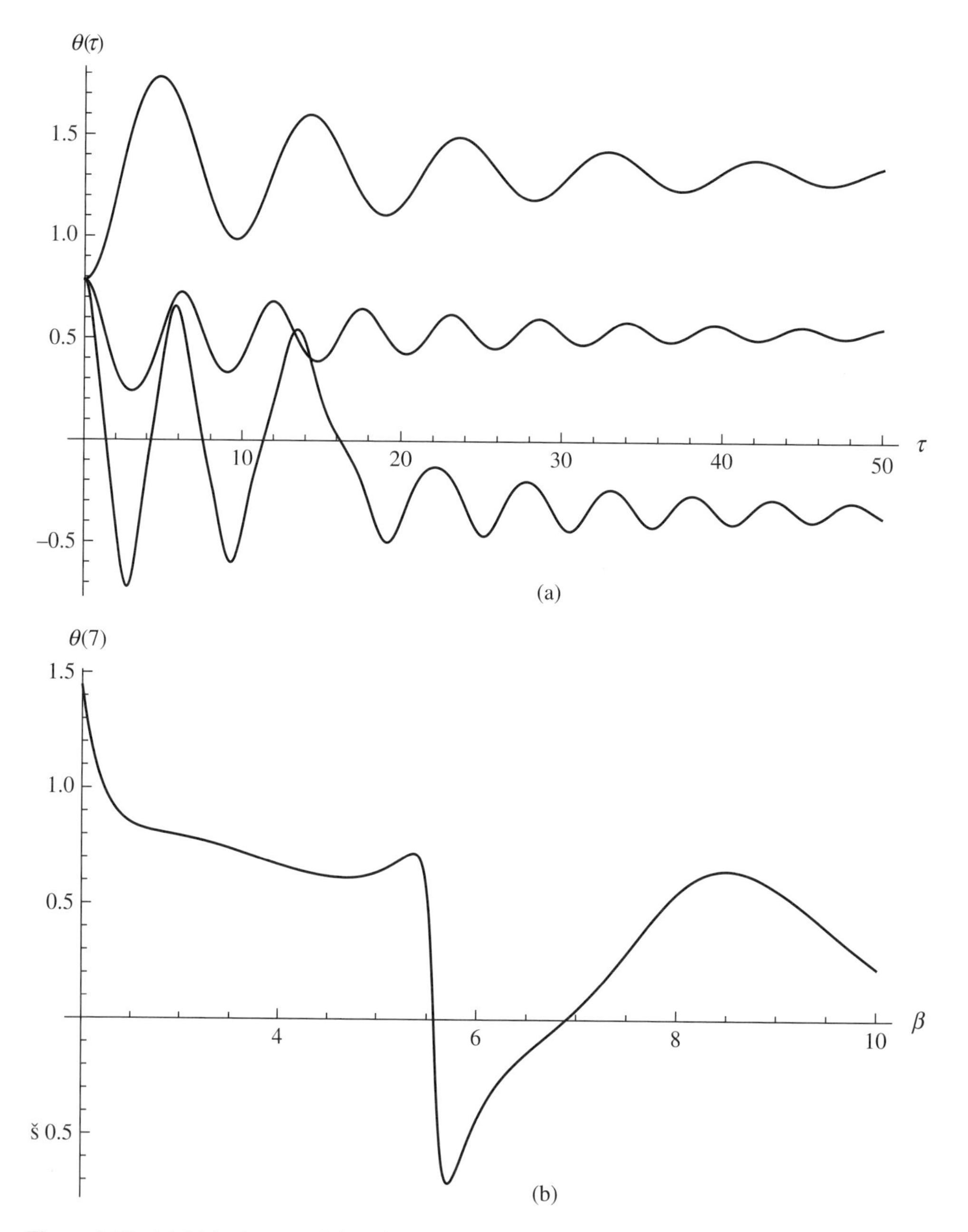

Figure 5.13 (a) $\theta(\tau)$ when $\alpha = 0.1$ and $\beta = 2$, 5, and 10 (the upper curve is for $\beta = 2$ and the lower curve for $\beta = 10$); (b) $\theta(7)$ as a function of β when $\alpha = 0.1$

Example 5.13

Heat Conduction in a Slab

The nondimensional equation for heat conduction in a slab that has a constant heat source Σ is given by

$$\frac{\partial \theta}{\partial \tau} = \frac{\partial^2 \theta}{\partial \xi^2} + \Sigma$$

where θ is proportional to the temperature in the slab. The initial condition is assumed to be

$$\theta(\xi, 0) = 1 - \alpha\xi$$

and the boundary conditions are given by

$$\left.\frac{\partial \theta}{\partial \xi}\right|_{\xi=0} = Bi\theta(0, \tau)$$

$$\theta(1, \tau) = 1 - \alpha$$

where α and Bi are constants.

We shall obtain a solution for this system for $0 \le \tau \le 0.5$, $Bi = 0.1$, $\alpha = 0.45$, and $\Sigma = 1.0$ and plot the results as a function of time for $\xi = 0.05, 0.3, 0.4, 0.7,$ and 0.9.

Before proceeding, however, it is noted that when the system implements **NDSolveValue** to solve partial differential equations, it checks to ensure that the initial conditions and the boundary conditions are consistent. From the initial condition in our case, it is seen that $\theta(0{,}0) = 1$ and $\theta(1{,}0) = 0.55$. From the second boundary condition, it is seen that $\theta(1{,}0) = 0.55$. Hence, we have consistency. However, from the first boundary condition, $\partial\theta(0{,}0)/\partial\xi - Bi\theta(0{,}0) \neq \theta(0{,}0)$. In other words, some adjustment must be made to this boundary condition so that at $\xi = 0$ and $\tau = 0$ the boundary condition and the initial condition are equal. Mathematica suggests using a rapidly, but smoothly decaying quantity such as $e^{-1000\tau}$. Thus, we modify the first boundary condition as follows

$$\left(1 - e^{-1000\tau}\right) \left.\frac{\partial \theta}{\partial \xi}\right|_{\xi=0} - Bi\theta(0, \tau) = -Bie^{-1000\tau}$$

which, when $\tau = 0$, is now consistent with the initial condition evaluated at $\xi = 0$.

With this modified boundary condition, the program is

```
Bi=0.1; θ1=0.55; Σ=1.;
temp=NDSolveValue[{D[θ[x,t],t]==D[θ[x,t],x,x]+Σ,
  θ[x,0]==1-0.45 x,θ[1,t]==θ1,
  Derivative[1,0][θ][0,t] (1-Exp[-1000 t])-Bi θ[0,t]==
  -Bi Exp[-1000 t]},θ,{x,0,1},{t,0,0.5}];
Plot[Table[temp[x,t],{x,{0.05,0.3,0.4,0.7,0.9}}],{t,0,0.5},
  AxesLabel->{"τ","θ(ξ,τ)"}]
```

which generates the results shown in Figure 5.14.

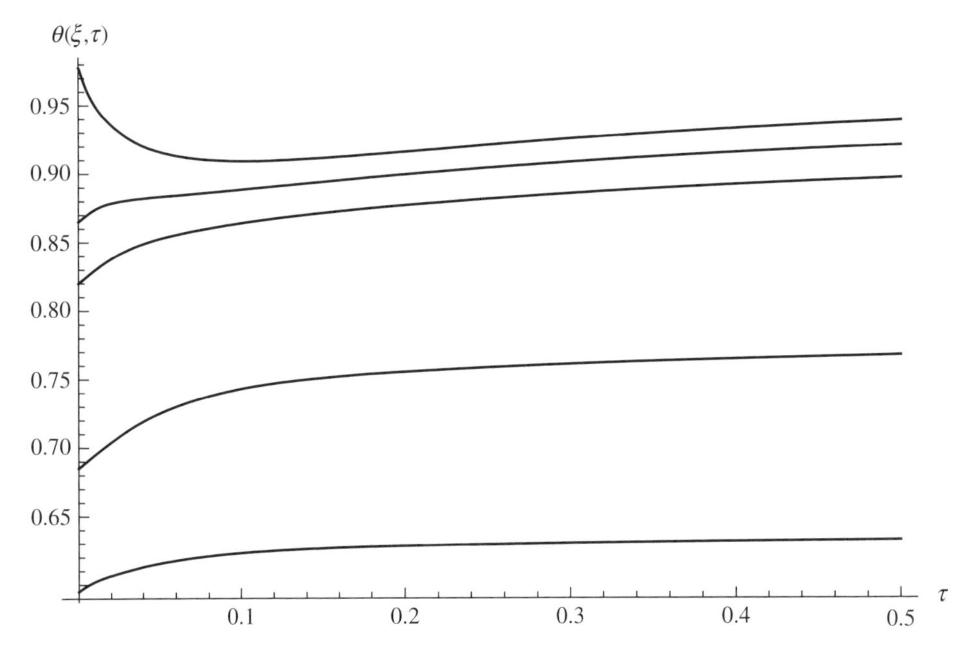

Figure 5.14 Graph of $\theta(\xi,\tau)$ in Example 5.13 for $\xi = 0.05$, 0.3, 0.4, 0.7, and 0.9: $\xi = 0.05$ is the top curve and $\xi = 0.9$ is the bottom curve

Example 5.14

Air Entrainment by Liquid Jets

The governing equations that model the injection of high-pressure water into a vertical water stream exiting into air through a contraction section is given by [3]

$$\frac{\partial a}{\partial t} + \frac{\partial}{\partial z}(au) = 0$$

$$\frac{\partial u}{\partial t} + u\frac{\partial u}{\partial z} = \frac{1}{\text{Fr}} - \frac{1}{\text{We}}\frac{\partial \kappa}{\partial z}$$

where $u = u(z,t)$ is the average velocity, $a = a(z,t)$ is the cross-sectional area of the jet a distance z from the exit, Fr is the Froude number, and We is the Weber number. The quantity κ is the local curvature given by

$$\kappa = \frac{1 + r'^2 - rr''}{r\left(1 + r'^2\right)^{3/2}}$$

where $r = r(z,t) = \sqrt{a(z,t)/\pi}$ is the local radius of curvature and the prime denotes the derivative with respect to z.

We assume that the boundary conditions are

$$u(0,t) = 1, \quad a(0,1) = 1, \quad a'(0,t) = 0$$

where the prime indicates the derivative with respect to z and that the initial conditions are

$$u(z,0) = 1, \quad a(z,0) = 1$$

The differentiation of κ with respect to z results in a term containing the second derivative of a with respect to z; therefore, the need for the derivative boundary condition for a.

This system of nonlinear equations with the given boundary conditions and initial conditions is solved with the following program, which also plots the results in Figure 5.15 as a surface using **Plot3D**. See Table 6.19 for a description of **Plot3D**.

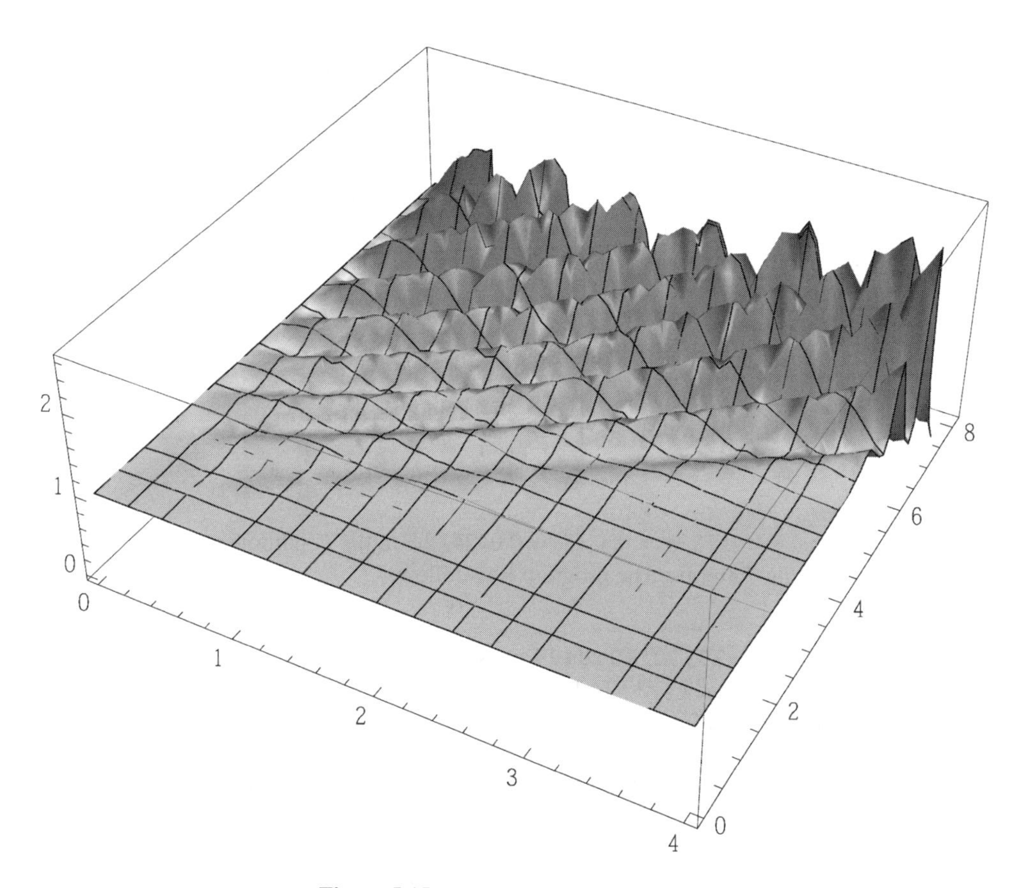

Figure 5.15 Results from Example 5.14

```
Fr=100;  We=50;
r=Sqrt[a[z,t]/π];
k=Simplify[D[(1+D[r,z]^2-r D[r,z,z])/
  (r (1+D[r,z]^2)^1.5)]];
eq1=D[a[z,t],t]+D[a[z,t] u[z,t],z];
eq2=D[u[z,t],t]+u[z,t] D[u[z,t],z];
{aa,uu}=NDSolveValue[{eq1==0,eq2==1/Fr-k/We,
  u[0,t]==1,a[0,t]==1,Derivative[1,0][a][0,t]==0,
  u[z,0]==1,a[z,0]==1},{a,u},{z,0,4},{t,0,3 π},
  PrecisionGoal->2];
Plot3D[aa[z,t],{z,0,4},{t,0,3 π},PlotRange->All]
```

In the above program, we have set the option **PrecisionGoal** to 2 to permit a larger relative error in the computations.

Example 5.15

Second-Order Differential Equation: Periodic Inhomogeneous Term

In this example, the procedure that one can use to obtain the solution to a differential equation when its inhomogeneous term is a Fourier series is illustrated. Consider the following equation, in which the inhomogeneous expression describes a periodically occurring pulse of duration α and period $2\pi/\Omega_o$

$$\frac{d^2y}{d\tau^2} + 2\zeta\frac{dy}{d\tau} + y = \alpha\left[1 + 2\sum_{m=1}^{\infty}\frac{\sin(m\pi\alpha)}{m\pi\alpha}\cos(m\Omega_o\tau)\right]$$

It is assumed that $\zeta = 0.15$, $\alpha = 0.4$, $\Omega_o = 0.0424$, the initial conditions are zero, and the series is summed using 70 terms. The program is as follows

```
ζ=0.15; α=0.4; M=70; Ωo=0.0424; ww={};
Do[yy=NDSolveValue[{y''[t]+2 ζ y'[t]+y[t]==
  If[m==0,a,2 α Sin[m π α]/(m π α) Cos[m Ωo t]],y[0]==0,
  y'[0]==0},y,{t,0,380}];AppendTo[ww,yy],{m,1,M}]
Plot[Total[Table[ww[[m]][t],{m,1,M}]],{t,100,380},
  PlotRange->{{100,380},All},AxesLabel->{"τ","y(τ)"}]
```

The results are shown in Figure 5.16, where the transient portion of the solution is ignored by plotting the results for $\tau \geq 100$.

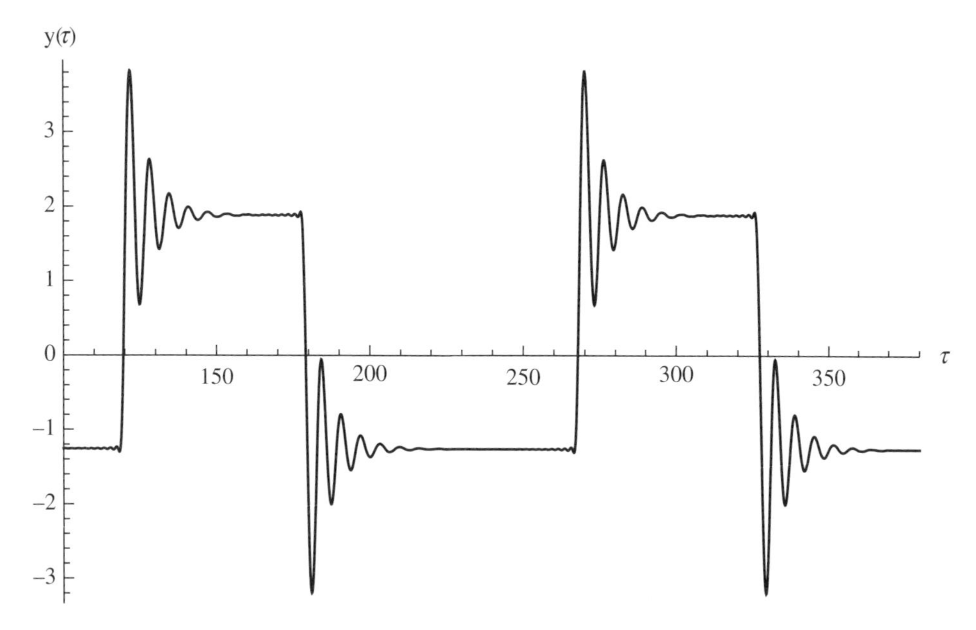

Figure 5.16 Solution to a second-order differential equation with a periodic inhomogeneous term that is represented by a Fourier series

Example 5.16

Deflection of a Uniformly Loaded Solid Circular Plate

We shall show how to deal with a singularity of the differential equation by determining the displacement of a uniformly loaded solid circular plate of uniform thickness that is clamped along its outer boundary. If the load is expressed in terms of the nondimensional parameter q, then the governing equation in nondimensional form is [4, pp. 54-6]

$$\frac{1}{\eta}\frac{d}{d\eta}\left\{\eta\frac{d}{d\eta}\left[\frac{1}{\eta}\frac{d}{d\eta}\left(\eta\frac{dw}{d\eta}\right)\right]\right\}=q$$

The boundary conditions at $\eta = 1$, the outer boundary, are that $w(1) = w'(1) = 0$, where the prime denotes the derivative with respect to η. Since the differential equation has a singularity at $\eta = 0$, we shall replace this boundary with a small hole of radius δ. Since the plate is uniformly loaded, the displacement will be symmetric about the origin. Therefore, one boundary conditions is $w'(\delta) = 0$. Based on some physical and mathematical arguments, the second boundary condition is that the nondimensional shear force $v_r(\delta)$ is zero. This quantity is given by

$$v_r(\delta) = -\frac{d}{d\eta}\left\{\frac{1}{\eta}\frac{d}{d\eta}\left(\eta\frac{dw}{d\eta}\right)\right\}_{\eta=\delta}=0$$

The quantity of interest is the radial stress, which is proportional to the nondimensional moment m_r, where

$$m_r = -\left(\frac{d^2w}{d\eta^2} + \frac{\nu}{\eta}\frac{dw}{d\eta}\right)$$

where ν is Poisson's ratio.

Using these relations, we shall determine for $q = 1$ and $\nu = 0.3$ the maximum displacement, which occurs near $\eta = \delta$, and the maximum value of m_r, which occurs at $\eta = 1$. If we select $\delta = 10^{-6}$, then the program to determine these quantities is

```
δ=10^(-6); q=1.; ν=0.3;
plate=Quiet[NDSolveValue[
 {D[r D[1/r D[r D[w[r],r],r],r],r]/r==q, w'[δ]==0,
  (w'''[δ]+1/δ w''[δ]-1/δ^2 w'[δ])==0,w[1]==0,
 w'[1]==0},w,{r,δ,1},InterpolationOrder->All]];
Print["wmax = ",plate[δ]]
Print["mr,max = ",-plate''[1]-ν plate'[1]]
```

which displays

```
wmax = 0.015625
mr,max = -0.124999
```

These values agree with published results.

5.4 Numerical Solutions of Equations: `NSolve[]`

The function **NSolve** obtains the numerical solution to an equation or a system of equations. For one equation, the form of **NSolve** is

```
NSolve[eqn,var]
```

where **eqn** is the equation and **var** is the independent variable. If there are L solution values, then the output of **NSolve** is a list consisting of L rows, where each row contains one value.

For a system of M equations in terms of M variables, we use

```
NSolve[{eqn1,eqn2, ... ,eqnM},{var1,var2, ... ,varM}]
```

If there are $L \times M$ solution values, then the output of **NSolve** is a list consisting of L rows with each row containing M values, one for each variable.

Example 5.17

Roots of a Polynomial #1

Consider the polynomial

$$0.1x^4 - (2 + 7j)x + 3.2 = 0$$

The roots are obtained with

```
r=x/.NSolve[0.1 x^4-(2+7 I) x+3.2==0,x]
```

which displays

```
{-3.45051+2.54576i},-0.434488-4.00786i,
 0.12126-0.422731i},3.76373+1.88484i}}
```

Hence, `r[[1]] = -3.45051 + 2.54576 I`, `r[[2]] = -0.434488 - 4.00786 I`, etc.

Example 5.18

Roots of Polynomials #2

Consider the polynomials

$$\frac{1}{4}x^2 + y^2 = 1$$

$$y - 4x^2 = -3$$

The solution is determined from

```
r={x,y}/.NSolve[{x^2/4+y^2==1,y-4x^2==-3},{x,y}]
```

which gives

```
{{-0.983702,0.870679},{0.718822,-0.933179},
 {-0.718822,-0.933179},{0.983702,0.870679}}
```

Thus, $r_1 = (x_1, y_1) =$ `(r[[1,1]],r[[1,2]]) = (-0.983702, 0.870679)`, $r_2 = (x_2, y_2) =$ `(r[[2,1]],r[[2,2]])`, etc.

Example 5.19

Roots of Polynomial #3

Consider the polynomial

$$2.1x^{1.6} + (0.6 + j)x^{-2.7} + 1.9 = 0$$

We shall determine its roots and then verify their correctness by direct substitution. Thus,

```
f[x_]:=2.1 x^1.6+(0.6+I)x^(-2.7)+1.9
r=x/.NSolve[f[x]==0,x]
Table[Chop[f[r[[n]]]],{n,1,Length[r]}]
```

which displays the roots

```
{-0.438802-1.07734 i,-0.648326+0.851714 i,
 0.274169+0.783737 i,0.583752-0.43059 i}
```

and their verification

```
{0,0,0,0}
```

5.5 Roots of Transcendental Equations: FindRoot[]

The function **FindRoot** determines numerically one root of the equation $f(x) = g(x)$ starting its search process in the vicinity of x_o, which is an initial guess provided by the user. The function $g(x)$ can be zero. It can also determine numerically one root of a system of M simultaneous equations $f_1(x) = g_1(x), f_2(x) = g_2(x), \ldots, f_M(x) = g_M(x)$. Some or all of the functions $g_k(x)$ can be zero. The function **FindRoot** for one equation is

```
FindRoot[f==g,sea]
```

and for a system of equations is

```
FindRoot[{f1==g1,f2==g2, ... ,fM==gM},{sea1,sea2, ... ,seaM}]
```

where two common forms of **sea** are **{x,xo}** and **{x,xo,xmin,xmax}**. In the latter form, **xmin** and **xmax**, respectively, are the minimum and maximum values of the limits of the search region. If the search for the root attempts to go outside these limits, the search process stops.

Important option: Consider the case where the equation is defined by a function **h[x]** such that a root is found using the following syntax

```
h[x_]:= ...
FindRoot[h[x]==0,sea]
```

As part of implementing **FindRoot**, Mathematica converts the variables to symbols before evaluating them numerically. In certain circumstances, it is necessary to avoid this solution procedure by restricting this variable to be a numerical quantity only. In this situation, the independent variable of the user-defined function is modified as follows

```
h[x_?NumberQ]:=...
```

This modification only allows **h[x]** to be evaluated when the argument is a number. For examples of its usages, see Sections 8.4.2 and 11.5.2.

Example 5.20

Natural Frequencies of a Beam Clamped at Both Ends

Consider the following characteristic equation from which the natural frequency coefficients of a beam clamped at both ends can be determined

$$f(\Omega) = \cos\Omega\cosh\Omega - 1 = 0$$

If an initial guess of $\Omega_o = 1.3\pi$ is assumed, then the lowest natural frequency can be determined from

```
r=Ω/.FindRoot[Cos[Ω] Cosh[Ω]==1,{Ω,1.3 π}]
```

which yields

```
4.73004
```

Since, in general, one doesn't always know what the search region should be, the function can be plotted and the approximate values of the zero crossings noted. This is especially important in an equation such as the one given above, which has an infinite number of such values. In this case, a poorly chosen initial guess could cause **FindRoot** to find the next highest or next lowest root instead of the desired one. In addition, for these types of functions it is often desirable to determine several of the lowest roots (natural frequency coefficients in this case). One way to do this without plotting the expression is to utilize the fact that a zero crossing always entails a sign change. Hence, one can compute the value of the expression for a crudely spaced set of values of $\Omega = \{\Omega_1, \Omega_2, \ldots, \Omega_K\}$ and then sequentially examine the sign of the product $f(\Omega_k)f(\Omega_{k+1})$. A zero crossing takes place when this product is negative. If the sign change occurs at $k = n$, then the initial guess can be taken as $\Omega_o = (\Omega_n + \Omega_{n+1})/2$.

We shall now illustrate one way in which to implement this procedure to find the lowest four roots of $f(\Omega)$ as follows

```
w[Ω_]:=Cos[Ω] Cosh[Ω]-1
Ωe=Range[0.25,25.,0.25];
wt=w[Ωe];
nf={}; (* To use AppendTo the list must be defined *)
```

```
Do[If[wt[[n]] wt[[n+1]]<0,
 AppendTo[nf,
  x/.FindRoot[w[x],{x,(Ωe[[n]]+Ωe[[n+1]])/2}]];
 If[Length[nf]==4,Break[]]],{n,1,Length[wt]-1,1}];
nf
```

The function **Break** is used to exit the **Do** command.

The execution of this program gives

```
{4.73004,7.8532,10.9956,14.1372}
```

Thus, Ω_1 = `nf[[1]] = 4.73004`, Ω_2 = `nf[[2]] = 7.8532`, Ω_3 = `nf[[3]] = 10.9956`, and Ω_4 = `nf[[4]] = 14.1372`.

Example 5.21

Root of a Series Expression

We shall determine the value of t_o that satisfies

$$\theta(e^{-\pi^2 t_o}) = \frac{\pi}{2\sqrt{5}}$$

where

$$\theta(q) = 2q^{1/4} \sum_{k=0}^{10} \frac{(-1)^k}{2k+1} q^{k(k+1)}$$

For an initial guess, we choose $t_o = 0.2$ and the program is

```
θ[q_]:=2 q^(1/4) Total[Table[(-1.)^k q^(k (k+1))/(2 k+1),
 {k,0,10,1}]]
r=to/.FindRoot[θ[Exp[-to π^2]]-π/(2 Sqrt[5]),{to,0.2}]
```

Its execution yields **0.424011**.

5.6 Minimum and Maximum: FindMinimum[] and FindMaximum[]

The minimum and maximum, respectively, of an expression that is a function of one variable can be found with

```
xmn=FindMinimum[{expr,con},{x,xo,xl,xu}]
xmx=FindMaximum[{expr,con},{x,xo,xl,xu}]
```

where **expr** is a function of **x**, **xo** is the initial guess for the search for the minimum or maximum, **xl** is the lower bound for the search region, and **xu** is the upper bound for the search region. Both **xl** and **xu** are optional. The parameter **con** represents any constraints placed on **expr**. The output of these commands is of the form

```
{min,{x->xminval}}
{max,{x->xmaxval}}
```

where **min/max** is the minimum/maximum value of **expr** in the vicinity of **xo** and occurs at **x = xminval/xmaxval**.

The value of **min** is equal to **xmn[[1]]** and the value of **xminval** is equal to **x/.xmn[[2]]**. Similarly for the maximum values, **max** is equal to **xmx[[1]]** and **xmaxval** is equal to **x/.xmx[[2]]**.

The minimum and maximum, respectively, of an expression that is a function of two or more variables can be found with

```
FindMinimum[{expr,con},{{x,xo,xl,xu},{y,yo,yl,yu}, ...}]
FindMaximum[{expr,con},{{x,xo,xl,xu},{y,yo,yl,yu}, ...}]
```

where **expr** are expressions that are functions of **x**, **y**, ... , **xo**, **yo**, ... are the initial guesses for the search for the minimum or maximum, **xl**, **yl**, ... , are the lower bounds for the search region, and **xu**, **yu**, ... , are the upper bounds for the search region. Both **xl**, **yl**, ... , and **xu**, **yu**, ... , are optional. The parameter **con** represents any constraints placed on **expr**. The output of these commands is of the form

```
{maxmin,{x->xval,y->yval, ...}}
```

where **maxmin** is the maximum/minimum value of **expr** in the vicinity of **xo**, **yo**, ... , and occurs at **x = xval**, **y = yval**, ...

Example 5.22

Determination of a Local Maximum and a Minimum

Consider the function

$$f(x) = \frac{1}{(x-0.4)^2+0.03} + \frac{1}{(x-1)^2+0.01} - 6$$

which is plotted in Figure 5.17.

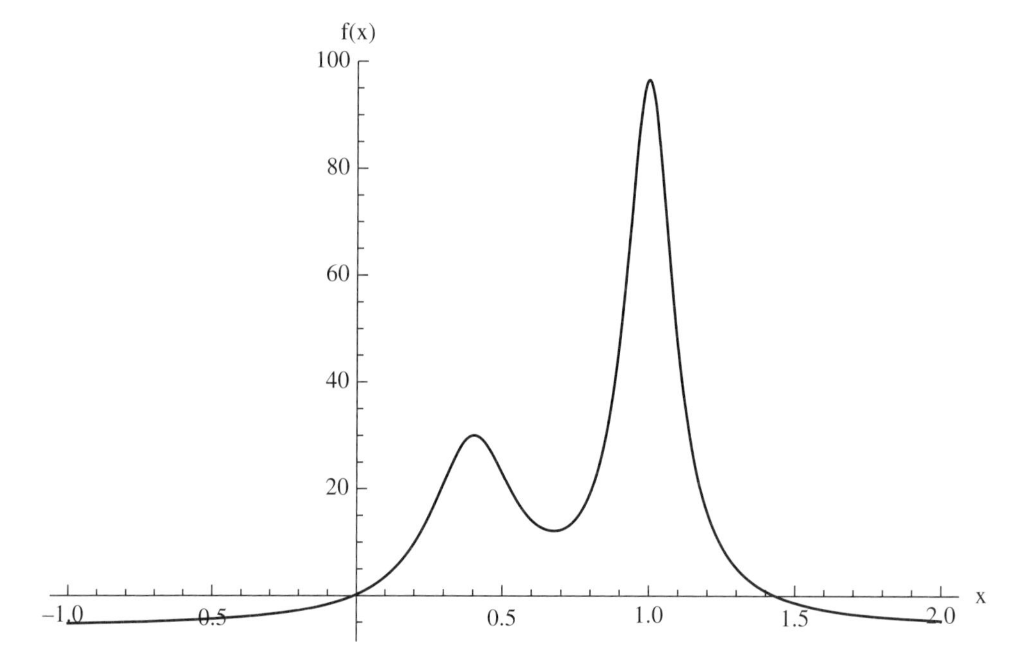

Figure 5.17 Function used in Example 5.22

Based on the graph of $f(x)$ given in Figure 5.17, we shall determine its maximum value in the vicinity of $x_o = 1$ and its minimum value in the vicinity of $x_o = 0.7$. The procedure to obtain these minimum and maximum values is

```
expr[x_]:=1/((x-0.4)^2+0.03)+1/((x-1)^2+0.01)-6;
mn=FindMinimum[expr[x],{x,0.7,0.5,0.9}]
mx=FindMaximum[expr[y],{y,1.0}]
```

which, upon execution, displays

```
{12.1158,{x->0.675927}}
{96.5657,{y->0.999605}}
```

Thus, the minimum occurs at x = `x/.mn[[2]] = 0.675927` where $f(0.675927)$ = `mn[[1]] = 12.1158` and the maximum occurs at y = `y/.mx[[2]] = 0.999605` where $f(0.999605)$ = `mx[[1]] = 96.5657`.

Example 5.23

Determination of a Maximum of a Surface

Consider the function

$$g(x,y) = \frac{1}{\sqrt{\left(0.8 - x^2\right)^2 + (0.5x)^2}\sqrt{\left(1.2 - y^2\right)^2 + (0.6y)^2}}$$

which is plotted in Figure 5.18. We shall determine its maximum value in the vicinity of $x_o = y_o = 1$. Then,

```
g[x_,y_]:=1/(Sqrt[(0.8-x^2)^2+(0.5 x)^2] Sqrt[(1.2-y^3)^2+
  (0.6 y)^2])
p=Quiet[FindMaximum[g[x,y],{x,1,0,2},{y,1,0,2}]]
```

The **Quiet** command suppresses the output of informational messages from Mathematica functions. The execution of these instructions gives

```
{3.71333,{x->0.821584,y->1.02698}}
```

Thus, the maximum occurs at x_{max} = **x/.p[[2,1]] = 0.821584** and y_{max} = **y/.p[[2,2]] = 1.02698** and, therefore, $g(x_{max}, y_{max})$ = **p[[1]] = 3.71333**.

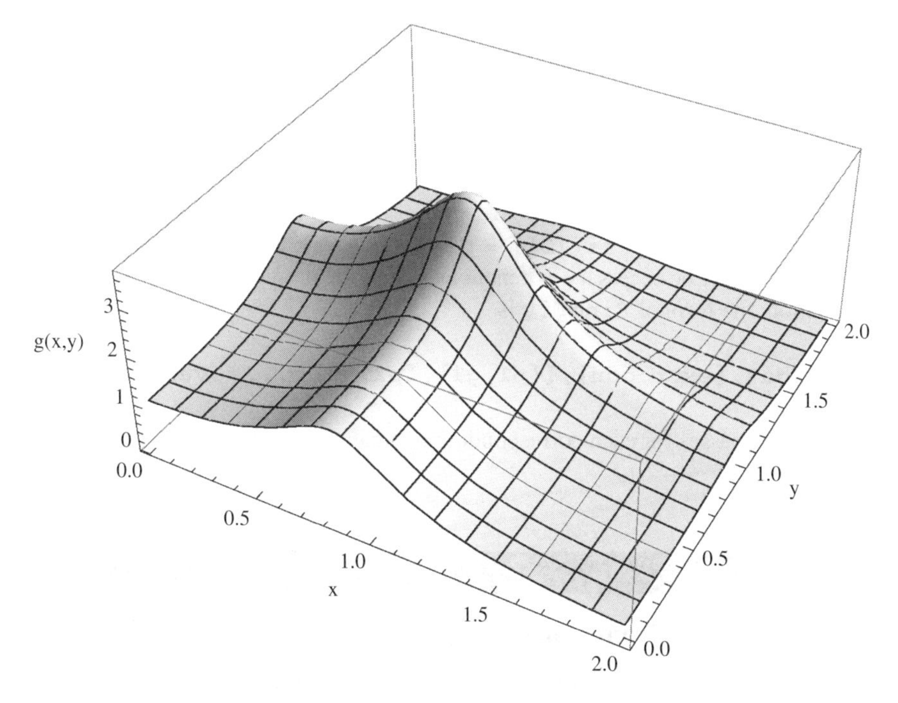

Figure 5.18 Surface used in Example 5.23

5.7 Fitting of Data: `Interpolation[]` and `FindFit[]`

There are several Mathematica functions available for fitting curves and surfaces to data. The two that we shall discuss are **`Interpolation`** and **`FindFit`**. The form for **`Interpolation`** is as follows. Consider a series of M amplitudes $f_1, f_2, \ldots$, that correspond to a set of values $x_1, x_2, \ldots$, such that $f_1 = f(x_1), f_2, = f(x_2), \ldots$. Then,

```
f=Interpolation[{{x1,f1},{x2,f2}, ...}]
```

constructs an interpolation of the function $f(x)$ over the range $x_1 \leq x \leq x_M$. This interpolation can then be used as if it were an ordinary function; that is, one could integrate it, find its roots, differentiate it, and so on.

The command **`FindFit`** is used to determine the parameters in an expression $f(x, y, \ldots, p_1, p_2, \ldots)$ that give the best fit to a set of data. The independent variables are $x, y, \ldots$, and the parameters are $p_1, p_2, \ldots$. The number of parameters and the number of variables are independent of each other. These parameters can be subject to constraints. The form of the expression for **`FindFit`** is

```
param=FindFit[dat,{expr,con},par,var]
```

where **`dat`** are the data in the form **`dat`** = **`{{x1,y1, ... ,f1},{x2,y2, ... , f2}, ...}`**, **`expr`** is an expression that is a function of the parameters **`par`** = **`{p1,p2, ...}`** and the independent variables **`var`** = **`{x,y, ...}`**, and **`con`** are the constraints on the parameters. If no constraints are specified, then this quantity is omitted.

To access the parameters, we use

```
para={p1,p2, ...}/.param
```

so that **`para[[1]] = p1`**, **`para[[2]] = p2, ...`** .

In some cases, it may be necessary or beneficial to provide initial guesses for the parameters. For this situation, the form of **`FindFit`** is

```
FindFit[dat,{expr,con},{{p1,g1},{p2,g2}, ...},var]
```

where **`gN`** are the guesses for each corresponding parameter. Not every parameter needs to be given an initial guess.

Example 5.24

Interpolation Function from Some Data

We shall illustrate **`Interpolation`** by creating a set of data from the following function

$$f(t) = e^{-at} \sin t \quad 0 \leq t \leq 25$$

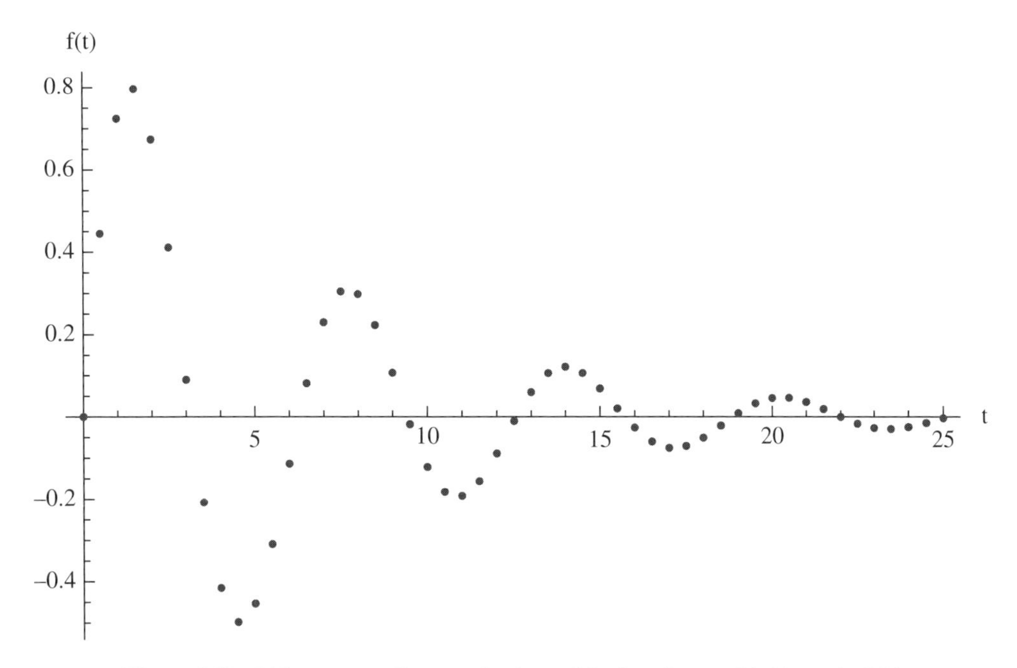

Figure 5.19 Fifty-one equally spaced values of the function used in Example 5.24

where $0 < a < 1$. These data will then be used to create two interpolation functions, one from 26 equally spaced values and the other from 51 equally spaced values. The sampled function using 51 data values is shown in Figure 5.19. Using these two interpolation functions, we shall obtain their derivatives, their integrals, and one of their roots and compare in tabular form these same quantities when determined from $f(t)$ directly. The integral is evaluated over the range $0 \le t \le 25$, the derivative is evaluated at $t = 2.0$, and the root is sought in the vicinity of $t = 3$.

The procedure to perform these operations is

```
fn[t_]:=Exp[-0.15 t] Sin[t]
tab={};
dt={1.,0.5};
Table[r=Range[0,25,dt[[n]]];
  dat=Table[{t,fn[t]},{t,r}];
  f=Interpolation[dat];
  AppendTo[tab,{Length[r],NIntegrate[f[t],{t,0,25}],f'[2],
   t/.FindRoot[f[t],{t,3}]}],{n,{1,Length[dt]}}];
AppendTo[tab,{"exact",NIntegrate[fn[t],{t,0,25}],
  D[fn[t],t]/.t->2.,t/.FindRoot[fn[t],{t,3}]}];
TableForm[tab,TableHeadings->{None,
  {"No. pts","Integral","Derivative","Root"}}]
```

The execution of this program displays

No. pts	Integral	Derivative	Root
26	0.9735	-0.357455	3.14945
51	0.955863	-0.403772	3.14233
exact	0.955654	-0.409333	3.14159

Example 5.25

Function's Parameters from a Fit to Some Data

We shall first create a set of data using the function given in Example 5.24. Then, the function that we shall fit to these data is

$$g = ae^{-bt}\sin(ct)$$

where a, b, and c, are the parameters to be determined and t is the independent variable. Then,

```
fn[t_]:=Exp[-0.15 t] Sin[t]
dat=Table[{t,fn[t]},{t,Range[0,25,1]}];
coe={a,b,c}/.FindFit[dat,a Exp[-b t] Sin[c t],{a,b,c},t]
```

gives

```
{1.,0.15,1.}
```

Thus, $a =$ `coe[[1]] = 1`, $b =$ `coe[[2]] = 0.15`, and $c =$ `coe[[3]] = 1`, which equal the parameters used initially to create the data.

Example 5.26

Parametric Solution to a Nonlinear Differential Equation

We shall use the parameters appearing in a nonlinear differential equation and in one of its initial conditions to fit data obtained from the measured response of a system that can be represented by such an equation. The governing equation is

$$\frac{d^2y}{dt^2} + 2\zeta\frac{dy}{dt} + y + \mu y^3 = 0$$

and the initial conditions are $y(0) = 0$ and $dy(0)/dt = v$. To demonstrate the procedure, we shall create some data by solving this equation for $\zeta = 0.25$, $\mu = 3$, and $v = 8$ and then use these data values in **FindFit**. The data values are obtained by uniformly sampling the solution from $0 \leq t \leq 15$ every $\Delta t = 0.25$ and saving them with **Reap** and **Sow**. The function to be fitted is that created by **ParametricNDSolveValue**. The program to perform these operations is

```
ζ=0.25; μ=3.; v=8.;
{oden,pts}=Reap[NDSolveValue[{y''[t]+2 ζ y'[t]+y[t]+
  μ y[t]^3==0,y[0]==0,y'[0]==v,WhenEvent[Mod[t,0.25]==0,
  Sow[{t,y[t]}]]},y,{t,0,15.}]];
Clear[ζ,μ,v]
fat=ParametricNDSolveValue[{y''[t]+2 ζ y'[t]+y[t]+
  μ y[t]^3==0,y[0]==0,y'[0]==v},y,{t,0,15.},{ζ,μ,v}];
const=FindFit[Flatten[pts,1],fat[ζ,μ,v][t],
  {{ζ,0.5},{μ,2.},{v,5.}},t]
```

Upon execution, the following is displayed

```
{ζ->0.25,μ->3.,v->8.}
```

which are the same values that were used to create the data. It is mentioned that the solution is sensitive to the initial guesses for the parameters.

5.8 Discrete Fourier Transforms and Correlation: Fourier[], InverseFourier[], and ListCorrelate[]

The discrete Fourier transform of a function $f(t)$ that is sampled every time interval Δt over the interval $0 \leq t \leq T$ is obtained from the expression

$$F_n = \frac{1}{M^{(1-a)/2}} \sum_{m=1}^{M} f_m e^{2\pi jb(m-1)(n-1)/M} \tag{5.1}$$

where $f_m = f(m\Delta t)$, $F_n = F(n\Delta f)$ are the transformed values, $\Delta f = 1/T$, $T = M\Delta t$, and M is the number of samples. For convenience, we shall assume that M is an even integer. In Eq. (5.1), F_n has the units of magnitude. The parameters a and b are used to accommodate the various definitions of the Fourier transform: for our purposes, we shall use $a = 1$ and $b = -1$. The value of Δt must satisfy $\Delta t < 1/(\alpha f_h)$, where $\alpha > 2$ and f_h is the highest frequency in $f(t)$. The discrete inverse Fourier transform is obtained from the expression

$$f_n = \frac{1}{M^{(1+a)/2}} \sum_{m=1}^{M} F_m e^{-2\pi jb(m-1)(n-1)/M} \tag{5.2}$$

To use the discrete Fourier transform to approximate the continuous Fourier transform, denoted G_n, we must multiply the discrete value F_n by Δt; that is, $G_n = \Delta t F_n$. Conversely, it is found that $g_n = f_n/\Delta t$. These relationships are valid when $f(t) = 0$ for $t < 0$. Furthermore, it can be shown that the F_n are symmetrical about $M/2$. Therefore, when G_n is displayed as the amplitude in a frequency band Δf centered at $n\Delta f$, it is displayed as $|G_n| = 2\Delta t\Delta f|F_n| = 2|F_n|/M$, where $n = 1, 2, \ldots, M/2$.

In this notation, it can be shown that the total energy in the signal using the time domain samples is

$$E = \Delta t \sum_{n=1}^{M} f_n^2 \tag{5.3}$$

This value is equal to the total power density obtained from the transformed quantities, which is given by

$$E = \frac{2\Delta t}{M} \sum_{n=1}^{M/2} |F_n|^2 \tag{5.4}$$

Consequently, the power density P_n in the frequency band centered at $n\Delta f$ is

$$P_n = \frac{2\Delta t}{M} |F_n|^2 \quad n = 1, 2, \ldots, M/2$$

The numerical implementations of Eqs. (5.1) and (5.2), respectively, are

```
Fn=Fourier[fn,FourierParameters->{a,b}]
```

and

```
fn=InverseFourier[Fn,FourierParameters->{a,b}]
```

In signal processing, auto- and cross-correlation on lists of signal amplitudes can be determined with

```
r12=Listcorrelate[list1,list2,{kL,kR}]
```

where `list1` and `list2` are lists of the same length and for signal processing `kL` = 1 and `kR` = 1 is adequate. When `list1` = `list2`, we have autocorrelation; that is, `r11`. Frequently, `r12` is normalized as `r12`/(`r11[[1]] r22[[1]]`)$^{1/2}$ and `r11` is normalized as `r11/r11[[1]]`, where the index 1 indicates the autocorrelation function at $t = 0$.

We now illustrate the use of these commands. Additional applications are given in Examples 10.3 and 10.4.

Example 5.27

Spectral Analysis of a Sine Wave

We shall examine the sampling of m periods of a sine wave of frequency f_o and a maximum magnitude of 2.5; that is,

$$f_n = f(n\Delta t) = 2.5\sin(2\pi f_o n\Delta t) \quad 1 \le n \le \alpha m$$

where α is the number of samples per period. Hence, is it seen that $\Delta t = 1/(\alpha f_o)$ (since $f_h = f_o$) and $M = m\alpha$. It is assumed that $f_o = 10$ Hz, $\alpha = 50$ and $m = 4$. We shall display the values of the sampled signal, plot the magnitude of the components of the frequency spectrum given by $|G_n|$, and take the inverse Fourier transform to recover the original signal. In addition, we shall compute the total energy in the waveform using Eqs. (5.3) and (5.4).

In the program that follows, **datp**, **datf**, and **ifdt** are the coordinate pairs that are placed in a format required by **ListPlot** and **ListLinePlot**. In addition, **dat** are the values of the sampled waveform, **ft** are the complex values of the Fourier transform of the sampled values, and **ift** are the values of the inverse of **ft**. In plotting $|G_n|$, only the first 20 data values are used. The option **Filling** is used to connect the data values with a straight line that ends on the x-axis; this option is illustrated in Table 6.15. The results are shown in Figure 5.20.

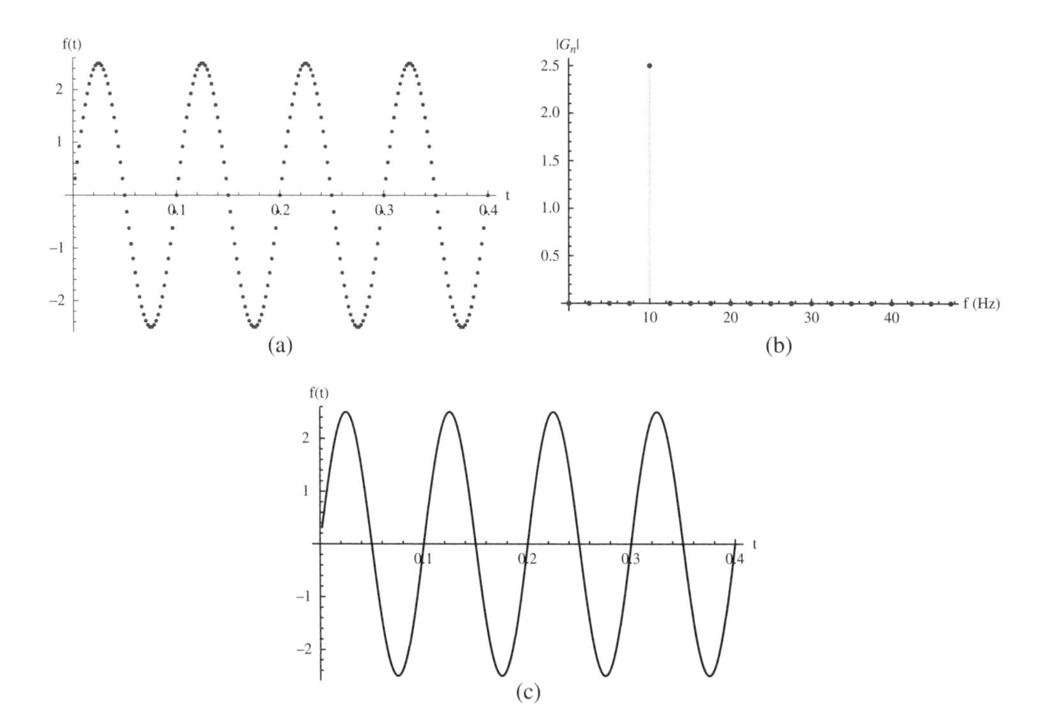

Figure 5.20 (a) Sampled waveform, (b) spectral plot of the sampled waveform, and (c) inverse Fourier transform

```
m=4.0; α=50.0; nn=m α; fo=10.0; dt=1/(α fo);
dat=Table[2.5 Sin[2 π fo n dt],{n,nn}];
datp=Table[{k dt,dat[[k]]},{k,1,nn}];
ListPlot[datp,AxesLabel->{"t","f(t)"}]
ft=Fourier[dat,FourierParameters->{1,-1}];
datf=Table[{(k-1)/(nn dt),2. Abs[ft[[k]]]/(m α)},
 {k,1,nn/2-1}];
ListPlot[datf[[1;;20]],PlotRange->All,Filling->Axis,
 AxesLabel->{"f (Hz)","|Gn|"}]
ift=InverseFourier[ft,FourierParameters->{1,-1}];
ifdt=Table[{k dt,ift[[k]]},{k,1,nn}];
ListLinePlot[ifdt,AxesLabel->{"t","f(t)"}]
daf2=Table[Abs[ft[[k]]]^2,{k,1,nn/2-1}];
Print["E from time data = ",dt Total[dat^2]]
Print["E from transformed data = ",2 dt/nn Total[daf2]]
```

The following values of the total energy are displayed

```
E from time data = 1.25
E from transformed data = 1.25
```

Example 5.28

Spectral Analysis of a Sine Wave of Finite Duration

We again consider the sign wave of Example 5.27, except it is now sampled over a time $T = Km$, where $K \geq 1$ is an integer and m is the number of periods of the sine wave. We assume that $f_o = 10$ Hz, $\alpha = 50$, $m = 4$, and $K = 8$. It is mentioned that as K increases, the frequency resolution increases; that is, Δf becomes smaller. We shall obtain a plot of the amplitudes $|G_n|$ comprising the components of its amplitude spectrum and display them over the range $0 \leq f \leq 20$ Hz. The maximum number of amplitudes to be plotted is chosen to be the integer value of $20Km/f_o$. Thus,

```
kk=8; m=4.0; α=50.0; nn=kk m α; fo=10.0; dt=1/(α fo);
dat=Table[Piecewise[{{Sin[2 π fo n dt],1<=n<=α m},
 {0,α m<=n<=nn}}],{n,nn}];
ft=Fourier[dat,FourierParameters->{1,-1}];
datf=Table[{(k-1)/(nn dt),2. Abs[ft[[k]]]/nn},{k,1,nn/2}];
last=Round[20 kk m/fo];
ListPlot[datf[[1;;last]],Filling->Axis,
 AxesLabel->{"f (Hz)","|Gn|"}]
```

produces the result shown in Figure 5.21.

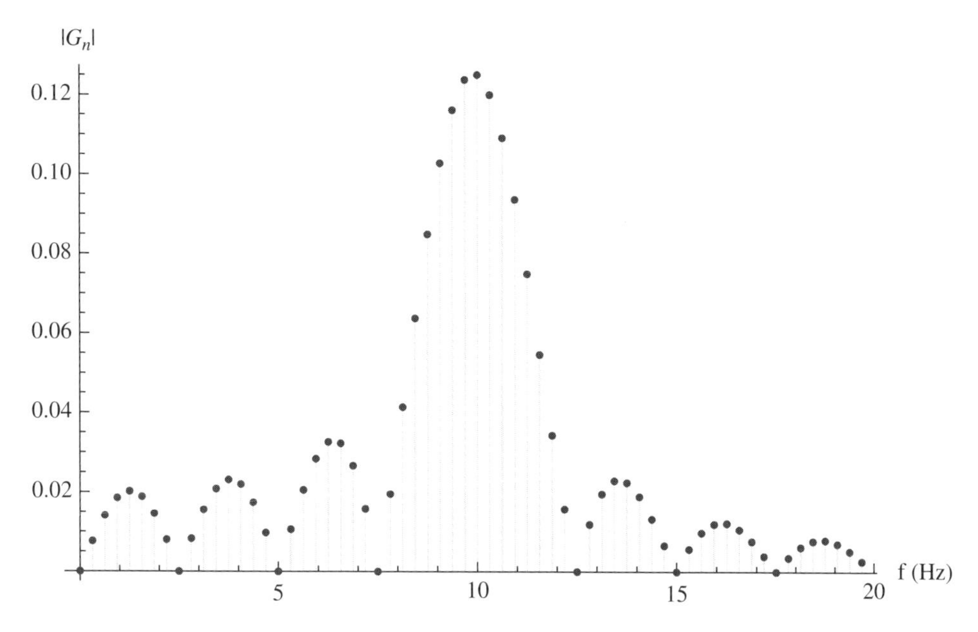

Figure 5.21 Amplitude spectral plot of a 10 Hz sine wave with a duration of four periods that is sampled over a time interval of thirty-two periods

Example 5.29

Cross-Correlation of a Signal with Noise

We shall obtain the normalized cross-correlation of a sine wave with a sine wave with noise. The amplitude of the noise will not exceed 20% of the sine wave amplitude. We use a procedure employed in the previous two examples to generate the sampled waveform and we use **RandomReal** to create the noise. The signal with noise is displayed along with the cross-correlation of the signals. The program that produces Figure 5.22 is as follows.

```
m=4.0; α=50.0; nn=m α; fo=10.0; dt=1/(α fo);
dat=Table[Sin[2 π fo n dt],{n,nn}];
datn=dat+0.2 RandomReal[{-1,1},nn];
datp=Table[{k dt,datn[[k]]},{k,1,nn}];
ListLinePlot[datp]
datauto=ListCorrelate[dat,dat,{1,1}];
datnauto=ListCorrelate[datn,datn,{1,1}];
datcross=ListCorrelate[dat,datn,{1,1}];
datplot=Table[{k dt,datcross[[k]]/
  Sqrt[datauto[[1]] datnauto[[1]]]},{k,1,nn}];
ListLinePlot[datplot]
```

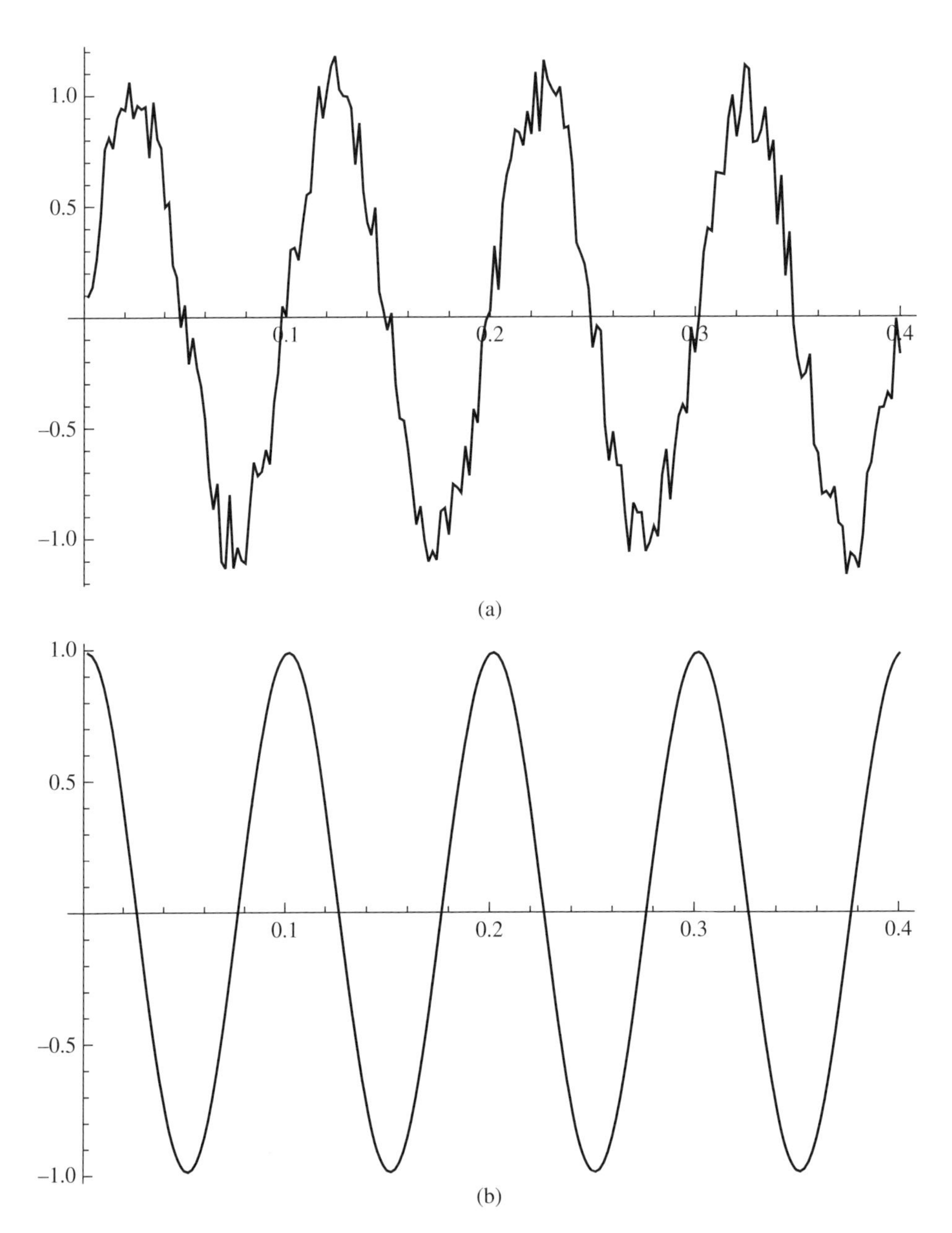

Figure 5.22 (a) Sine wave with noise and (b) cross correlation of a sine wave with a sine wave with noise

5.9 Functions Introduced in Chapter 5

The functions introduced in this chapter are listed in Table 5.1.

Table 5.1 Commands introduced in Chapter 5

Command	Purpose
`Break`	Exit a `Do` or `While` command
`FindFit`	Finds the numerical values of the parameters appearing in an expression that will provide the best fit to a set of data
`FindMaximum`	Searches for a numerical approximation to a local maximum of an expression in a specified region
`FindMinimum`	Searches for a numerical approximation to a local minimum of an expression in a specified region
`FindRoot`	Searches for a numerical approximation to a root of an expression in a specified region
`Fourier`	Determines the discrete Fourier transform of a list of numerical values
`Interpolation`	Constructs a function that approximates a given set of data
`InverseFourier`	Determines the discrete inverse Fourier transform on a list of numerical values
`ListCorreleate`	Determines the cross-correlation of two lists of numerical values
`NDSolveValue`	Finds the numerical solution to one or a set of ordinary differential equations or to a set of partial differential equations with two independent variables
`NIntegrate`	Obtains a numerical approximation to a definite integral or definite multiple integral
`NSolve`	Attempts to obtain a numerical solution to an equation or a system of equations
`ParametricNDSolveValue`	Finds the numerical solution to one or a set of ordinary differential equations or to a partial differential equation with two independent variables as a function of one or more parameters
`Quiet`	Suppresses the display of any messages generated by the implementation of a Mathematica function
`RandomReal`	Generates a pseudorandom real number

References

[1] S. Chatterjee, A. K. Mallik, and A. Ghosh, "On impact dampers for non-linear vibrating systems", *Journal of Sound and Vibration*, 1995, 187(3), pp. 403–420.

[2] F. C. Moon, "Experiments on chaotic motions of a forced nonlinear oscillator: strange attractors," *ASME Journal of Applied Mechanics*, 1980, 47(3), pp. 638–644.

[3] Y. Zhu, H. N. Oğuz, and A. Prosperetti, "On the mechanism of air entrainment by liquid jets at a free surface," *Journal of Fluid Mechanics*, 2000, 404, pp. 151–177.

[4] S. Timoshenko and S. Woinowsky-Krieger, *Theory of Plates and Shells*, 2nd edn, McGraw Hill, New York, 1959.

[5] F. P. Incropera and D. P. Dewitt, *Introduction to Heat Transfer*, 4th edn, John Wiley & Sons, New York, 2002.

[6] T. Dahlberg, "Procedure to calculate deflections of curved beams," *Journal of Engineering Education*, 2004, 20(3), pp. 503–513.

[7] S. Kostić, I. Franović, K. Todorović, and N. Vasović, "Friction memory effect in complex dynamics of earthquake model", *Nonlinear Dynamics*, 2013, 73(3), pp. 1933–1943.

[8] B. S. Shvartsman, "Analysis of large deflections of a curved cantilever subjected to a tip-concentrated follower force," *International Journal Nonlinear Mechanics*, 2013, 50, pp. 75–80.
[9] A. Belendez, A. Hernandez, T. Belendez, M. L. Alvarez, S. Gallego, M. Ortuno, and C. Neipp, "Application of the harmonic balance method to a nonlinear oscillator typified by a mass attached to a stretched string,", *J. Sound Vibration*, 2007, 302, pp. 1018–1029.
[10] A. S. dePaula, M. A. Savi, and F. H. I. Pereira-Pinto, "Chaos and transient chaos in an experimental nonlinear pendulum," *Journal of Sound and Vibration*, 2006, 295, pp. 585–595.
[11] Y. Starosvetsky and O.V. Gendelman, "Interactions of nonlinear energy sink with a two degrees of freedom system: Internal resonance," *Journal of Sound and Vibration*, 2010, 329, pp. 1836–1852.
[12] J. J. Thomsen, *Vibrations and Stability*, 2nd edn, Springer, Berlin, 2003.
[13] A. A. Al-Qaisia and M. N. Hamdan, "Subharmonic resonance and transition to chaos of nonlinear oscillators with a combined softening and hardening nonlinearities," *Journal of Sound and Vibration*, 2006, 305, pp. 772–782.
[14] S. S. Rao, *Engineering Optimization*, John Wiley & Sons, New York, 1996, p.706.
[15] I. G. Currie, *Fundamentals of Fluid Mechanics*, McGraw-Hill, New York, 1993, p. 181.
[16] B. R. Munson, D. F. Young, and T. H. Okiishi, *Fundamentals of Fluid Mechanics*, 4th edn, John Wiley & Sons, New York, 2002, p. 641.
[17] J. H. Duncan, Chapter 11, Fluid Mechanics, in E. B. Magrab, et al., *An Engineer's Guide to MATLAB*, 3rd edn, Prentice Hall, Upper Saddle River, New Jersey, 2011, p. 618.
[18] E. E. Lundquist and E. Z. Stowell, "Critical compressive stress for flat rectangular plates supported along all edges and elastically restrained against rotation along the unloaded edges," NACA Report 733, 1942.

Exercises

Section 5.2

5.1 Show numerically that

$$\int_{2}^{\sqrt{6}} \frac{du}{\sqrt{6-u^2}} = \cot^{-1}\sqrt{2}$$

5.2 The total emissive power of a black body is determined from [5, p. 676]

$$E_b = \int_{0}^{\infty} \frac{c_1 d\lambda}{\lambda^5\left(e^{c_2/(\lambda T)}-1\right)} = \sigma T^4$$

where $c_1 = 3.742 \times 10^8$ W μm^4·m^{-2}, $c_2 = 1.4388 \times 10^4$ μm·K, and $\sigma = 5.670 \times 10^{-8}$ W·m^{-2}·K^{-4}. Show that for $T = 500$ K the difference between the exact solution (right-hand side) and the approximate solution (numerical solution to the integral) is 0.006272%.

5.3 In determining the deflection of curved beams, the following relationship is obtained [6]

$$\delta(\beta) = I_1(\beta) + 1.3 I_2(\beta)$$

where δ is the nondimensional deflection, $\beta \geq 0$, and

$$I_1(\beta) = \int_0^1 \frac{\left[-\beta^2\xi - (1-\xi)\right]^2}{\sqrt{1+y^2}} d\xi$$

$$I_2(\beta) = \int_0^1 \frac{\left[\beta\sqrt{1-\xi^2} + (1-\xi)y\right]^2}{\sqrt{1+y^2}} d\xi$$

$$y = \frac{-\beta\xi}{\sqrt{1-\xi^2}}$$

Find the value of $\delta(1.5)$.

5.4 Show numerically for $z = 0.5$ and $y = 0.15$ that

$$\frac{1}{\pi}\int_0^\pi e^{y\cos\theta}\cos(z\sin\theta)d\theta = J_0\left(\sqrt{z^2-y^2}\right)$$

where $J_0(x)$ is the Bessel function of the first kind of order zero.

Section 5.3

5.5 Determine the solution to the following system of nonlinear ordinary differential equations

$$\frac{dy_1}{dt} = y_2$$

$$\frac{dy_2}{dt} = -\left(\frac{2}{L}\frac{dL}{dt} + \gamma L\right)y_2 - \frac{1}{L}\sin(y_1)$$

over the range $0 \leq t \leq 30$ where

$$L = 1 + \varepsilon\sin^7(\omega t + 9\pi/8)$$

$$\frac{dL}{dt} = 7\varepsilon\omega\sin^6(\omega t + 9\pi/8)\cos(\omega t + 9\pi/8)$$

The initial conditions are $y_1(0) = -1$ and $y_2(0) = 1$ and the constants have the following values: $\varepsilon = 0.16$, $\gamma = 0.4$, and $\omega = 0.97$. Plot $y_1(t)$ versus $y_2(t)$ using `ParametricPlot`.

5.6 The nondimensional axial velocity U_z of blood in a large artery undergoing sinusoidal oscillations can be can be determined from

$$\frac{\partial U_z}{\partial\tau} = -\sin\tau + \beta\left(\frac{1}{\eta}\frac{\partial}{\partial\eta}\left(\eta\frac{\partial U_z}{\partial\eta}\right)\right)$$

where $0 \le \eta \le 1$, τ is a nondimensional time, and β is a constant. The initial condition is

$$U_z(\eta, 0) = 0$$

and the boundary conditions are

$$U_z(1, \tau) = 0$$
$$\left.\frac{\partial U_z}{\partial \eta}\right|_{\eta=0} = 0$$

Solve this system and plot U_z as a function of η and τ, $0 \le \tau \le 10$, using **Plot3D** when $\beta = 0.05$. This equation has a discontinuity at $\eta = 0$ so that the technique given in Example 5.16 will have to be used.

5.7 The nondimensional equation for heat conduction in a slab is given by

$$\frac{\partial \theta}{\partial \tau} = \frac{\partial^2 \theta}{\partial \xi^2}$$

where $\theta = \theta(\xi,\tau)$ is proportional to the temperature in the slab. We are interested in the region $0 \le \tau \le 4.0$ and $0 \le \xi \le 1$. The initial condition is $\theta(\xi,0) = 0$ and the boundary conditions are

$$\theta(0, \tau) = \tau e^{-\tau}$$
$$\theta(1, \tau) = 0$$

Solve this system and use **Plot3D** to display θ.

5.8 The equilibrium equations in terms of nondimensional quantities that govern the large deflections of a solid circular plate of constant thickness are given by [4, p. 402]

$$\frac{d^2u}{d\eta^2} + \frac{1}{\eta}\frac{du}{d\eta} - \frac{u}{\eta^2} = -\frac{1-\nu}{2\eta}\left(\frac{dw}{d\eta}\right)^2 - \frac{dw}{d\eta}\frac{d^2w}{d\eta^2}$$

$$\frac{d^2w}{d\eta^2} + \frac{1}{\eta}\frac{dw}{d\eta} - \frac{w}{\eta^2} = h_a\frac{dw}{d\eta}\left[\frac{du}{d\eta} + \nu\frac{u}{\eta} + \frac{1}{2}\left(\frac{dw}{d\eta}\right)^2\right] + q_o\frac{\eta}{2}$$

If the plate is clamped along the outer boundary, the boundary conditions are

$$u(1) = 0, \quad w(1) = 0, \quad w'(1) = 0$$

At the center of the plate, symmetry is used to require that

$$u(0.001) = 0, \quad w'(0.001) = 0$$

where, to avoid the discontinuity at $\eta = 0$, the center of the plate is approximated by a hole of radius $\eta = 0.001$.

The displacement of a solid circular plate using linear theory that is clamped at its outer edge and subjected to a constant load is given by

$$w(\eta) = \frac{q_o}{64}\left(1 - 2\eta^2 + \eta^4\right)$$

Use these equations to plot the linear and nonlinear transverse displacements of the plate when $\nu = 0.3$, $h_a = 0.001$, and $q_o = 10000$.

5.9 The following three coupled nonlinear equations in terms of nondimensional quantities are given [7]

$$\frac{d\theta}{dt} = -v\left[\theta + (1+\varepsilon)\ln v\right]$$

$$\frac{du}{dt} = v - 1$$

$$\frac{dv}{dt} = -\gamma^2\left[u + (\theta + \ln v)/\xi\right]$$

Obtain the solution to these equations when $u(0) = v(0) = 1$ and $\theta(0) = 0$ and for $\varepsilon = 0.5$, $\gamma = 0.8$, and $\xi = 0.5$. Display the results two ways: as a plot of $v(t)$ versus t using `Plot` and as phase portrait of $u(t)$ versus $v(t)$ using `ParametricPlot`. In both displays, let $0 \le t \le 1000$.

5.10 Consider a curved cantilever beam of initial radius of curvature R, length L, Young's modulus E, and moment of inertia I. A load P is applied to the free end of the beam at an angle α, where $\alpha = 0$ indicates a load that is parallel to the axis of the beam and an angle $\alpha = \pi/2$ indicates a load perpendicular to axis of the beam. If s is the arc length of the beam and $\varphi(s)$ is the slope of the centroidal axis of the beam, then the equation from which the deflection of the beam can be determined is obtained from [8]

$$\theta''(\eta) + p_o\theta(\eta) = 0$$

where $\eta = s/L$, the prime denotes the derivative with respect to η, and

$$\varphi(\eta) = \theta(\eta) - \theta(L)$$

$$p_o = \frac{PL^2}{EI}$$

The boundary conditions are given by

$$\theta(0) = \alpha \quad \text{and} \quad \theta'(0) = -\frac{L}{R}$$

The nondimensional Cartesian coordinates of a point η on the beam are given by

$$x(\eta) = \int_{\eta}^{1} \cos(\varphi(\rho))d\rho$$

$$y(\eta) = \int_{\eta}^{1} \sin(\varphi(\rho))d\rho$$

which describe the deformed shape of the beam.

Determine the shape of the beam for the case of $\alpha = 0$ and $p_o = \{0.02, 3, 9, 12\}$ and for the case of $\alpha = \pi/2$ and $p_o = \{0.02, 5, 15, 25, 35\}$. The results should look like those shown in Figure 5.23. Assume that the initial shape of the beam is a semicircle so that $L/R = \pi$. The sign of the y-coordinate has been changed to obtain the curves shown in this figure. The topmost curve belongs to the largest value of p_o.

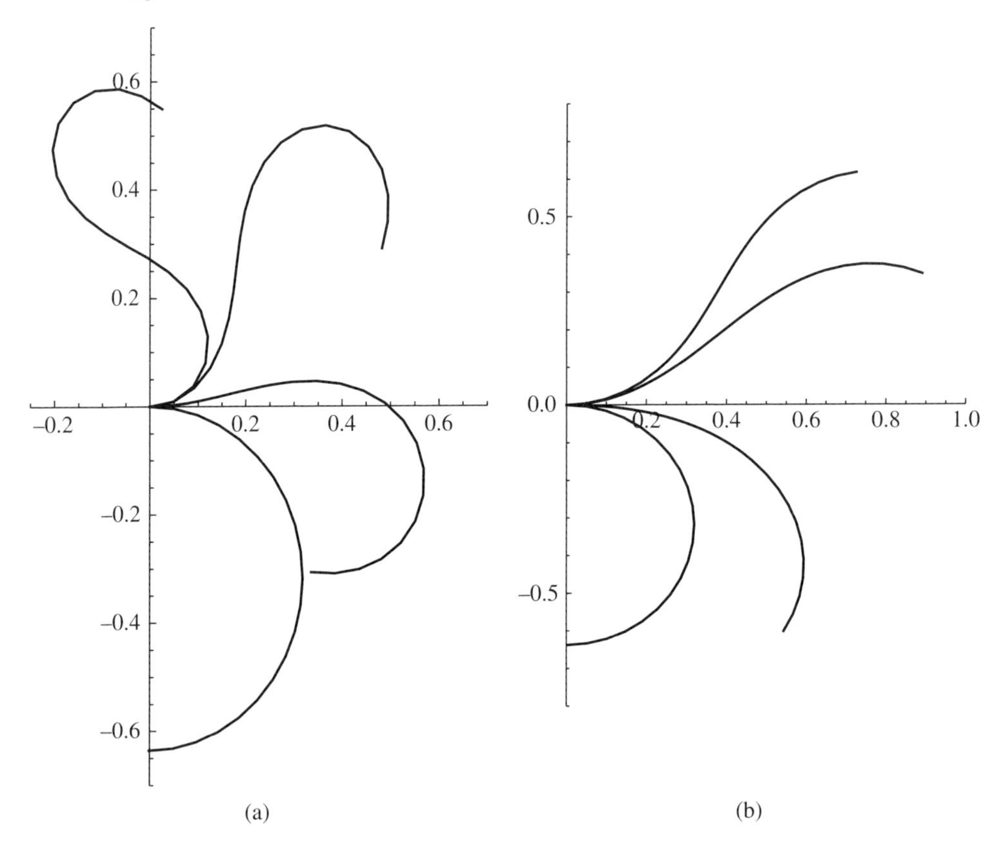

Figure 5.23 Solution to Exercise 5.10: (a) $\alpha = 0$ and $p_o = \{0.02, 3, 9, 12\}$ (b) $\alpha = \pi/2$ and $p_o = \{0.02, 5, 15, 25, 35\}$

5.11 Given the following nonlinear equation [9]

$$\frac{d^2y}{d\tau^2} + y - \frac{\lambda y}{\sqrt{1+y^2}} = 0$$

where $0 < \lambda \le 1$. The initial conditions are

$$y(0) = A, \quad \dot{y}(0) = 0$$

where the over dot indicates the derivative with respect to τ. The solution to this system is periodic with period T. Determine the frequency $\omega = 2\pi/T$ for the following set of parameters: $\lambda = \{0.1, 0.9, 1\}$ and $A = \{0.1, 1, 10\}$. Display the results in tabular form with nine rows of values and the headings λ, A, and ω.

5.12 Given the following nonlinear equation [10]

$$\frac{d^2\varphi}{d\tau^2} + \zeta\frac{d\varphi}{d\tau} + \varphi + \frac{2\mu}{\pi}\tan^{-1}\left(q\frac{d\varphi}{d\tau}\right) + \omega_r \sin\varphi = f\left(\sqrt{1+\alpha^2 - 2\alpha\cos\Omega\tau} - (1-\alpha)\right)$$

where all quantities are nondimensional. For $\omega_r = 2.4073$, $\zeta = 0.033673$, $\mu = 0.0447031$, $q = 10^6$, $\alpha = 0.375$, $f = 3.3333$, and $\Omega = 1.38648$, obtain the plot shown in Figure 5.24. In this case, the sampling interval is $2\pi/\Omega$ and let $0 \le \tau \le 60{,}000$. In **`NDSolveValue`**, set **`MaxSteps`** to **`Infinity`**.

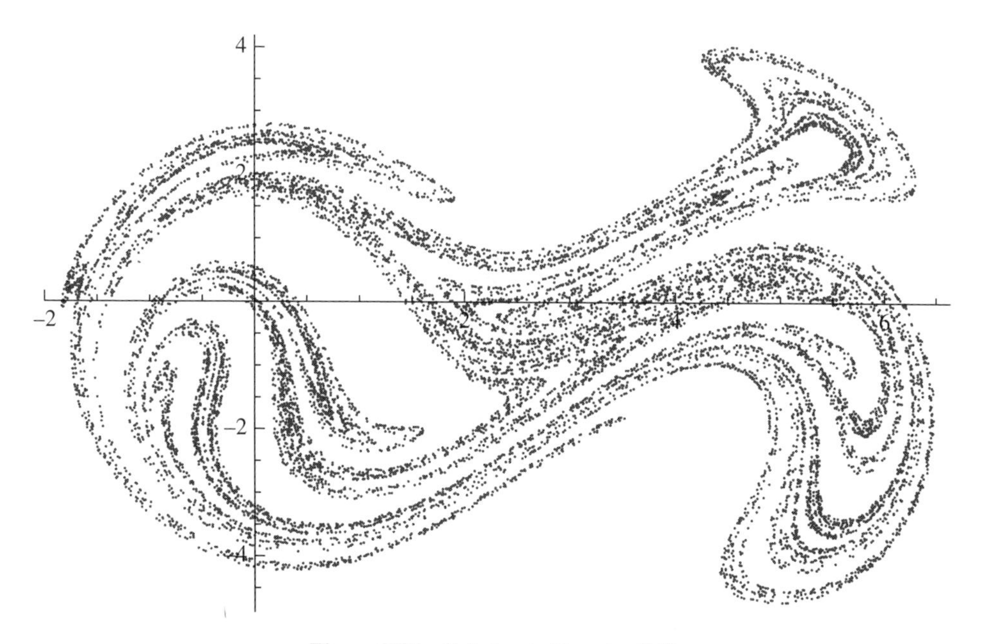

Figure 5.24 Solution to Exercise 5.12

5.13 Given the following three coupled nonlinear nondimensional equations [11]

$$\frac{d^2y_1}{d\tau^2} + y_1 + \varepsilon k_1 \left(y_1 - y_2\right) = 0$$

$$\frac{d^2y_2}{d\tau^2} + y_2 + \varepsilon k_1 \left(y_2 - y_1\right) + \varepsilon\lambda \left(\frac{dy_2}{d\tau} - \frac{dv}{d\tau}\right) + \varepsilon k_a \left(y_2 - v\right)^3 = 0$$

$$\frac{d^2v}{d\tau^2} + \varepsilon\lambda \left(\frac{dv}{d\tau} - \frac{dy_2}{d\tau}\right) + \varepsilon k_a \left(v - y_2\right)^3 = 0$$

where $\varepsilon \ll 1$. For $\lambda = 0.2$, $k_a = 4/3$, $k_1 = 1$, $\varepsilon = 0.01$, and $0 \le \tau \le 2000$, plot $y_2(\tau) - v(\tau)$ for the following three sets of initial conditions: (1) $\dot{y}_1(0) = 1$ and $\dot{y}_2(0) = 0$; (2) $\dot{y}_1(0) = 1.5$ and $\dot{y}_2(0) = 0$; and $\dot{y}_1(0) = 1.5$ and $\dot{y}_2(0) = 1.5$. The over dot indicates the derivative with respect to τ. In all three sets of initial conditions $y_1(0) = y_2(0) = v(0) = \dot{v}(0) = 0$. In using, **NDSolveValue** set the option **MaxSteps** to 50,000.

5.14 Given the following equation [12, p. 72]

$$\frac{d^2\theta}{d\tau^2} + 2\zeta\frac{d\theta}{d\tau} + \left(1 - \Omega^2 \cos(\Omega\tau)\right) \sin\theta = 0$$

with the initial conditions $\theta(0) = \theta_o$ and $\dot{\theta}(0) = r_o$, where the over dot indicates the derivative with respect to τ. For $\zeta = 0.04$, $\theta_o = -\pi/2$, $r_o = 0.5$, $q = 0.15$, $\Omega = 2.0$, and $0 \le \tau \le 300$, obtain two different plots of the results: (a) a phase portrait of $\theta(\tau)$ versus $\dot{\theta}(\tau)$ and (b) $\theta(\tau)$ versus τ.

5.15 Given the following equation [12, p. 132]

$$\left(1 + m_r u^2\right) \frac{d^2u}{d\tau^2} + m_r u \left(\frac{du}{d\tau}\right)^2 + u + \beta u^3 = 0$$

with the initial conditions $u(0) = u_o$ and $\dot{u}(0) = 0$, where the over dot indicates the derivative with respect to τ. For $m_r = 0.55$, $\beta = 1$, $u_o = 6$, and $0 \le \tau \le 15$, obtain two different plots of the results: (a) a phase portrait of $\theta(\tau)$ versus $\dot{\theta}(\tau)$ and (b) $\theta(\tau)$ versus τ.

5.16 Given the following pair of coupled nonlinear equations [12, p. 274]

$$\frac{d^2f}{d\tau^2} + 2\beta\frac{df}{d\tau} + \left(1 - m\omega^2 u\right) f = 0$$

$$\frac{d^2u}{d\tau^2} + 2\beta\omega\frac{du}{d\tau} + \omega^2 u + \kappa\left(f\frac{d^2f}{d\tau^2} + \left(\frac{df}{d\tau}\right)^2\right) = \frac{q}{m}\cos(\Omega\tau)$$

with zero initial conditions. Obtain a Poincare plot (see Example 5.11) when $q = 0.2$, $\omega = 2.44$, $\beta = 0.03$, $m = 3.2$, $\kappa = 2.63$, $\Omega = 2.4$ and $0 \le \tau \le 30{,}000$.

Section 5.4

5.17 Given the following nonlinear algebraic equations [13]

$$A_3\left(-\frac{5\varepsilon_1\Omega^2}{18}\left(2A_1^2+A_3^2+B_3^2\right)+\frac{3\varepsilon_2}{4}\left(2A_1^2+A_3^2+B_3^2\right)-\frac{\Omega^2}{9}+1\right)+$$

$$A_1\left(A_3^2-B_3^2\right)\left(\frac{\varepsilon_1\Omega^2}{6}+\frac{3\varepsilon_2}{4}\right)+\frac{B_3 d\Omega}{3}=0$$

$$B_3\left(-\frac{\varepsilon_1\Omega^2}{18}\left(10A_1^2+A_3^2+B_3^2\right)+\frac{3\varepsilon_2}{4}\left(2A_1^2+A_3^2+B_3^2\right)-\frac{\Omega^2}{9}+1\right)+$$

$$A_1A_3B_3\left(\frac{\varepsilon_1\Omega^2}{3}-\frac{3\varepsilon_2}{2}\right)-\frac{A_3 d\Omega}{3}=0$$

where

$$A_1=-\frac{9P}{8\Omega^2}$$

Solve for A_3 and B_3 when $P = 10.0$, $\varepsilon_1 = 0.2$, $\varepsilon_2 = 1.0$, $d = 0.02$, and $\Omega = 3.8$.

Section 5.5

5.18 Determine the value of p for which

$$\left(1+k_3\right)h^3+\left(2\left(1+k_3\right)+k_1k_3-\left(1+k_3\right)y\right)h^2+k_1k_3\left(1+k_2-y\right)h-yk_1k_2k_3=0$$

when $h = 10^{-p}$, $k_1 = 2.5 \times 10^{-4}$, $k_1 = 5.6 \times 10^{-11}$, $k_3 = 1.7 \times 10^{-3}$, and $y = 1$ and $y = 2$.

5.19 Determine the value of x that satisfies

$$x=\left(1+x^{-1}\right)^x$$

5.20 When $\alpha = 0.9$, determine the value of λ that satisfies

$$\frac{\alpha}{\lambda}=\tanh\lambda$$

5.21 Given the following equation for the temperature $T = T(t)$ of a systems that represents a spherical thermocouple with convection conditions and includes radiation exchange with its surrounding walls [5, p. 250.]

$$c_2\frac{dT}{dt}=-\left(T-T_\infty+c_1\left(T^4-T_{sur}^4\right)\right)$$

In this equation, $c_1 = 1.27575 \times 10^{-10}$ K^{-3} and $c_2 = 0.991667$ s.

If it is assumed that $T_\infty = 473.15$ K, $T_{sur} = 673.15$ K, and $T(0) = 298.15$ K, then determine the time t_s at which $T(t_s) = 490.85$ K.

5.22 Find the value of λ that satisfies

$$\frac{1}{\sqrt{\lambda}} + 2\log_{10}\left[\frac{2.51}{\text{Re}\sqrt{\lambda}} + \frac{0.27}{d/k}\right] = 0$$

when $d/k = 200$ and $\text{Re} = 8000$.

5.23 Create a table with the appropriate headings that display the lowest three values of Ω_{mn} that satisfy

$$J_m(\Omega_{mn})I_{m+1}(\Omega_{mn}) + I_m(\Omega_{mn})J_{m+1}(\Omega_{mn}) = 0$$

for $m = 0$, 1, and 2 and $n = 1, 2, 3$. The quantity J_m is the Bessel function of the first kind of order m and I_m is the modified Bessel function of the first kind of order m. The use of the technique to find the initial guesses for a function with multiple roots shown in Example 5.20 should prove useful.

5.24 Consider the following differential equation as a function n, $n = 0, 1, \ldots$.

$$\frac{d^2\theta_n}{dx^2} + \frac{2}{x}\frac{d\theta_n}{dx} + \theta_n^n = 0$$

The boundary conditions are $\theta(0) = 1$ and $d\theta(0)/dx = 1$. This system has an analytical solution for $n = 0$, which is

$$\theta_0(x) = 1 - \frac{x^2}{6}$$

and for $n = 1$, which is

$$\theta_1(x) = \frac{\sin x}{x}$$

Use **`ParametricNDSolveValue`** and **`FindRoot`** to determine the smallest value of x, $2 < x < 6$, that satisfies $\theta(x) = 0$ for $n = 0$, 1, and 2. To avoid the discontinuity at $x = 0$, specific the initial conditions at $x = 0.001$. Compare your results with those obtained from the analytical solutions.

Section 5.6

5.25 The following expression can be used to determine the normalized contact shear stress between two spheres of the same material that are kept in contact by an external force

$$\tau_s = \frac{3}{4\left(1+\eta^2\right)} - \frac{1+\nu}{2}\left(1 - \eta\tan^{-1}\frac{1}{\eta}\right)$$

A positive η is the radial distance in one of the spheres and a negative η is the radial distance from the plane of contact in the other sphere. Determine the maximum value of τ_s in each sphere and the values of η at which these maxima occur. Assume that $\nu = 0.3$ in each sphere.

5.26 The following expression can be used to determine the normalized contact shear stress between two cylinders of the same material that are kept in contact by an external force

$$\tau_c = \frac{1}{2\sqrt{1+\xi^2}} + \xi - \sqrt{1+\xi^2}\left(1 - \frac{1}{2\left(1+\xi^2\right)}\right)$$

Determine the maximum magnitude of τ_c for $\xi > 0$ when $\nu = 0.3$ and the value of ξ where the maximum occurs.

5.27 Minimize the following function [14]

$$f = 2x_1 + x_2 + \sqrt{2}x_3$$

subject to the constraints that $x_j > 0, j = 1, 2, 3$, and

$$g_1 = 1 - \frac{\sqrt{3}x_2 + 1.932x_3}{d(x_1, x_2, x_3)} \geq 0$$

$$g_2 = 1 - \frac{0.634x_1 + 2.828x_3}{d(x_1, x_2, x_3)} \geq 0$$

$$g_3 = 1 - \frac{0.5x_1 - 2x_2}{d(x_1, x_2, x_3)} \geq 0$$

$$g_4 = 1 + \frac{0.5x_1 - 2x_2}{d(x_1, x_2, x_3)} \geq 0$$

$$d(x_1, x_2, x_3) = 1.5x\,x_2 + \sqrt{2}x_2x_3 + 1.319x_1x_3$$

5.28 The nondimensional propagation speed of a surface wave that includes the effects of surface tension is given by [15]

$$\hat{c} = \sqrt{\hat{\lambda}\left(1 + \hat{\sigma}/\hat{\lambda}^2\right)\tanh\left(1/\hat{\lambda}\right)}$$

For $\hat{\sigma} = 0.1$, determine the minimum value of $\hat{c}$ and the value of $\hat{\lambda}$ at which the minimum occurs.

5.29 Given the following function of α, $\alpha > 0$

$$g(\alpha) = (2 + \sin(10\alpha)) \int_0^2 x^{\alpha} \sin\left(\frac{\alpha}{2-x}\right) dx$$

Determine the maximum and minimum values of $g(\alpha)$ and the value of α where these extrema occur. Use **Quiet** and expect the execution time to be relatively lengthy because of the highly oscillatory nature of the integrand near $x = 2$.

5.30 The flow rate in a pipe with a circular cross section is proportional to [16]

$$Q = \frac{(\theta - \sin\theta)^{5/3}}{\theta^{2/3}} \qquad 0 \le \theta \le 2\pi$$

Find the maximum value of Q and the angle at which it occurs.

5.31 The force required to hold open a planar gate of a reservoir at an angle θ is given by [17]

$$f_g = \frac{c_1 h^3(\theta)}{\sin\alpha(\theta)\cos^2\theta} + \frac{c_2 \sin\theta}{\sin\alpha(\theta)}$$

where

$$\alpha(\theta) = \theta + \cos^{-1}\left(\frac{\cos\theta}{\sqrt{2}}\right)$$

$$h(\theta) = \frac{-1}{\tan\theta}\left(1 - \sqrt{1 + c_3\tan\theta}\right)$$

For $c_1 = 204{,}000$, $c_2 = 50{,}000$, and $c_3 = 4$, determine the value of θ for which f_g is a minimum.

5.32 Given the following expression from which the buckling load coefficient Δ of a rectangular plate subjected to a nondimensional in-plane compressive force β is approximated [18]

$$\Delta = \frac{A}{B}$$

where

$$A = \beta^2\left(\frac{\pi^2}{120\gamma^2} + \frac{\gamma^2}{\pi^2} + \frac{1}{6}\right) + \left(1 + \frac{\beta}{2}\right)\left(\gamma + \frac{1}{\gamma}\right)^2\left(\frac{1}{2} + \frac{\beta}{4} - \frac{4\beta}{\pi^2}\right) + \frac{2\beta\gamma^2}{\pi^2}$$

$$B = \frac{\pi^2\beta^2}{120} - \frac{4\beta}{\pi^2}\left(1 + \frac{\beta}{2}\right) + \frac{1}{2}\left(1 + \frac{\beta}{2}\right)^2$$

and γ is proportional to the ratio of the lengths of the sides of the plate. For $\beta = \{0, 1, 5, 10, 100\}$, determine the values of γ for which Δ is a minimum. Place your results in tabular form with the labeled headings: β, γ_{min}, and Δ_{min}.

Section 5.7

5.33 An experiment produces the data given in Table 5.2. The theory indicates that a model that should fit these data is of the form

$$y = \tan^{-1}\left[\frac{a\cot x\sin^2 x - b\cot x}{c + d\cos(2x)}\right]$$

Table 5.2 Data for Exercise 5.33

x	y
0.01	0.0
0.1141	0.09821
0.2181	0.1843
0.3222	0.2671
0.4262	0.3384
0.5303	0.426
0.6343	0.5316
0.7384	0.5845
0.8424	0.6527
0.9465	0.6865
1.051	0.8015
1.155	0.8265
1.259	0.7696
1.363	0.7057
1.467	0.4338
1.571	0.0

Determine the constants a, b, c, and d. Use the two argument form of $\tan^{-1}$ and use the following starting values for each search region for the constants: $a = c = d = 2$ and $b = 0.1$. Create a table of the difference in the values obtained from the fitted curve to whose given by the data. Show that with different starting values for the search, one obtains very different values for a, b, c, and d, but the values in the table of differences between the original data and the fitted curve remain the same.

Section 5.8

5.34 A signal is composed of m rectangular pulses of duration t_d and period T_o as shown in Figure 5.25. Obtain a spectral plot similar to that shown in Figure 5.21 when $T_o = 4$, $m = 8$, the sampling interval $\Delta t = T_o/50$, and $t_d = 1$. The signal is sampled over the interval $0 \le t \le mT_o$. Plot the first 100 spectral values.

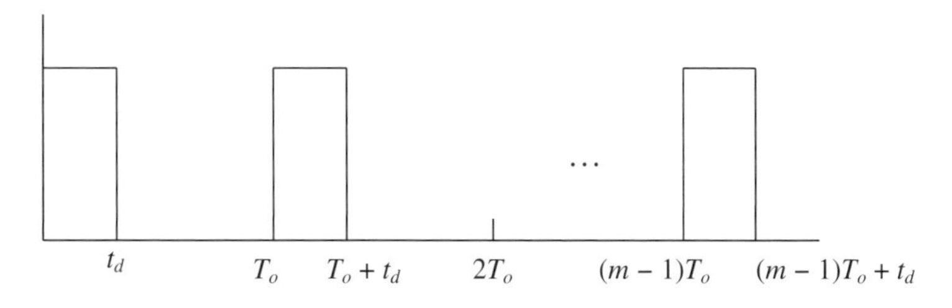

Figure 5.25

6

Graphics

6.1 Introduction

There are many different ways to display numerical results and these different ways depend on whether the data are discrete, are obtained from expressions, require logarithmic compression, are to be represented in 2D or 3D, or are specific to a given application area such as statistics (histograms, bar charts, and pie charts), computational geometry (Voronoi diagrams and convex hulls), wavelet analysis, or controls (Bode, Nichols, and root locus plots). To accommodate these wide-ranging needs, Mathematica provides a large set of high-level plotting commands that require little user involvement to generate a basic figure; that is, one with minimum annotation. However, for each plot command, one is given the means to control virtually all aspects of the graph, thereby providing a high degree of flexibility in enhancing a figure. We shall discuss a subset of these plot commands and introduce instructions that can be used to modify, enhance, and individualize the graph's curves and the overall figure.

Mathematica also provides a set of straightforward commands for creating an interactive environment for the presentation of numbers, expressions, and graphics by using **`Manipulate`**. This command is introduced in Chapter 7.

The introduction and usage of the various 2D and 3D plotting commands and their enhancement are introduced, primarily, via tables. The usage of these enhancements is then illustrated by examples from engineering topics.

6.2 2D Graphics

6.2.1 Basic Plotting

The basic forms of the 2D plotting commands that will be introduced in this chapter are presented in Table 6.1. They are divided into four categories: those that display the numerical evaluation of expressions using linear axes and those using logarithmic axes; and those that display lists of data using linear axes and those using logarithmic axes. It is mentioned that each plotting command that numerically evaluates an expression uses an internal procedure to determine the number and spacing of the values of the independent variable. These values

An Engineer's Guide to Mathematica®, First Edition. Edward B. Magrab.

Companion Website: www.wiley.com/go/magrab

Table 6.1 2D plotting commands for plotting one or more expressions or one or more lists of values. Optional enhancements, which are denoted **enh**, are comma-separated options given in subsequent tables

Plotting type	Mathematica function*
Linear axes	
Basic	`Plot[{f,g, ...},{x,xs,xe},enh]`
Parametric	`ParametricPlot[{f,g},{x,xs,xe}},enh]`
Polar	`PolarPlot[r,{th,ths,the}},enh]`
Contour	`ContourPlot[h,{x,xs,xe},{y,ys,ye}},enh]`
Region	`RegionPlot[reg`†`,{x,xs,xe},{y,ys,ye}},enh]`
Plotting expressions with logarithmic axes	
y-axis logarithmic, x-axis linear	`LogPlot[{f,g, ...},{x,xs,xe}},enh]`
y-axis linear, x-axis logarithmic	`LogLinearPlot[{f,g, ...},{x,xs,xe}},enh]`
y-axis logarithmic, x-axis logarithmic	`LogLogPlot[{f,g, ...},{x,xs,xe}},enh]`
Plotting of lists of coordinate pairs with linear axes	
Basic: plot points only	`ListPlot[{list1,list2, ...}},enh]`
Basic: connect data values	`ListLinePlot[{list1,list2, ...}},enh]`
Polar	`ListPolarPlot[{listr1,listr2, ...}},enh]`
Contour	`ListContourPlot[lst,enh]`
Plotting lists of coordinate pairs with logarithmic axes	
y-axis logarithmic, x-axis linear	`ListLogPlot[{list1,list2, ...}},enh]`
y-axis linear, x-axis logarithmic	`ListLogLinearPlot[{list1,list2, ...}},enh]`
y-axis logarithmic, x-axis logarithmic	`ListLogLogPlot[{list1,list2, ...}},enh]`

* **f** = $f(x)$, **g** = $g(x)$; **r** = $r(\theta)$; **h** = $h(x,y)$;
list1, **list2**, ..., are lists of Cartesian coordinate pairs **{{x1,y1},{x2,y2}, ...}**;
listr1, **listr2**, ..., are lists of polar coordinate pairs **{{th1,r1},{th2,r2}, ...}**;
xs or **ys** or **ths** = start value;
xe or **ye** or **the** = end value;
lst = {{x1,y1,f1},{x2,y2,f2}, ...}

† **reg** = is a set of relational equations each of which is separated by **&&**; **reg** is plotted only for those values of **x** and **y** that satisfy the set of relational equations.

are determined in a recursive fashion and the number of recursions that are used is, by default, determined automatically. However, if one wants to specify the number of recursions to n recursions, where n is a positive integer, this is done with the option **MaxRecursion->n** which is one of the optional arguments in the plotting command. Examples of the use of the 2D plot commands in Table 6.1 are shown in Tables 6.2 to 6.4.

All figures can have their displayed size adjusted by employing the option

```
...[..., ImageSize->sz, ...]
```

where typically **sz** is either **Tiny**, **Small**, **Medium**, or **Large**.

Table 6.2 Examples of basic plotting expressions with linear axes

Plot command	Expressions plotted and the plotting program	Figure created

`Plot`	$f(t) = e^{-0.15t} \sin t \quad f'(t) = \frac{df}{dt}$
	`f[t_]:=Exp[-0.15 t] Sin[t]`
	`Plot[{f[t],f'[t]},{t,0,25},`
	`  PlotRange->All]`
		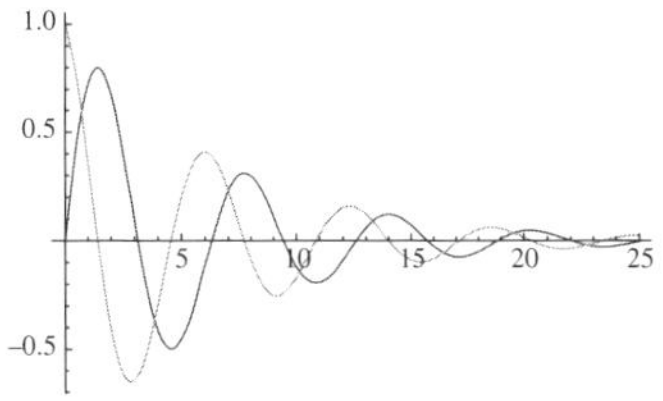

`ParametricPlot`	$f(t) = e^{-0.15t} \sin t \quad f'(t) = \frac{df}{dt}$
	`f[t_]:=Exp[-0.15 t] Sin[t]`
	`ParametricPlot[{f[t],f'[t]},`
	`  {t,0,25},PlotRange->All]`
		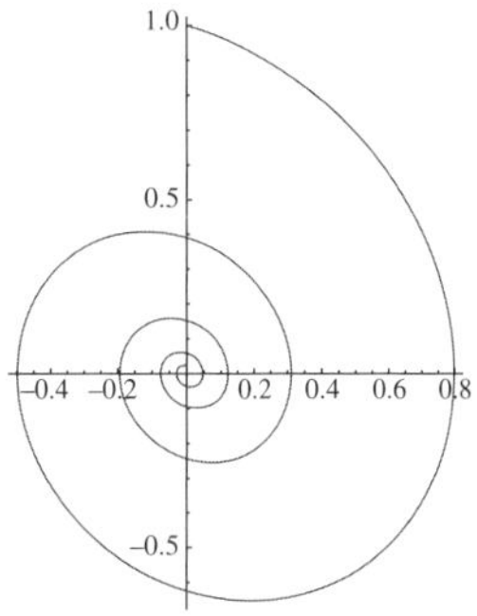

`PolarPlot`	$r(\theta) = \lvert J_0(2\pi(0.6 - \cos\theta))\rvert$
	`r[θ_]:=Abs[BesselJ[0,`
	`  2 π (0.6-Cos[θ])]]`
	`PolarPlot[r[θ],{ θ,0,2 π}]`
		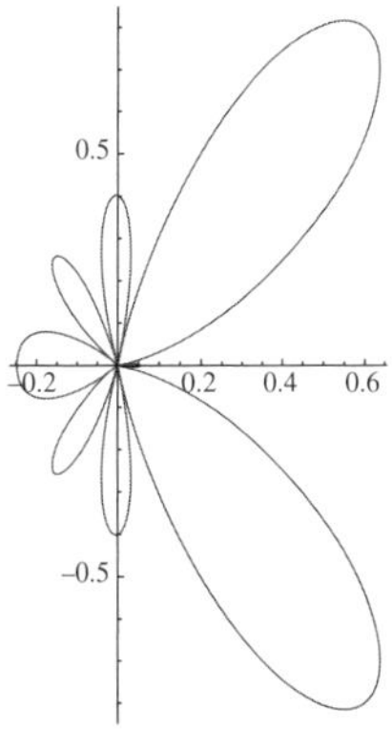

`ContourPlot`	$w(x, y) = \sin(2\pi x)\sin(3\pi y)$
	`w[x_,y_]:=Sin[2 π x]*`
	`  Sin[3 π y]`
	`ContourPlot[w[x,y],`
	`  {x,0,1},{y,0,1}]`
		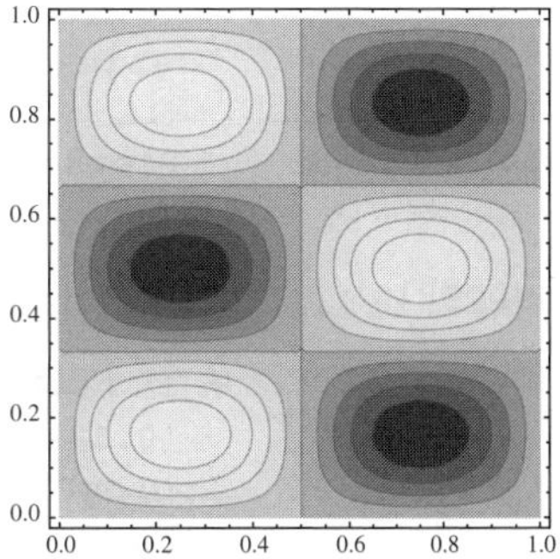

(*continued*)

Table 6.2 (*Continued*)

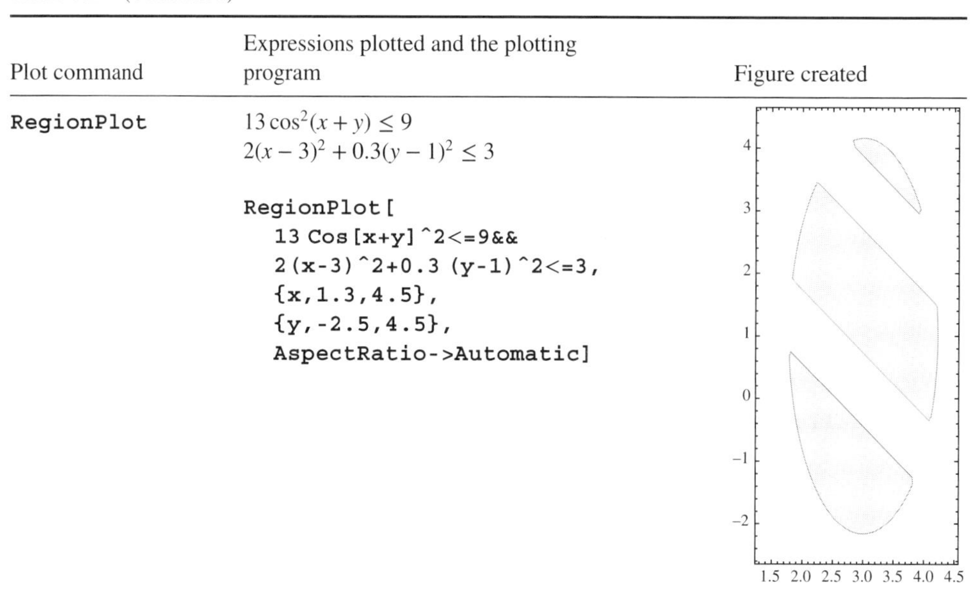

Plot command	Expressions plotted and the plotting program	Figure created

`RegionPlot`	$13\cos^2(x+y) \le 9$
	$2(x-3)^2 + 0.3(y-1)^2 \le 3$
	
	`RegionPlot[`
	`13 Cos[x+y]^2<=9&&`
	`2(x-3)^2+0.3 (y-1)^2<=3,`
	`{x,1.3,4.5},`
	`{y,-2.5,4.5},`
	`AspectRatio->Automatic]`	

Table 6.3 Examples of plotting expressions with logarithmic axes

Plot command	Expressions plotted and the plotting program	Figure created

`LogPlot`	$H(\Omega) = \{(1-\Omega^2)^2 + (0.1\Omega)^2\}^{-1/2}$
	
	`h[Ω_]:=1/Sqrt[(1-Ω^2)^2+`
	`(0.1 Ω)^2]`
	`LogPlot[h[Ω],{Ω,0.01,10}]`	

`LogLinearPlot`	$H(\Omega) = \{(1-\Omega^2)^2 + (0.1\Omega)^2\}^{-1/2}$
	
	`h[Ω_]:=1/Sqrt[(1-Ω^2)^2+`
	`(0.1 Ω)^2]`
	`LogLinearPlot[h[Ω],`
	`{Ω,0.01,10},`
	`PlotRange->All]`	

`LogLogPlot`	$H(\Omega) = \{(1-\Omega^2)^2 + (0.1\Omega)^2\}^{-1/2}$
	
	`h[Ω_]:=1/Sqrt[(1-Ω^2)^2+`
	`(0.1 Ω)^2]`
	`LogLogPlot[h[Ω],`
	`{Ω,0.01,10}]`	

Table 6.4 Examples of plotting of lists with linear axes. See also Table 6.15

Plot command	Expressions used to generate discrete data and the plotting program	Figure created

`ListPlot`	$f(t) = e^{-0.15t} \sin t \quad f'(t) = \frac{df}{dt}$
	
	`f[t_]:=Exp[-0.15 t] Sin[t]`
	`tau=Range[0,25,0.5];`
	`lst1=Table[{t,f[t]},`
	`  {t,tau}];`
	`lst2=Table[{t,f'[t]},`
	`  {t,tau}];`
	`ListPlot[{lst1,lst2},`
	`   PlotRange->All]`	

`ListLinePlot`	$f(t) = e^{-0.15t} \sin t \quad f'(t) = \frac{df}{dt}$
	
	`f[t_]:=Exp[-0.15 t] Sin[t]`
	`tau=Range[0,25,0.5];`
	`lst1=Table[{t,f[t]},`
	`  {t,tau}];`
	`lst2=Table[{t,f'[t]},`
	`  {t,tau}];`
	`ListLinePlot[{lst1,lst2},`
	`   PlotRange->All]`	

6.2.2 *Basic Graph Enhancements*

To enhance a figure, the arguments of the basic plotting commands given in Table 6.1 can be augmented with a very large set of optional instructions. We shall introduce several of them in this section and several others in the following sections.

Color

The named color attributes available for equations, text, curves, fill, regions, points, background, and graphics primitives are listed in Table 6.5. In addition to these named colors, each of the main colors can be made lighter by using **Lighter** or darker by using **Darker** as indicated in this table.

There is also available a very large number of color schemes for use by such commands as **ContourPlot**, **ListContourPlot**, **ParamentricPlot**, **RegionPlot**, **Plot3D**, **SphericalPlot3D**, and the **Filling** option in **Plot**, **ListPlot**, and others. The color schemes are accessed with the *Color Schemes* palette obtained from the *Palettes* menu as shown in Figure 6.1 in either of two ways. The first way is to use

```
ColorFunction->ColorData["Color scheme name"]
```

Table 6.5 Named color attributes available for equations, text, curves, fill, regions, graphics primitives, background, and points

Option	Main color	Lighter color	Lighter color*	Darker color†
Colors	`Red`	`LightRed`	`Lighter[`*Main color*`,p]`	`Darker[`*Main color*`,q]`
	`Green`	`LightGreen`		
	`Blue`	`LightBlue`		
	`Black`	`LightGray`		
	`White`	-		
	`Gray`	-		
	`Cyan`	`LightCyan`		
	`Magenta`	`LightMagenta`		
	`Yellow`	`LightYellow`		
	`Brown`	`LightBrown`		
	`Orange`	`LightOrange`		
	`Pink`	`LightPink`		
	`Purple`	`LightPurple`		

*$0 \leq$ `p` ≤ 1: `p` $= 0$ gives *Main color* and `p` $= 1$ gives white
†$0 \leq$ `q` ≤ 1: `q` $= 0$ gives *Main color* and `q` $= 1$ gives black

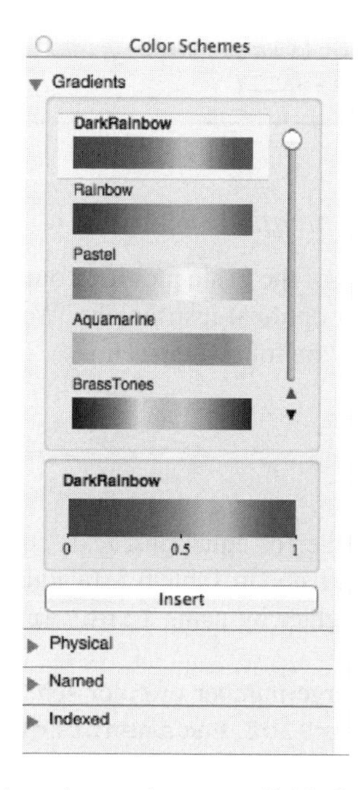

Figure 6.1 Selection of color schemes that are available from the *Color Schemes* palette

where "*Color scheme name*" (quotation marks required) can be entered manually from any of the choices appearing in the palettes in any of the four headings: *Gradient*, *Physical*, *Named*, and *Indexed*. Shown in Figure 6.1 are the first few choices appearing under *Gradient*. The other choices become visible by scrolling down with the slider. The second way is to use the palette directly by typing the plotting option as

```
ColorFunction->
```

Leaving the cursor after the ">", one goes to the palette menu and clicks on the desired selection and then clicks on *Insert*. When this is done for the palette selection shown in Figure 6.1, the following will appear

```
ColorFunction->ColorData["DarkRainbow"]
```

Curve Characteristics

Mathematica automatically selects the properties of the curves that it displays such that each curve is a different color, but each line is solid and has the same thickness. This automation can be overridden and line colors, line thickness, and the line type – solid or dashed – can be selected for each curve. The instruction to perform these changes is **`PlotStyle`** and illustrations of its usage are given in Table 6.6.

Text Characteristics

The characteristics that text can have are listed in Table 6.7. These text enhancement attributes can be employed in the text creation portions of labeling the axes (**`AxisLabel`**), in creating a legend (**`PlotLegend`**), in giving a plot a title (**`PlotLabel`**), in framing the figure (**`FrameLabel`**), in identifying individual curves (**`Epilog`**, **`Inset`**, and **`Tooltip`**), and in placing text annotation anywhere within a figure (**`Epilog`** and **`Inset`**).

Text can include subscripts and superscripts and other mathematical notation as described in Section 1.5.2, numerical values can have the number of digits specified as described in Section 1.9.2, and the format of equations can be specified as described in Section 1.9.2.

In creating text for labels, **`Style`** is used to specify the attributes of the text or an equation. The general means of employing **`Style`** is shown in Figure 6.2. Several examples of how **`Style`** is used with text and equations are given in Table 6.8. Additional examples of the use of **`Style`** are given subsequent examples. Text can also be framed and its background color can be selected using **`Frame`** and **`Background`** as indicated in Figure 6.2 and shown in Table 6.8.

Axes Characteristics

There are two ways in which axes can appear. The first way is for the axes to appear with tick marks and with tick labels. This is the default way in which axes are displayed. The second way is for the axes to appear without tick marks and without labels and instead the entire figure is framed and the tick marks appear on the inside of the frame and the tick labels appear

Table 6.6 Curve enhancements using `PlotStyle`

`PlotStyle->`{{*curve* 1 *attribute* #1, *curve* 1 *attribute* #2, ... },
{*curve* 2 *attribute* #1, *curve* 2 *attribute* #2, ... }, ... }*

Option	*curve k attribute #m*
Dashed lines	`Dashing[arg]` `arg` = `Tiny`, `Small`, `Medium`, or `Large`; `arg` = `{}` specifies a solid line
Line thickness	`Thickness[arg]` `arg` = `Tiny`, `Small`, `Medium`, or `Large` or decimal number $0 \leq p \leq 0.1$†
Color	From Table 6.5

Usage

Default

```
h[t_,p_]:=Exp[-0.15 t] Sin[t-p]
Plot[{h[t,0],h[t,π/3.]},
  {t,0,15}]
```

(Note: Line colors are different)

Modified

```
h[t_,p_]:=Exp[-0.15 t] Sin[t-p]
Plot[{h[t,0],h[t,π/3.]},{t,0,15},
  PlotStyle->{{Dashing[{}],
    Thickness[0.015],Black},
    {Dashing[Small],
    Thickness[Large],Red}}]
```

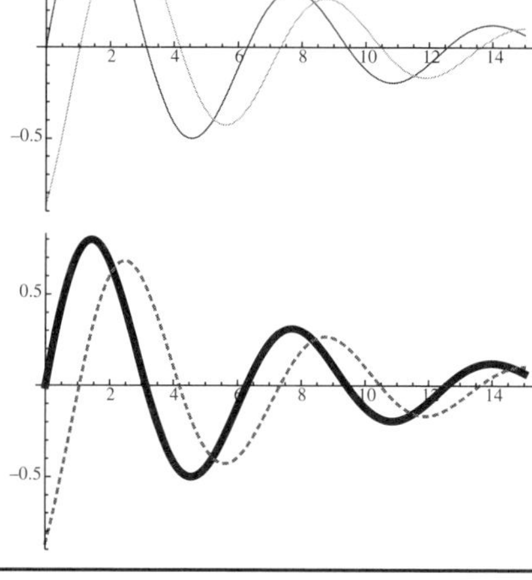

*Order of attributes is arbitrary.
†Typical region for p.

Table 6.7 Options for text. See Table 6.8 for typical usage.

Option	Instruction

Font size	`Large`
	`Small`
	`Tiny`
	n (For text, equals number of points: default is 12)
	p (For points, decimal number where typically $0.01 \leq p \leq 0.1$)

Font attribute	`Bold` (Can be used with `Italic` and `Underlined`) `Italic` (Can be used with `Bold` and `Underlined`) `Underlined` (Can be used with `Bold` and `Italic`) `Plain` (None of the above: default)

Font style	`"Times"`
	`"Courier"` (default)
	`"Helvetica"`

Create one expression

If an equation → **expr=TraditionalForm[...]**
If a numerical value → **expr=NumberForm[...,n]**
If text → **expr=" Text "**

Modify one expression

Framed

```
modexpr=Style[Framed[expr],color,font size,
            font attribute,FontFamily->font style,
            Background->color]
```

No Frame

```
modexpr=Style[expr,color,font size,font attribute,
   FontFamily->font style,Background->color]
```

Concatenation

```
labelR=Row[{modexpr1,modexpr2, ... }]
labelC=Column[{modexpr1,modexpr2, ... }]
labelRC=Column[{labelR1,labelR2,...}]
```

Application

modexpr$_{sub}$ (* from *Basic Math Assistant Typesetting* template $\square_\square$ *)

modexpr^{sup} (* from *Basic Math Assistant Typesetting* template $\square^\square$ *)

modexpr$^{sup}_{sub}$ (* from *Basic Math Assistant Typesetting* template $\square^\square_\square$ *)

```
PlotLabel->label
PlotLegend->{label,...}
Inset[label,location]
FrameLabel->{label,...}
AxesLabel->{label,...}
```

Figure 6.2 General approach to creating labels using `Style`: *`color`* is taken from Table 6.5, and *`font size`*, *`font attribute`*, and *`font style`* are taken from Table 6.7. See Table 6.8 for examples of usage

on the outside of the frame. This option is applied by using the option `Frame->True`. The characteristics of the axes can be changed as shown in Table 6.9 and those for frames as shown in Table 6.10.

6.2.3 *Common 2D Shapes:* `Graphics[]`

There are several commands that create common 2D shapes. These commands can be used to create separate figures or to enhance a figure. They can also be used to create plot markers as

Table 6.8 Examples of text enhancement using Tables 2.1, 6.5, and 6.7, Figure 6.2, and Sections 1.5.2 and 1.9.2

Mathematica statements	Output
modexpr=Style[Framed["Text"],Blue,12, Bold,Italic,FontFamily->"Helvetica", Background->Yellow]	***Text***
expr=TraditionalForm[Sqrt[x+y^2]]; modexpr=Style[expr,Blue,14,Bold, FontFamily->"Times"]	$\sqrt{x+y^2}$
expr=NumberForm[Sqrt[5.],6]; modexpr=Style[Framed[expr],14,Bold, FontFamily->"Times"]	**0.2307**
n1=TraditionalForm[Sqrt[1+Cos[k θ]]]; n2=NumberForm[Sin[π/5.]^2,4]; label=Row[{$\overset{\wedge}{\text{Style["A", Italic]}}_3$,"=", n1," and ", $\text{Style["h",Italic]}^2_{\text{Style["max",Italic]}}$,n2}]	$\hat{A}_3 = \sqrt{\cos(\theta k)+1}$ and $h^2_{max} = 0.3455$
n1=TraditionalForm[Sqrt[1+Cos[k θ]]]; n2=NumberForm[Sin[π/5.]^2,4]; label=Column[{Row[{$\overset{\wedge}{\text{Style["A", Italic]}}_3$, " = ",n1}],Row[{ $\text{Style["h",Italic]}^2_{\text{Style["max",Italic]}}$, " = ",n2}]}]	$\hat{A}_3 = \sqrt{\cos(\theta k)+1}$ $h^2_{max} = 0.3455$
m=1; n=2; Ω=3.512; str=StringJoin[ToString[n],ToString[m]] Row[{Ω^2_{str}," = ",NumberForm[Ω^2,4]}]	$\Omega^2_{21} = 12.33$
n=3; ex=Row[{"-",n}]; label="α = 10"^{ex}	$\alpha = 10^{-3}$

shown in Table 6.15. A list of these shapes is given in Table 6.11. They are created with the following command

```
Graphics[{s1,s2, ... c1,c2, ... },opt]
```

where **opt** are options such as **Axes** and **Frame** that can be used to further modify the figure and

```
sK={col1K,EdgeForm[{col2K,thkK,dashK}],shape,Opacity[n]}
cK={colK,thk,dash,cla}
```

Table 6.9 Options for figure axes

	Axes->*Instruction*	
	Option	Instruction
Axes drawn	Draw axes (Default)	**True** or omit the option
	Omit axes	**False**
	Draw only *x*-axis	**{True,False}**
	Draw only *y*-axis	**{False,True}**
	PlotRange->*Instruction*	
	Option	Instruction
Axes limits	Use all values	**All**
	Use most points, omitting outliers	**Automatic** (default)
	Set specific limits	**{{xmin,xmax},{ymin,ymax}}**
	AxesStyle->{{*x-axis attribute* #1, *x-axis attribute* #2, . . . **}**, **{***y-axis attribute* #1, *y-axis attribute* #2, . . . **}}***	
	Option	*x*-axis or *y*-axis attribute
Axes characteristics	Change color of axis, ticks, and tick labels	Color from Table 6.5
	Size of tick labels	Font size from Table 6.7
	Tick label style	Font attribute from Table 6.7
	Use default attributes	**None** or omit the option
	Axis thickness	**Thickness[...]** (see Table 6.6)
	Axis dashed	**Dashing[...]** (see Table 6.6)
	AxesLabel->{*x-axis instruction*, *y-axis instruction*}†	
	Option	*x*-axis or *y*-axis instruction
Axes labels (appear at end of each axis)	No label (Default)	**{None,None}** or omit the option
	Labels	**{*labelx*,*labely*}**†
	Label only *x*-axis	**{*labelx*,None}**†
	Label only *y*-axis	**{None,*labely*}**†
	Ticks->{*x-axis instruction*, *y-axis instruction*}	
	Option	*x*-axis or *y*-axis instruction
Tick mark characteristics and labels	System decides locations and their labels (default)	**Automatic**
	No tick marks	**None** or omit the option
	Specify locations and system determines labels	**{*loc*}** = list of *x* or *y* coordinates
	Specify locations and their respective labels	**{{*loc1*,*lab1*},{*loc2*,*lab2*}, ...}**‡

*If only one attribute is selected and it is to be applied to both axes, then all braces may be omitted
† ***labelx*** and ***labely*** are, in general, given by **label** shown in Figure 6.2.
‡ **locN** is either the value of the *x*-coordinate along the *x*-axis or the value of the *y*-coordinate along the *y*-axis

Table 6.10 Options for figure frames

Option	Instruction
Frame->*Instruction*	
No frame (default)	**False** or **None** or omit the option
Draw frame on all four sides	**True**
Draw any combination of portions of frame	**{{True** or **False,True** or **False}, {True** or **False,True** or **False}}** where the locations are {{left, right},{bottom, top}}
FrameLabel->*Instruction*	
No labels (default)	**None** or omit the option
Label at bottom edge	***label*****
Label at bottom and left edges	***{bottom,left}***†
Label all edges	***{{left,right},{bottom,top}}***†
FrameTicks->*Instruction*	
No tick marks	**None** or omit the option
Place tick marks automatically	**Automatic**
Place tick marks automatically on bottom and left edges	**True**
Place tick marks automatically on all edges	**All**
Specify locations and labels for each tick mark on each edge	***{{left,right}{bottom,top}}***† ***left*={{loc1,label1},{loc2,label2},...}**‡ or **None** or **Automatic** Similarly for ***right***, ***bottom***, and ***top***

* ***label*** or ***labelN*** is given by the general form shown in Figure 6.2
† ***bottom***, ***top***, ***left***, and ***right*** specify edge location based on its position in the list; when implemented, an appropriate label of the general form given in Figure 6.2 replaces each of these identifiers
‡ ***locN*** equals x-coordinate value for ***bottom*** and ***top*** and equals the y-coordinate value for ***left*** and ***right***

In these expressions,

colK, **col1K**, and **col2K** are colors from Table 6.5;
thkK , **thk = Thickness[...]** specifies the line thickness as shown in Table 6.6;
dashK, **dash = Dashing[...]** specifies a dashed line as shown in Table 6.6;
Opacity[n] is the degree of opaqueness, where n is a number from 0 to 1 with 1 being opaque and 0 being invisible: default is opaque.
shape = Disk (Ellipse), **Rectangle**, or **Polygon**
cla = Circle, **Line**, **Point**, or **Arrow**
EdgeForm controls the properties of the perimeter of the shape

With regard to **Point**, the options **dash** and **thk** are replaced with

```
PointSize[arg]
```

Table 6.11 Built-in 2D shapes

Shape	Mathematica function
Lines	`Line[{{x1,y1},{x2,y2}, ...}]`
Connected lines with the last point containing an arrowhead (unless **Arrowheads** indicates otherwise)	`Arrow[{{x1,y1},{x2,y2}, ...}]`
Arrowhead size or size of two arrowheads, one at each end of line	`Arrowheads[p]` for arrowhead at last point (typically, $0 < p \leq 0.1$) `Arrowheads[{-p1,p2}]` for arrowheads at first and last points: `p1` size of arrowhead of first point (minus sign required), `p2` size of arrowhead of last point. Note: must be used to draw two arrowheads and this specification must precede **Arrow**.
Points	`Point[{{x1,y1},{x2,y2}, ...}]`
Circle or arc of circle	`Circle[{x0,y0},r,{θ1,θ2}]` `{x0,y0}` = coordinates of circle center `r` = radius of circle (if omitted, `r` = 1) `{θ1,θ2}` = angles of an arc of the circle (if omitted, complete circle)
Disk or disk sector	`Disk[{x,y},r,{ θ1,θ2}]` `{x,y}` = coordinates of disk center `r` = radius of disk (if omitted, `r` = 1) `{θ1,θ2}` = angles of a sector of the disk (if omitted, complete disk)
Ellipse	`Disk[{x,y},{rx,ry}]` `{x,y}` = coordinates of ellipse center `rx` = semi-axis length parallel to x-axis `ry` = semi-axis length parallel to y-axis
Rectangle	`Rectangle[{xl,yl},{xu,yu}]` `{xl,yl}` = coordinates of lower left hand corner `{xu,yu}` = coordinates of upper right hand corner
Polygon	`Polygon[list]` `list` = list of coordinates of the vertices

where **arg** is either **Tiny**, **Small**, **Medium**, **Large**, or a decimal number that typically is in the range $0 < p \leq 0.1$.

The usage of these shapes is illustrated with the following statement

```
Graphics[{{Red,Rectangle[{0,0},{1,1}]},
  {Blue,Opacity[0.4],Rectangle[{0.5,0.5},{1.5,1}]},
  {Thick,Yellow,Dashing[Large],Circle[{0.5,0.5},0.5]},
  {Purple,Polygon[{{1,0},{1.5,0},{1.5,0.5}}]},
  {Dashing[Small],Arrow[{{1,0.5},{1.25,0.25}}]},
  {Black,PointSize[0.05],Point[{1,0.5}]}}]
```

the execution of which yields Figure 6.3.

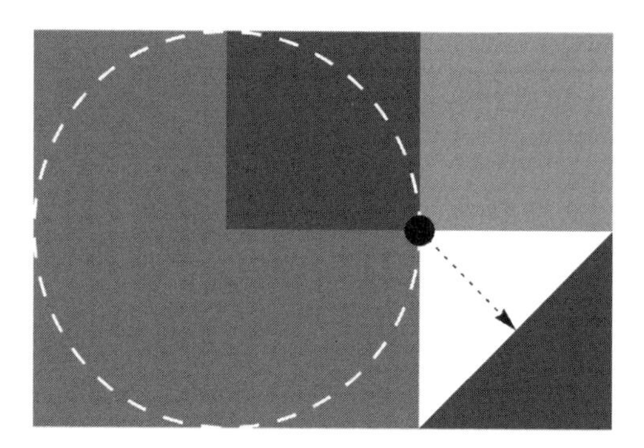

Figure 6.3 Application of several built-in 2D shapes

For another illustration, two of these built-in shapes are used to create n concentric circles and n concentric unfilled rectangles as follows. It is assumed that the rectangle and the circle are centered at (x_c,y_c), the width of the rectangle is a and its height is b, and the radius of the circle is r. The coordinates of the end points of the rectangle in terms of x_c and y_c are given by $(x_c - na/2, y_c - nb/2)$ for the lower left corner and by $(x_c + na/2, y_c + nb/2)$ for the upper right corner, where $n = 1, 2, \ldots,$ is the number of rectangles to be drawn. If it is assumed that $x_c = 1, y_c = 1, b = 0.5, a = r$ and r varies from 0.5 to 2 in increments of 0.5, then the program to draw these concentric shapes is

```
xc=1;  yc=1;  a=1;  b=0.5;
Graphics[{Table[{EdgeForm[Black],Opacity[0.],
  Rectangle[{xc-n a/2,yc-n b/2},{xc+a n/2,yc+b n/2}]},
   {n,1,4}],
  Table[Circle[{xc,yc},r],{r,0.5,2,0.5}]}]
```

where we have used **EdgeForm** to define the perimeter of the rectangle independently of the rectangle's opaqueness, which has been set to 0 (invisible). The execution of this program produces Figure 6.4.

6.2.4 *Additional Graph Enhancements*

In addition to the enhancements described above, there are several other enhancements that are often used to increase the visual impact of a figure. These are: filling regions under curves, drawing grid lines, placing additional objects within the figure, placing a figure title, and rotating the figure. Typical objects are text, graphics shapes, or another figure. These figure enhancements are listed in Table 6.12 and illustrated in Table 6.13.

Legends

The basic legend commands that are introduced are: **PlotLegends**, which creates legends; **LegendLabel**, which places a title over the legends; **LineLegend**, which selects the

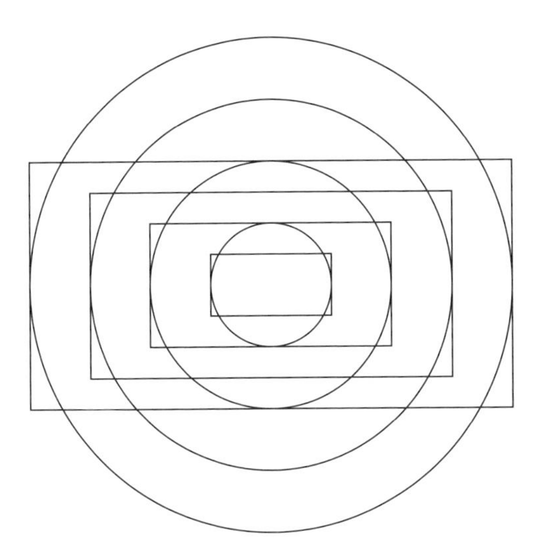

Figure 6.4 Concentric circles and concentric rectangles using built-in 2D shapes

Table 6.12 Additional figure enhancements: see Table 6.13 for typical usage

Enhancement	Instruction
Filling area under a curve, place a line under a point, or place a "volume" under a surface	**`Filling->loc`** `loc` = **`None`**; no filling (default) `loc` = **`Axis`**; fill to x-axis `loc` = **`Top`**; fill to top of plot `loc` = **`Bottom`**; fill to bottom of plot `loc` = **`{1->{2}}`**; fill between the first and second curves drawn
	`FillingStyle->typ` `typ` = **`Automatic`** (default) `typ` = **`{col,opa,thk,dash}`** `opa` = **`Opacity[n]`** ($0 \leq n \leq 1$) `col` = color from Table 6.5 `thk` = **`Thickness[...]`** from Table 6.6 `dash` = **`Dashing[...]`** from Table 6.6
Grid lines	**`GridLines->spec`** `spec` = **`None`** (default) `spec` = **`Automatic`** `spec` = **`{listx,listy}`**, where **`listx`** and **`listy`** specify locations of grid lines in each direction: **`listx`** or **`listy`** can equal **`None`** or can equal **`Automatic`**
Plot title	**`PlotLabel[label]`** *
Place additional objects on a figure	**`Epilog->{{obj1},{obj2},...}`** **`objN`** is typically a graphics primitive or an **`Inset`** object
Rotate a graphics object including an entire figure created by any plot command	**`Rotate[obj,th,{xo,yo}]`** **`obj`** = object to be rotated **`th`** = counterclockwise angle of rotation (radians) **`{xo,yo}`** = optional point about which rotation occurs

* **`label`** is, in general, of the form given in Figure 6.2.

Table 6.13 Examples of graph enhancement using the options listed in Table 6.12

```
h[t_,p_]:=Exp[-0.15 t] Sin[t-p];
gx=Table[g,{g,Range[1,15,1]}]; (* Coordinates of grid lines for
   x-axis *)
lab=Style["Two curves",Blue];
eq=Table[{n,h[n,0]},{n,2.π/3,14.π/3,π}]; (* See footnote *)
p1=Plot[{h[t,0],h[t,π/3.]},{t,0,15},PlotStyle->{{Thickness[Large],
  Red},{Thickness[0.01],Black}},options]
```

options: `PlotLabel->lab`

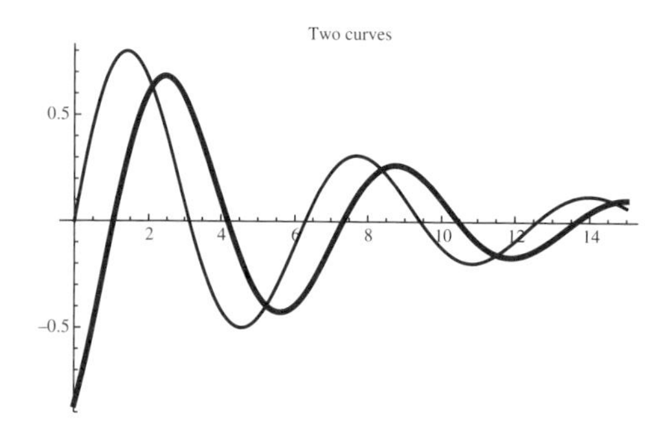

options: `PlotLabel->lab, Filling->{1->{2}}, FillingStyle->Orange, Frame->True`

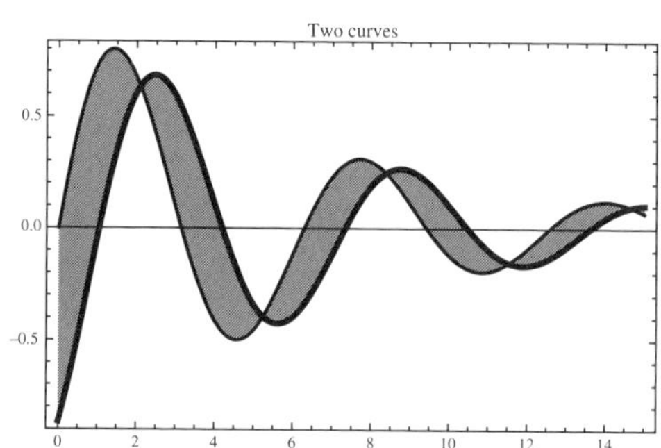

options: `PlotLabel->lab, Filling->{1->{2}}, FillingStyle->Orange, Frame->True, GridLines->{gx,Automatic}`

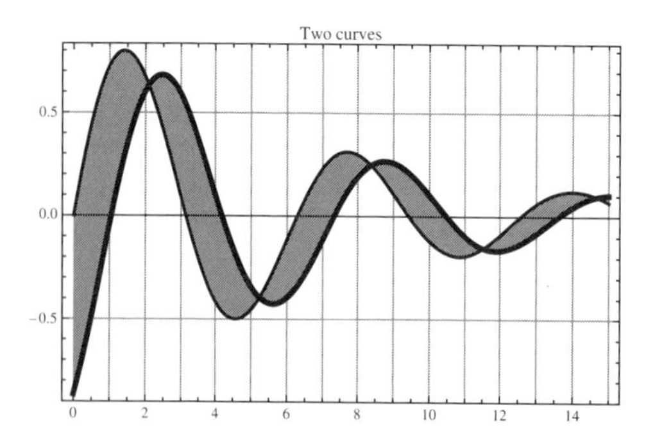

options: `PlotLabel->lab, Filling->{1->{2}}, FillingStyle->Orange, Frame->True, GridLines->{gx,Automatic}, Epilog->{PointSize[0.02], Yellow,Point[eq]}`

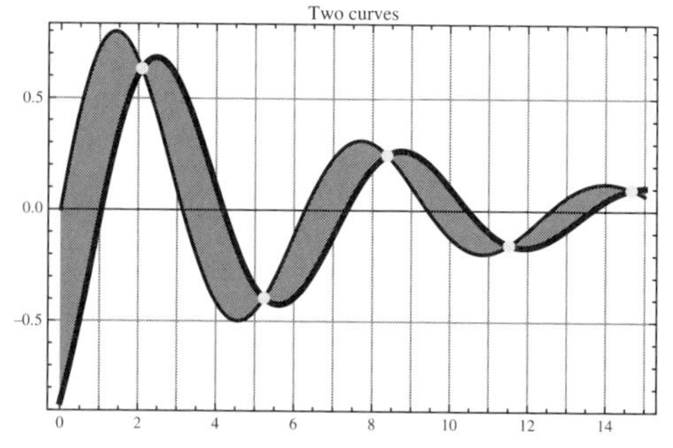

Table 6.13 *(Continued)*

`Rotate[p1,π/2]`
using in `p1` the options:
`PlotLabel->lab,`
`Filling->{1->{2}},`
`FillingStyle->Orange,`
`Frwme->True,`
`GridLines->{gx,Automatic},`
`Epilog->{PointSize[0.02],`
`Yellow,Point[eq]}`
	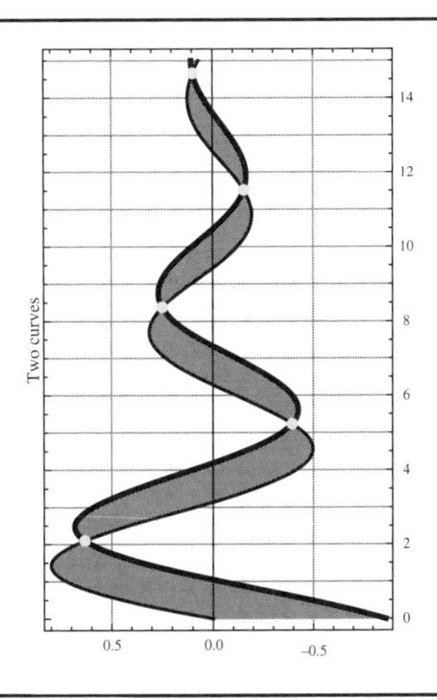

*Coordinates of curves' intersections

characteristics of the legend identifiers such as line colors and their labels; and **`Legend-Function`**, which specifies the characteristics of the box that contains the legends. To have additional flexibility when using legends, two additional commands are often employed. The first is

```
Placed[expr,{xp,yp}]
```

which places **`expr`** at the location **`{xp,yp}`**. The magnitude of **`{xp,yp}`** is a function of the command in which it is used. For graphs, it is typically a pair of positive numbers between 0 and 1. The second command is

```
Legended[expr,label]
```

which, when used with the appropriate legend commands, will associate **`label`** with **`expr`**. It is used when only selected curves are to be identified in a legend.

Examples of the use of these commands are given in Table 6.14 for **`Plot`**, **`ListPlot`**, **`ListLinePlot`**, and **`ContourPlot`**. In addition, to improve readability of the programs in Table 6.14, the command

```
Sequence[arg]
```

is used. The quantity **`arg`** represents a sequence of comma-separated arguments that can be placed into any function. In Table 6.14, it is used to specify the characteristics of line and point styles of the graphs and is implemented with **`Evaluate`**.

Table 6.14 Graph enhancement using legends

```
plt=Sequence[{BesselJ[0,x],Exp[-(x-4)^2/2]},{x,0,10}];
lst1=Table[{x,BesselJ[0,x]},{x,Range[0,10,0.15]}];
lst2=Table[{x,Exp[-(x-4)^2/2]},{x,Range[0,10,0.15]}];
pstyl=Sequence[PlotStyle->{{Dashing[{}],Black},{Dashing[Medium],Blue}}];
lstl=Sequence[{TraditionalForm[BesselJ[0,x]],TraditionalForm[Exp[-(x-4)^2/2]]}];
frd=Sequence[{RoundingRadius->10,Background->Yellow,FrameStyle->Blue}];
```

```
Plot[Evaluate[plt],Evaluate[pstyl],
  PlotLegends->Placed["Expressions",Right]]
```

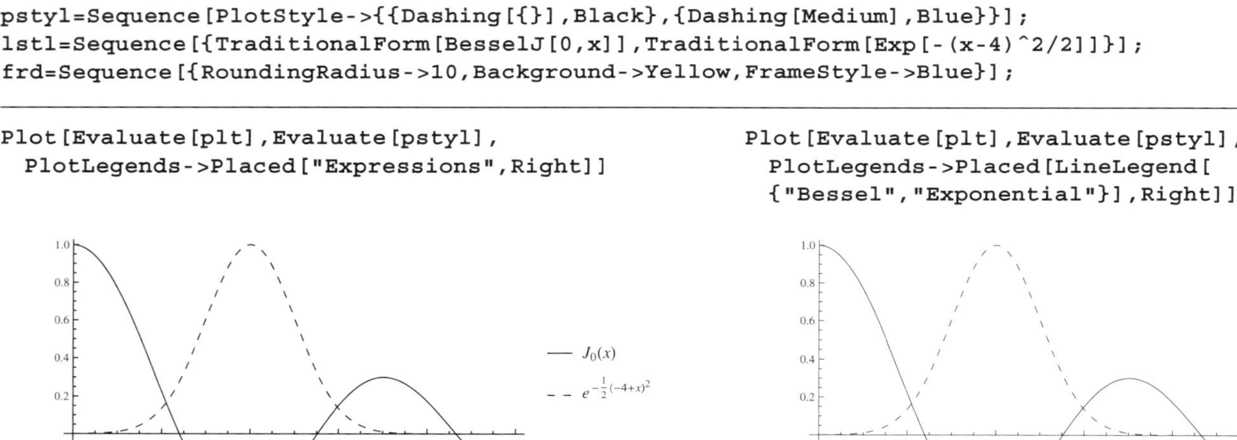


```
Plot[Evaluate[plt],Evaluate[pstyl],
  PlotLegends->Placed[LineLegend[
  {"Bessel","Exponential"}],Right]]
```

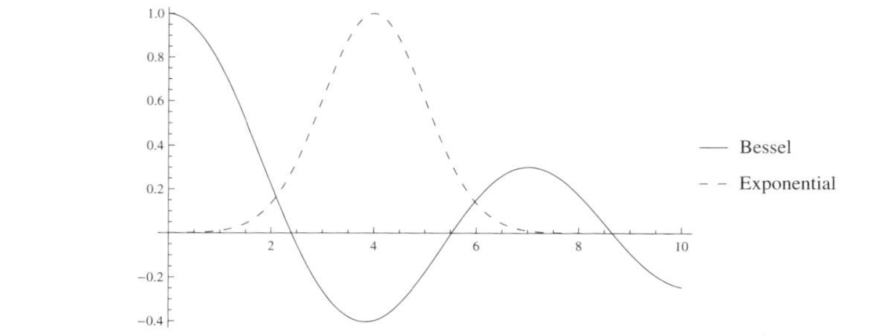


```
Plot[Evaluate[plt],Evaluate[pstyl],PlotLegends->
  Placed[LineLegend["Expressions",
  LegendLabel->"Functions"],Right]]
```

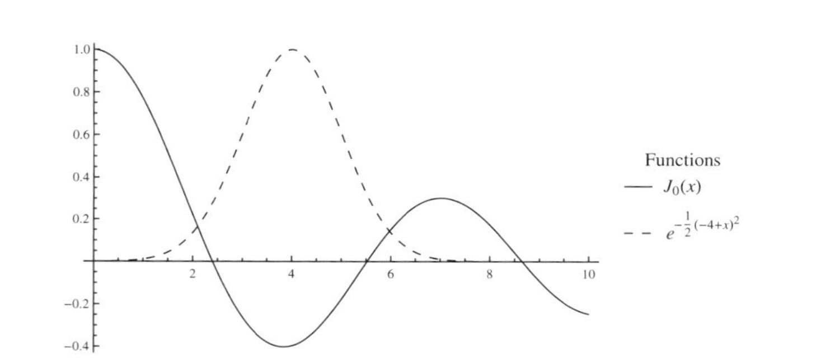


```
Plot[Evaluate[plt],Evaluate[pstyl],
  PlotLegends->Placed[LineLegend["Expressions",
  LegendLabel->"Functions",
  LegendFunction->Framed],Right]]
```

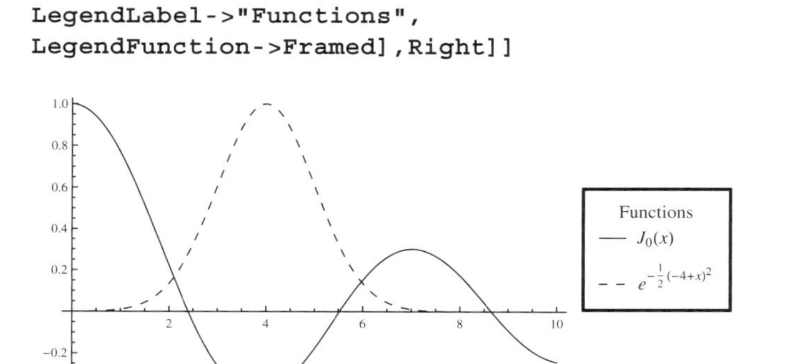

```
Plot[Evaluate[plt],Evaluate[pstyl],
  PlotLegends->Placed["Expressions",Top]]
```

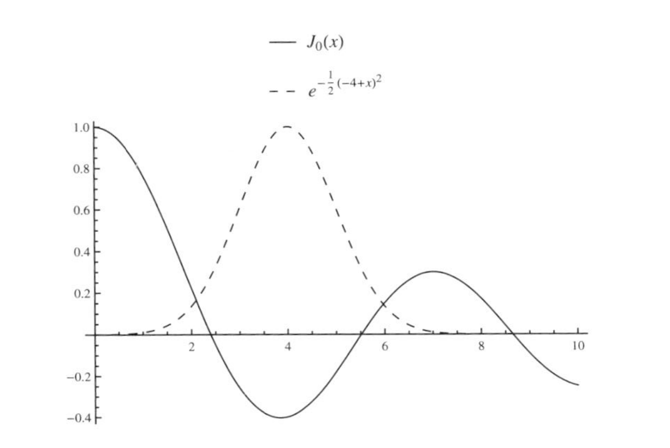


```
Plot[Evaluate[plt],Evaluate[pstyl],
  PlotLegends->Placed[LineLegend["Expressions",
   LegendLabel->"Functions",LegendFunction->
    (Framed[#,Evaluate[frd]] &)],Right]]
```

```
ListLinePlot[{lst1,lst2},PlotLegends->
  Placed[LineLegend[Sequence[lstl],
  LegendLabel->"Functions"],Right]]
```

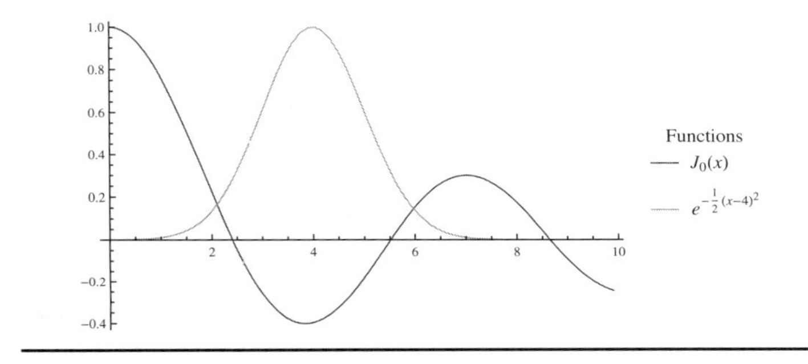

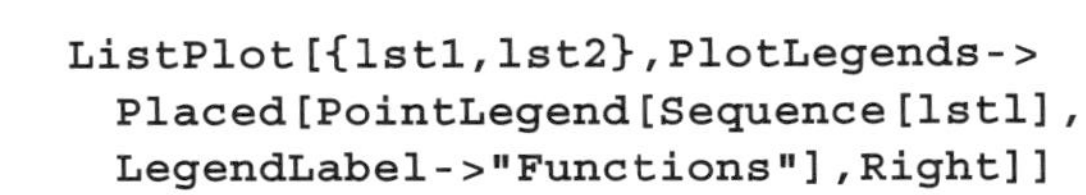

```
ListPlot[{lst1,lst2},PlotLegends->
  Placed[PointLegend[Sequence[lstl],
  LegendLabel->"Functions"],Right]]
```

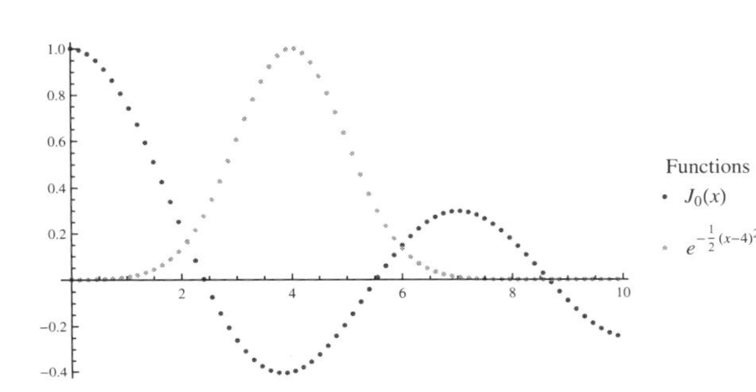

(continued)

Table 6.14 *(Continued)*

```
ListLinePlot[{lst1,Legended[lst2,
  TraditionalForm[Exp[-(x-4)^2/2]]]},PlotLegends->
  Placed[LineLegend[None,LegendLabel->"Function",
  LegendFunction->Framed],Right]]
```

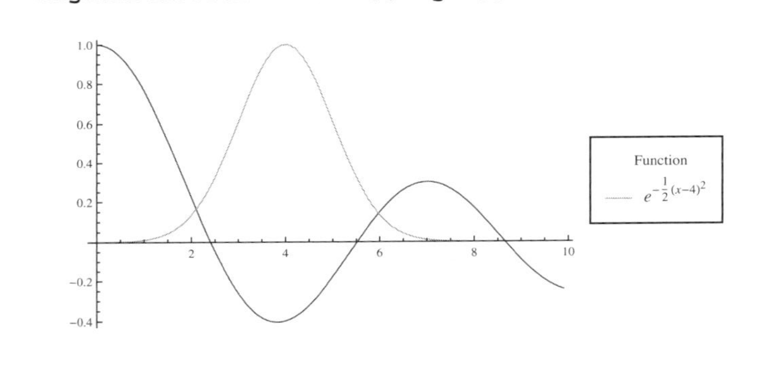


```
ContourPlot[Sin[2 x] Sin[y],{x,0,1},{y,0,1},
  ColorFunction->ColorData["Rainbow"],
  PlotLegends->Placed[BarLegend[Automatic],
    Right]]
```

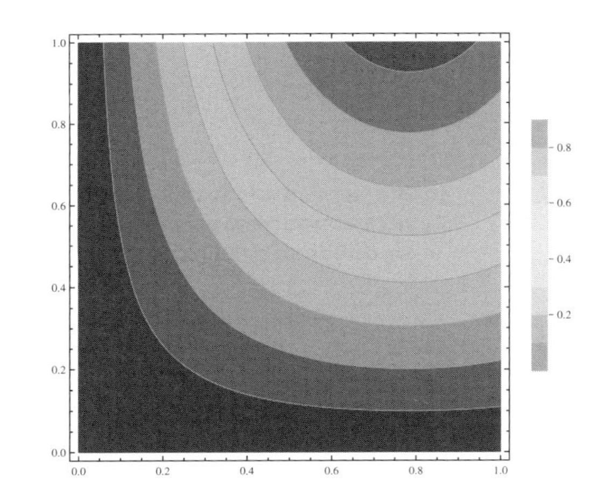

```
Plot[Evaluate[plt],Evaluate[pstyl],
  PlotLegends->Placed["Expressions",{0.8,0.8}]]
```

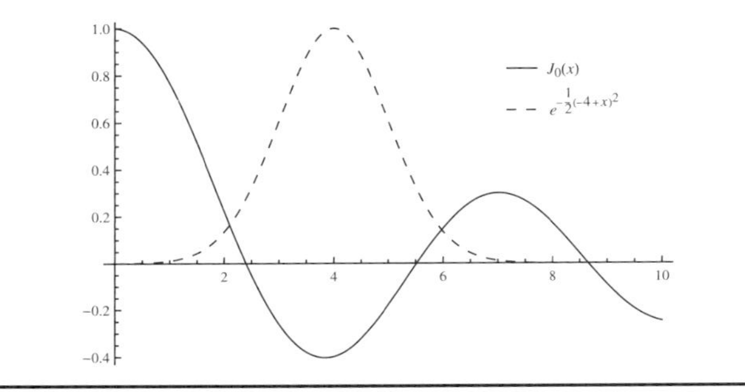

Table 6.15 Examples of using `PlotMarkers` and `Filling`

Objective	Program*	Output

Use default values	`lst1=Table[{n π/4.,Sin[n π/4.]},{n,0,16,0.5}];`
	`lst2=Table[{n π/4.,Exp[-n π/16.]},{n,0,16}];`
	`ListPlot[{lst1,lst2},PlotMarkers->Automatic]`
		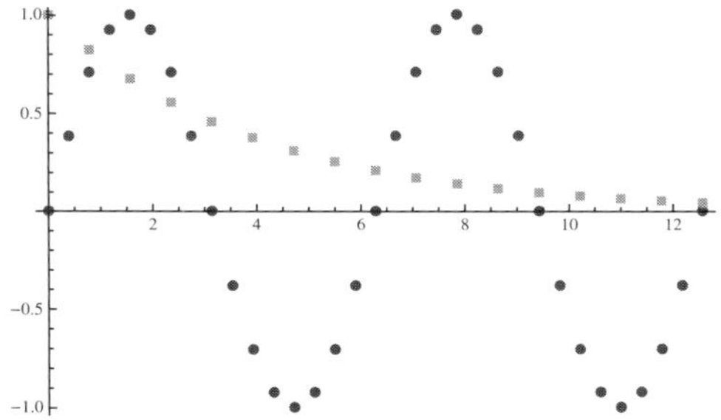

Use default values	`lst1=Table[{n π/4.,Sin[n π/4.]},{n,0,16,0.5}];`
	`lst2=Table[{n π/4.,Exp[-n π/16.]},{n,0,16}];`
	`ListLinePlot[{lst1,lst2},PlotMarkers->`
	`  Automatic]`
		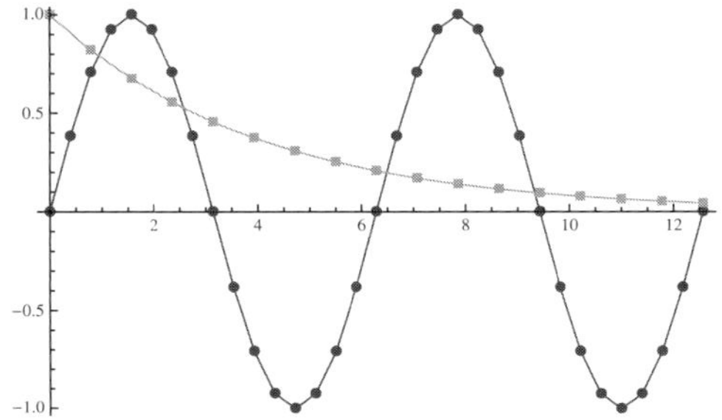

Select fill and marker characteristics	`lst1=Table[{n π/4.,Sin[n π/4.]},{n,0,16}];`
	`ListPlot[lst1,Filling->Axis,`
	`  FillingStyle->{Dashing[Medium],`
	`    Blue,Thickness[Large]},`
	`  PlotMarkers->{Graphics[{Red,Rectangle[]}],`
	`    0.03}]`
		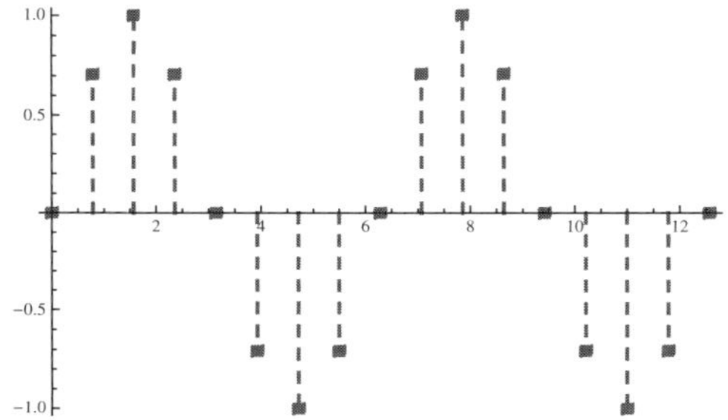

(continued)

Table 6.15 *(Continued)*

Objective	Program*	Output

Place a series of points whose locations differ from those used to plot curve	`lst1=Table[{n π/5.,Sin[n π/5.]},{n,0,20}];`
	`Plot[Sin[x],{x,0,4 π},},Filling->Axis,`
	`  Epilog->{Red,PointSize[0.025],Point[lst1]}]`
		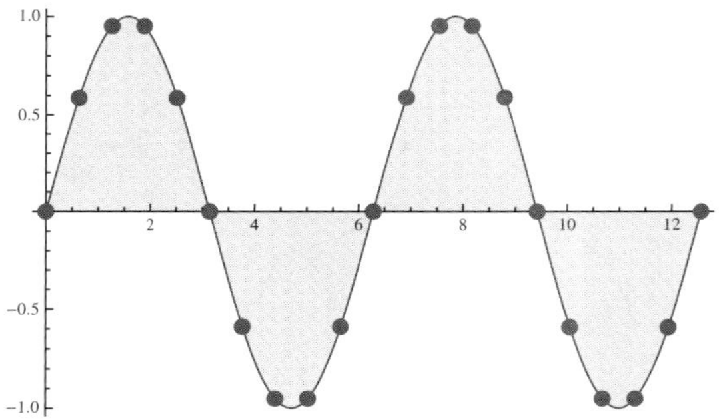

Place different markers at same *x* location	`lst1=Table[{n π/5.,Sin[n π/5.]},{n,0,20}];`
	`lst2=Table[{n π/5.,0.8 Sin[n π/5.]},{n,0,20}];`
	`Show[ListPlot[lst1,Filling->Axis,`
	`  FillingStyle->{Blue,Thickness[Medium]},`
	`  PlotMarkers->`
	`    {Graphics[{Red,Rectangle[]}],0.03}],`
	`  ListPlot[lst2,PlotMarkers->`
	`    {Graphics[{Black,Disk[]}],0.03}]]`
		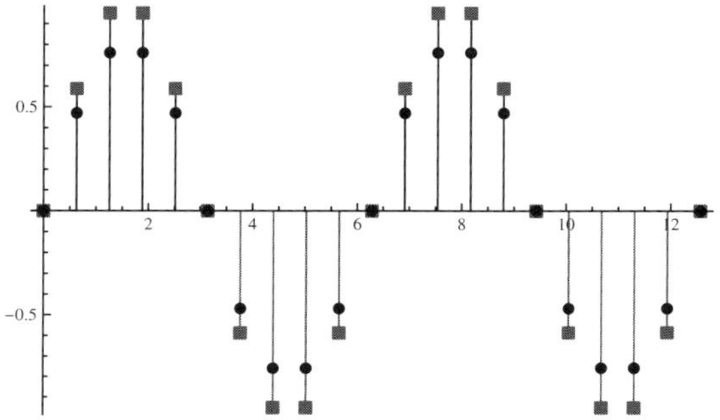

*Additional plot marker shapes, which include the shapes shown above, can be obtained by using **`Polygon`** and symbols from the palette shown in Figure 1.2b.

Plot Markers

Plot markers are a way to specify the shape and size of the symbol used to plot the data values that appear in the **`ListPlot`** family of plotting functions. The basic option is

```
PlotMarkers->Automatic
```

To specify a specific shape, color, and size of the plot marker, the following is used

```
PlotMarkers->{Graphics[{color,shape}],n}
```

where **`color`** is a the color selected from Table 6.5, **`shape`** is a graphic shape selected from Table 6.11, and **`n`** is a number, a typical value of which is 0.03. Examples of the usage of **`PlotMarkers`** are given in Table 6.15.

Placement of an Object within a Figure

One or more objects can be placed anywhere in a figure by using **`Epilog`**, **`Insert`**, and the definitions in Figure 6.5. The **`Inset`** command is

```
Inset[obj,loc]
```

where **`obj`** is a graphics object or text, **`loc`** = **`{xo,yo}`** is the location where the *center* of the object's coordinate system will be placed, **`sft={xs,ys}`** is the location with respect to the inserted figure's coordinate system, and **`sz={βΔx,αΔy}`**. When used with **`LogPlot`**, **`LogLinearPlot`**, and **`LogLogPlot`**, the following transformation has to be performed on the location coordinates:

```
LogPlot → loc = {x1,Log[y1]}
LogLinearPlot → loc = {Log[x1],y1}
LogLogPlot → loc = {Log[x1],Log[y1]}
```

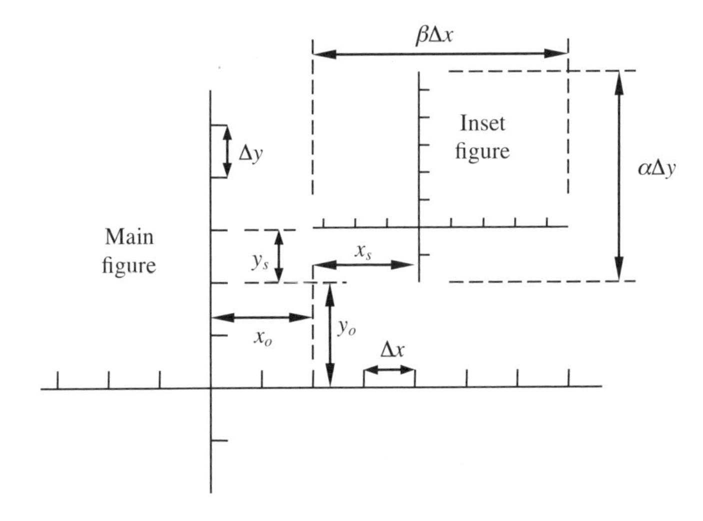

Figure 6.5 Definitions of the parameters used by **`Inset`** for placement of a figure within a figure

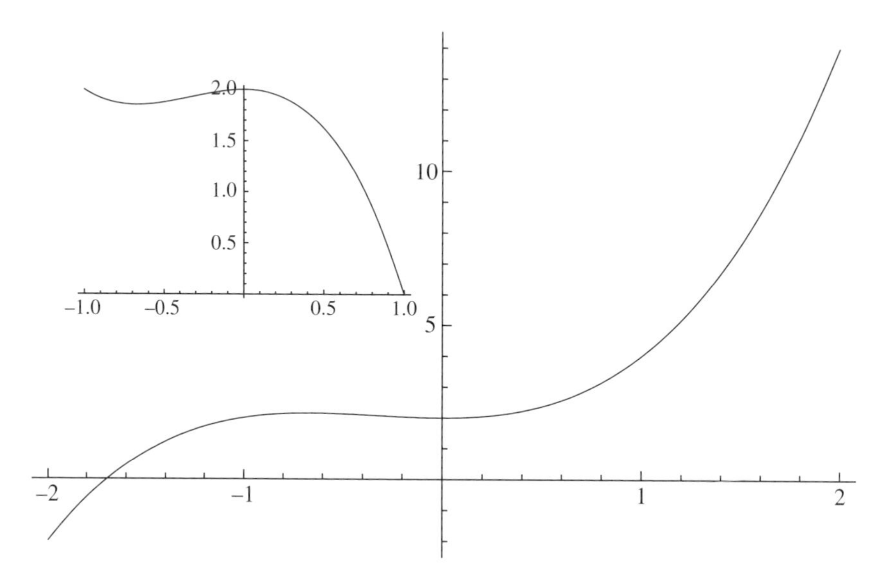

Figure 6.6 Illustration of **Inset** that placed a figure within a figure

As an example, consider following statements

```
p2=Plot[2-x^2-x^3,{x,-1,1}];
Plot[2+x^2+x^3,{x,-2,2},PlotRange->All,
  Epilog->Inset[p2,{-1,6},{0,0},{8,8}]]
```

which produces Figure 6.6.

These graph enhancement techniques are now illustrated with the following examples.

Example 6.1

Graph Annotation #1

We shall plot the numerical evaluation of the following equation and then label the result by identifying its maximum value, identifying a local minimum, filling the area between the curve and the x-axis, placing this equation in a figure title, labeling the axes, and altering the attributes of the axes.

$$h(x) = \frac{1}{(x-0.4)^2+0.03} + \frac{1}{(x-1)^2+0.01} - 6$$

To illustrate the enhancement options with their corresponding effects, we shall display intermediate output of the program below in Table 6.16. The intermediate results are created with the set of options as defined by **opt1** to **opt4**. The program is

Table 6.16 Output of Example 6.1 as a function of plot enhancements

```
Plot[h[x],{x,-1,2},PlotRange->{{-1,2},{-10,100}},options]
opt1=Sequence[PlotLabel->plab,Filling->Axis,FillingStyle->Orange];
opt2=Sequence[AxesLabel->{"x","h(x)"},LabelStyle->{14,Blue}];
opt3=Sequence[{PointSize[Large],Point[{xmin,hmin}]}];
opt4=Sequence[{Dashing[Medium],Line[{{0,hmax},{1.5,hmax}}]},
  {Arrowheads[0.03],Arrow[{{0.7,80},{xmax,hmax}}]},
  Inset[hminlab,{0.8,7.}],Inset[hmaxlab,{0.4,80}]];
```

options: none

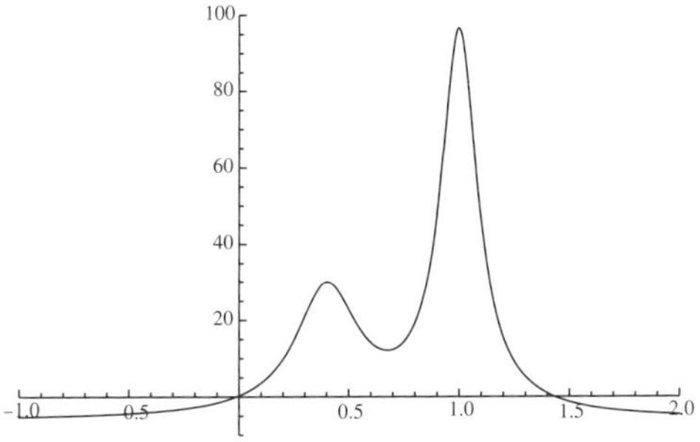

options: `Evaluate[opt1]`

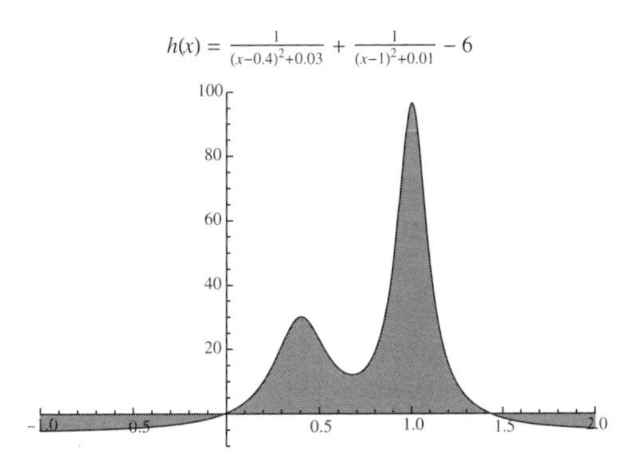

options: `Evaluate[opt1], Evaluate[opt2], Epilog->{Evaluate[opt3]}`

options: `Evaluate[opt2], Evaluate[opt2], Epilog->{Evaluate[opt3], Evaluate[opt4]}`

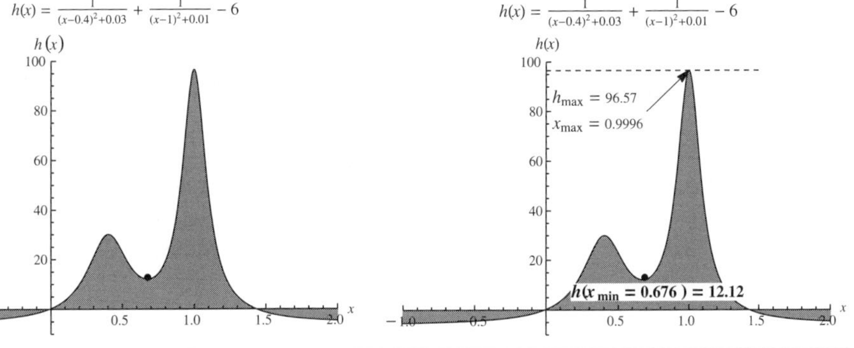


```
h[x_]:=1/((x-0.4)^2+0.03)+1/((x-1)^2+0.01)-6;
```

(* Find minimum *)

```
hmn=FindMinimum[h[x],{x,0.5,0.5,0.9}];
hmin=hmn[[1]];    xmin=x/.hmn[[2]];
```

```
(* Find maximum *)
hmx=FindMaximum[h[x],{x,1}];
hmax=hmx[[1]];    xmax=x/.hmx[[2]];
(* Create labels *)
plab=Column[{Style[Row[{"h(x) = ",TraditionalForm[
  1/((x-0.4)^2+0.03)+1/((x-1)^2+0.01)-6]}],Black,14]," "}];
hminlab=Style[Row[{"h(xmin = ",NumberForm[xmin,3],") = ",
  NumberForm[hmin,4]}],Background->White,14,Bold];
hmaxlab=Style[Column[{Row[{"hmax = ",NumberForm[hmax,4]}],
  Row[{"xmax = ",NumberForm[xmax,4]}]}],14];
(* Plot and annotate results *)
opt1=Sequence[PlotLabel->plab,Filling->Axis,
  FillingStyle->Orange];
opt2=Sequence[AxesLabel->{"x","h(x)"},LabelStyle->
  {14,Blue}];
opt3=Sequence[{PointSize[Large],Point[{xmin,hmin}]}];
opt4=Sequence[{Dashing[Medium],Line[{{0,hmax},{1.5,hmax}}]}
  ,{Arrowheads[0.03],Arrow[{{0.7,80},{xmax,hmax}}]},
  Inset[hminlab,{0.8,7.}],Inset[hmaxlab,{0.4,80}]];
Plot[h[x],{x,-1,2},PlotRange->{{-1,2},{-10,100}},
  Evaluate[opt1],Evaluate[opt2],Epilog->{Evaluate[opt3],
  Evaluate[opt4]}]
```

It is mentioned that the labels using subscripts were created with the appropriate templates from *Typesetting* in the *Basic Math Assistant* palette.

Example 6.2

Graph Annotation #2

The numerical evaluation of the following two equations shall be plotted on one figure. In addition, the default style and color of each curve will be altered, the right-hand y-axes will be labeled to correspond to the values of $\theta(\Omega)$, and a legend will be placed on the figure. The equations to be numerically evaluated and plotted are

$$h(\Omega) = [(1-\Omega^2)^2 + (2\zeta\Omega)^2]^{-1/2}$$

$$\theta(\Omega) = \tan^{-1}\frac{2\zeta\Omega}{1-\Omega^2}$$

where $0 \le \Omega \le 2.5$, $\zeta = 0.15$, and $\theta(\Omega)$ will be displayed in degrees.

To display these two quantities with widely differing magnitudes, the quantity $\theta(\Omega)$ has to be scaled by the maximum value of the y-axis as dictated by the maximum value of $h(\Omega)$. This maximum value is denoted r_g. The program is

```
h[Ω_,ζ_]:=1/Sqrt[(1-Ω^2)^2+(2 ζ Ω)^2]
θ[Ω_,ζ_]:=ArcTan[1-Ω^2,2 ζ Ω]
ζ=0.15; rg=3.5;
raxis={{0,0},{rg/4,45 Degree},{rg/2,90 Degree},
 {3 rg/4,135 Degree},{rg,180 Degree}};
Plot[{h[Ω,ζ],θ[Ω,ζ] rg/π},{Ω,0,2.5},
  PlotRange->{{0,2.5},{0,rg}},
  PlotStyle->{{Black},{Black,Dashing[Small]}},
  Epilog->{Arrow[{{1.4,2.8},{1.7,2.8}}],
   Arrow[{{0.7,2.8},{0.4,2.8}}]},
  Frame->{{True,True},{True,False}},
  FrameTicks->{{Automatic,raxis},{Automatic,None}},
  FrameLabel->{{"H(Ω)","θ(Ω)"},{"Ω",None}},
  PlotLegends->Placed[LineLegend[{"H(Ω)","θ(Ω)"},
   LegendFunction->Framed],{0.8,0.55}]]
```

which, upon execution, produces Figure 6.7.

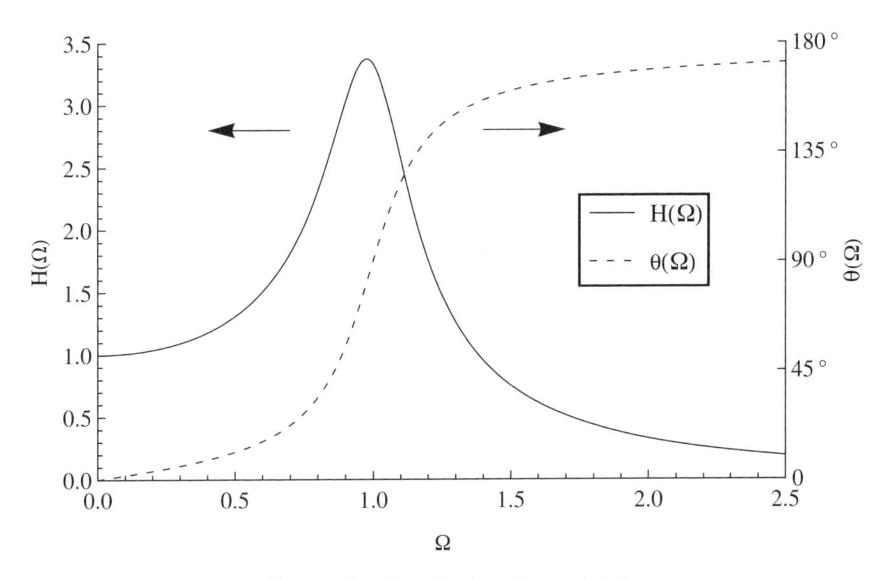

Figure 6.7 Results from Example 6.2

Example 6.3

Labeling a Family of Curves

We shall illustrate how to label each curve of a family of curves by considering the following expression for the percentage error in an electric circuit as a function of radian frequency ω

when one connects a device with an output resistance R_g to the input of amplifier that has an input resistance R_i and a shunt capacitance C_i. This expression is

$$\text{error} = 100\{1 - [(1+\alpha)^2 + (\alpha\omega\tau_i)^2]^{-1/2}\}\%$$

where $\alpha = R_g/R_i$ and $\tau_i = R_iC_i$. We shall plot the numerical evaluation of this equation for $\alpha = 5 \times 10^{-n}$, $n = 1, 2, \ldots, 5$, $10^{-2} \le \omega R_iC_i \le 10^6$, and label each curve accordingly. These curves are displayed best by using logarithmic axes. The quantity **tab** creates a table of **Inset** text commands containing the labels and their locations.

```
f[x_,a_]:=100 (1-1/Sqrt[(1+a)^2+(a x)^2]) (* x=ωRiCi *)
tab=Table[Inset[Row[{"Rg/Ri = ",NumberForm[0.5 10^(n),5]}],
  {Log[0.1],Log[f[0.01,0.9 10^(n)]]}],{n,-5.,0,1}];
LogLogPlot[Table[f[x,10^n],{n,-5,0,1}],{x,10^(-2),10^6},
  Epilog->tab,Frame->True,
  FrameLabel->{"ωRiCi","Percentage Error"},
  PlotLabel-> "Error Due to Impedance Mismatch"]
```

The results are given in Figure 6.8. As mentioned previously, the coordinates for the labels appearing in the **Inset** command had to be given as their natural logarithms. Also, the label was placed slightly above the line by multiplying its calculated value by 1.5, a somewhat arbitrary value.

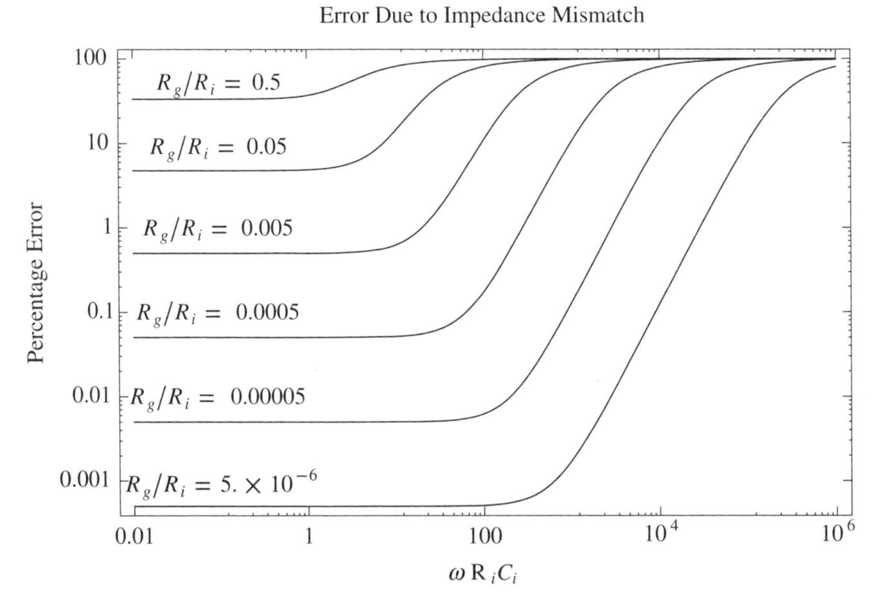

Figure 6.8 Results from Example 6.3

Example 6.4

Figure within a Figure

The amplitude response of a base-excited single degree-of-freedom system is given by

$$h_b(\Omega) = \frac{\sqrt{1 + (2\zeta\Omega)^2}}{\sqrt{\left(1 - \Omega^2\right)^2 + (2\zeta\Omega)^2}}$$

where $\Omega \geq 0$ and $0 < \zeta < 1$ are nondimensional quantities.

We shall annotate a plot of the numerical evaluation of this result for $0 \leq \Omega \leq 2.5$ and $\zeta = 0.15$ by adding an image of a spring–mass system. The spring is represented by a sine wave and is denoted as **p1** and when rotated 90° it is denoted as **p4**. The mass and the base, respectively, are created by **Rectangle** and are denoted as **p2** and **p3**. The program is

```
h[Ω_,ζ_]:=Sqrt[1+(2 ζ Ω)^2]/Sqrt[(1-Ω^2)^2+(2 ζ Ω)^2]
p1=Plot[Sin[x],{x,0,6 π},Axes->False,
 PlotStyle->{Thickness[0.07],Blue}];
p2=Graphics[{Red,Rectangle[{-0.25,-0.25},{0.25,0.25}]}];
p3=Graphics[{Magenta,Rectangle[{-0.25,-0.05},{0.25,0}]}];
p4=Rotate[p1,π/2];
Plot[h[x,0.15],{x,0,2.5},AxesLabel->{"Ω","hb(Ω)"},
 Epilog->{Inset[p4,{2,2},{0,0},{0.3,4}],
  Inset[p2,{2,2.5},{0,-0.07},{0.5,0.5}],
  Inset[p3,{2,1.5},{0,-0.15},{0.3,0.5}],
  Arrowheads[{-0.03,0.03}],Arrow[{{2.2,1.3},{2.2,2}}]}]
```

which, upon execution, gives Figure 6.9. The coordinates of each insertion's position were obtained after a few iterative adjustments to their positions.

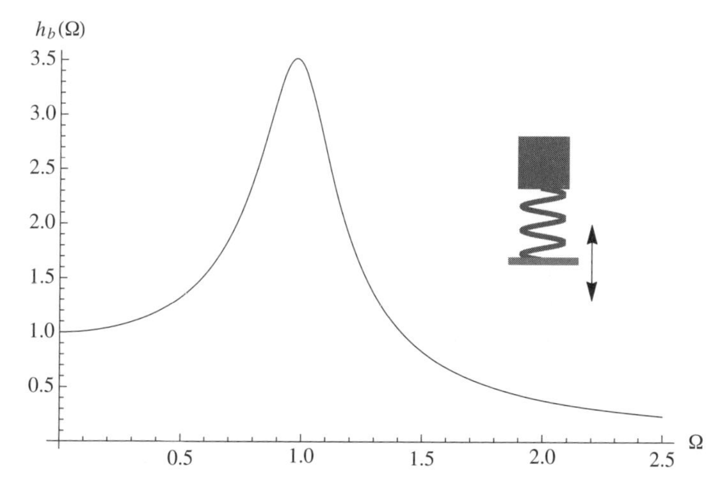

Figure 6.9 Insertion of a representation of a base-excited spring–mass system using graphics primitives

6.2.5 *Combining Figures:* **Show[]** *and* **GraphicsGrid[]**

The command **Show** allows one to combine several graphical entities into one figure, including the combination of those created with different plotting commands such as **Plot** and **ListPlot**. However, the major labeling and overall appearance of the final graphic will be determined by the first plotting or graphics command appearing in **Show**. Therefore, one should place the desired overall graphic options in this command.

Show combines several distinct graphs onto one graph. If one wants to keep several distinct graphs separate, but arrange them in a specific manner based in some sort of grid, then there are three other commands to perform this function. The first command is **GraphicsGrid**, which arranges the graphs on an $n \times m$ grid. The second is **GraphicsColumn**, which place the graphs in a column one below the other. The third command is **GraphicsRow**, which places the graphs adjacent to each other in a row. These commands and their respective output are illustrated in Table 6.17.

The use of **Show** is illustrated with the following example.

Table 6.17 Various ways of grouping individual figures into one figure entity

Expressions used in examples:

```
g1=Plot[2+x^2+x^3,{x,-1,1}];  g2=Plot[2-x^2+x^3,{x,-1,1}];
g3=Plot[2-x^2-x^3,{x,-1,1}];  g4=Plot[2+x^2-x^3,{x,-1,1}];
```

Command	Usage	Figure
`GraphicsRow[ {h1,h2,...}]`	`GraphicsRow[ {g1,g2}]`	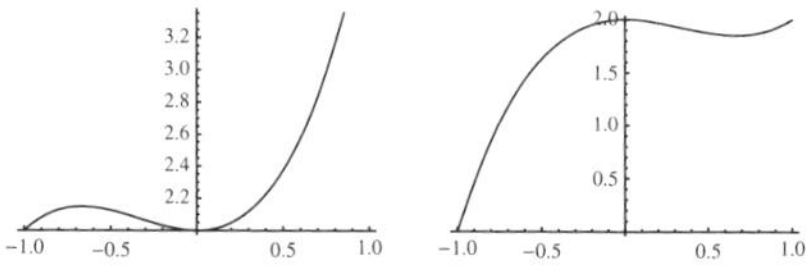
`GraphicsColumn[ {h1,h2,...}]`	`GraphicsColumn[ {g1,g2}]`	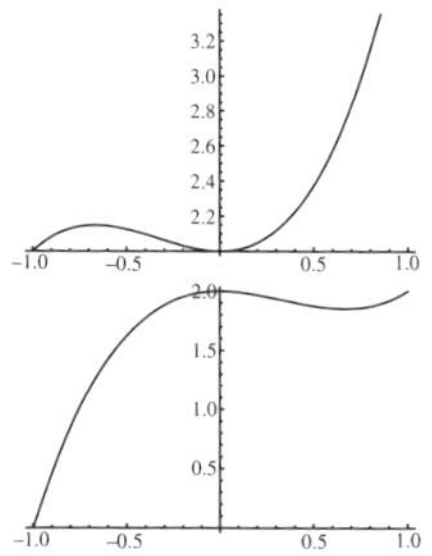
`GraphicsGrid[ {{h11,h12,...}, {h21,h22,...}, ...}]`	`GraphicsGrid[ {{g1,g2}, {g3,g4}}]`	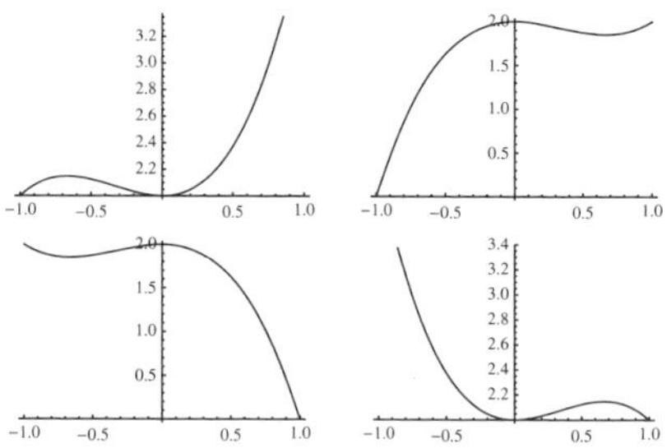

Table 6.17 (*Continued*)

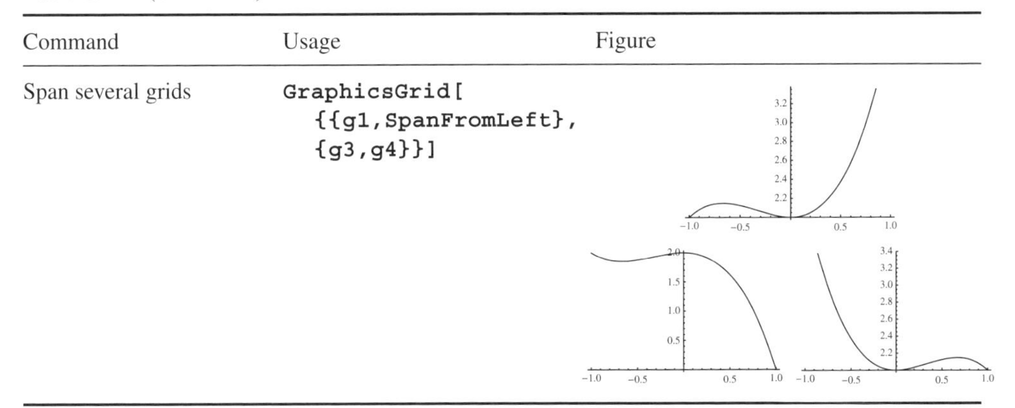

Command	Usage	Figure
Span several grids	`GraphicsGrid[ {{g1,SpanFromLeft}, {g3,g4}}]`	

Example 6.5

Sudoku Grid

We shall create a 9×9 Sudoku grid in which the initial array of numbers, which in the program below is denoted **`sud`**, are displayed in the center of their respective squares. The squares that contain these initial values will be yellow and the empty squares will be white. These squares are generated in the array **`aa`**. In addition, two thick vertical lines and two thick horizontal lines will be added to delineate nine 3×3 sub groupings. The program is given below and the results are shown in Figure 6.10.

				1				
1			9		2			4
2		9	6		8	5		3
		3				8		
4		5				1		6
		7				2		
8		1	2		5	3		7
3			4		1			8
				9				

Figure 6.10 Sudoku grid with yellow squares highlighting the initial values and thick lines delineating the 3×3 sub arrays

```
ge[p_,n_,m_]:=If[p!=0,Graphics[{Yellow,EdgeForm[Thin],
  Rectangle[{n-1,m-1},{n,m}],Text[Style[p,16,Black],
   {n-0.5,m-0.5}]}],
  Graphics[{White,EdgeForm[Thin],Rectangle[{n-1,m-1},
   {n,m}]}]]
sud={{0,0,0,0,9,0,0,0,0},{3,0,0,4,0,1,0,0,8,0},
  {8,0,1,2,0,5,3,0,7},{0,0,7,0,0,0,2,0,0},
  {4,0,5,0,0,0,1,0,6},{0,0,3,0,0,0,8,0,0},
  {2,0,9,6,0,8,5,0,3},{1,0,0,9,0,2,0,0,4},
  {0,0,0,0,1,0,0,0,0}};
aa=Table[Table[ge[sud[[m,n]],n,m],{n,1,9,1}],{m,1,9,1}];
Show[aa,Graphics[{{Thick,Line[{{3,0},{3,9}}]},
  {Thick,Line[{{6,0},{6,9}}]},
  {Thick,Line[{{0,3},{9,3}}]},
  {Thick,Line[{{0,6},{9,6}}]}}]]
```

Example 6.6

Flow Around a Cylinder

Consider the streamline pattern ψ of flow about a circular cylinder that has a uniform velocity in the positive x-direction and a cross-circulation. This pattern can be obtained by adding the potential functions for a uniform field that has a velocity U_o, a doublet of strength K at location (x_K,y_K), and vortex of strength Γ at location (x_Γ,y_Γ). The result of such an operation is

$$\psi = U_o y - \frac{\Gamma}{2\pi}\ln\sqrt{(x-x_\Gamma)^2+(y-y_\Gamma)^2} - \frac{K}{\sqrt{(x-x_K)^2+(y-y_K)^2}}\sin\left(\tan^{-1}\frac{y-y_K}{x-x_K}\right)$$

The cylinder is centered at (x_K,y_K) and its radius is given by $\sqrt{K/U_o}$.

It is assumed that $U_o = 5$, $K = 5$, $x_K = -1$, $y_K = -1$, $\Gamma = 8\pi$, $x_\Gamma = -1$, and $y_\Gamma = -1$. Then the streamlines are determined from

```
sf[x_,y_]:=(pK=-k Sin[ArcTan[x-xK,y-yK]]/
   Sqrt[(x-xK)^2+(y-yK)^2];
  pG=-gam Log[Sqrt[(x-xG)^2+(y-yG)^2]]/(2 π);
  uo y+pK+pG)
xK=-1.;  yK=-1.;  xG=-1.;  yG=-1.;  k=5.;  gam=8. π;  uo=5.;
ContourPlot[sf[x,y],{x,-3,1},{y,-3,1.5},
  FrameLabel->{"x","y"}]
```

which results in Figure 6.11a.

We would like to revise this figure in the following ways. The circular boundary of the figure does not display as a circle. This is because the figure was implemented with an aspect ratio of 1 but the axis lengths are different: the x-axis is 4 and the y-axis is 4.5. This distortion is corrected by changing the aspect ratio to 4.5/4. In addition, since we are interested in the

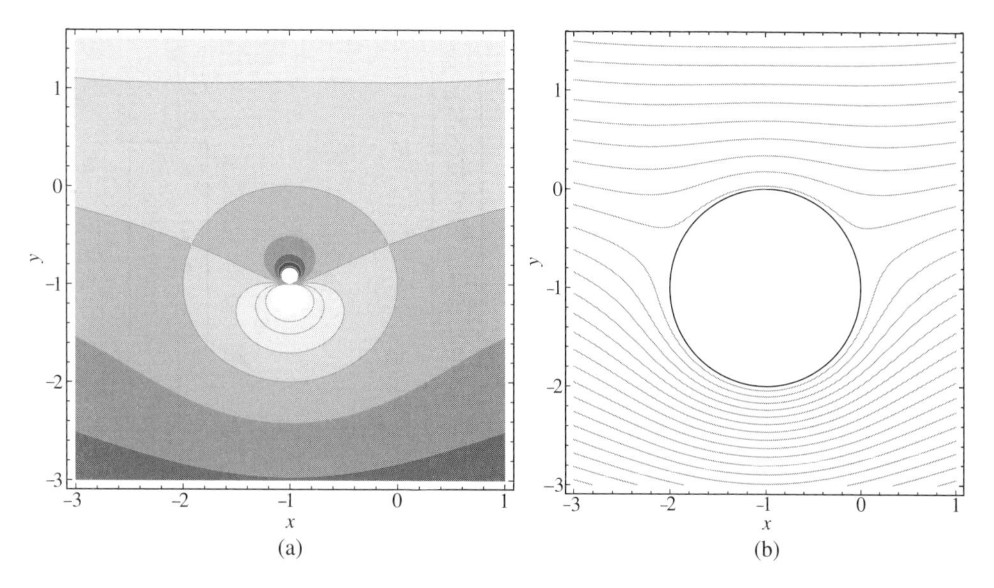

Figure 6.11 Example of (a) a default contour plot; (b) modified contour plot shown in (a)

flow about the exterior of the cylinder, we shall suppress the streamlines shown in the interior of the circle with a white disk that has a black perimeter. Lastly, we shall remove the colors appearing between the contour lines and increase number of contour lines to 50. These changes result in the following program

```
sf[x_,y_]:=(pK=-k Sin[ArcTan[x-xK,y-yK]]/
   Sqrt[(x-xK)^2+(y-yK)^2];
 pG=-gam Log[Sqrt[(x-xG)^2+(y-yG)^2]]/(2 π);
 uo y+pK+pG)
xK=-1.;  yK=-1.;  xG=-1.;  yG=-1.;  k=5.;  gam=8. π;  uo=5.;
Show[ContourPlot[sf[x,y],{x,-3,1},{y,-3,1.5},Contours->50,
   AspectRatio->4.5/4,ContourShading->False],
   Graphics[{White,EdgeForm[Black],
    Disk[{xK,yK},Sqrt[k/uo]]}]]
```

which produces Figure 6.11b.

6.2.6 `Tooltip[]`

Tooltip is a graph enhancement that allows one to use a mouse pointer to display information about curves or data points when the mouse pointer is over a curve or a data point. Its form is

```
Tooltip[expr,label]
```

where **`label`** will be displayed in a framed box. The use of **`Tooltip`** in plotting is shown in Table 6.18.

Table 6.18 Examples of `ToolTip` usage

Tooltip objective	Instructions	Figure*

Identify each curve	`c1=Tooltip[Sin[x],`
	`  TraditionalForm[Sin[x]]];`
	`c2=Tooltip[Cos[2 x],`
	`  TraditionalForm[Cos[2 x]]];`
	`Plot[{c1,c2},{x,0,5 π}]`
		cos(2*x*)
		sin(*x*)
		1.0 0.5 −0.5 −1.0 5 10 15

Display values of individual data points [unformatted]	`lst1=Table[Tooltip[{x,Sin[x]}],`
	`  {x,Range[0,5. π,5 π/51.]}];`
	`ListPlot[lst1]`
		{2.46399, 0.626924}
		1.0 0.5 −0.5 −1.0 15

Display values of individual data points [formatted]	`lst1=Table[Tooltip[`
	`  {x,Sin[x]},Column[{Row[{"x = ",`
	`  NumberForm[x,3]}],`
	`  Row[{"sin(x) = ",`
	`  NumberForm[Sin[x],3]}]}]],`
	`  {x,Range[0,5. π,5 π/51.]}];`
	`ListPlot[lst1]`
		x = 2.46
		sin(x) = 0.627
		1.0 0.5 −0.5 −1.0 15

Display values of individual data points and connect data points [unformatted]

```
ListLinePlot[Table[Tooltip[
  {x,Sin[x]},{x,Sin[x]}],
  {x,Range[0,5. π,5 π/102.]}],
  PlotMarkers->{Automatic,5}]†
```

Display values of individual data points, connect data points, and identify data set [unformatted]

```
lst1=Table[Tooltip[{x,Sin[x]},
  Column[{"Data set #1",
  {x,Sin[x]}}]],
  {x,Range[0,5. π,5 π/102.]}];
lst2=Table[Tooltip[{x,Cos[x]},
  Column[{"Data set #2",
  {x,Cos[x]}}]],
  {x,Range[0,5. π,5 π/102.]}];
ListLinePlot[{lst1,lst2},
  PlotMarkers->{Automatic,5}]
```

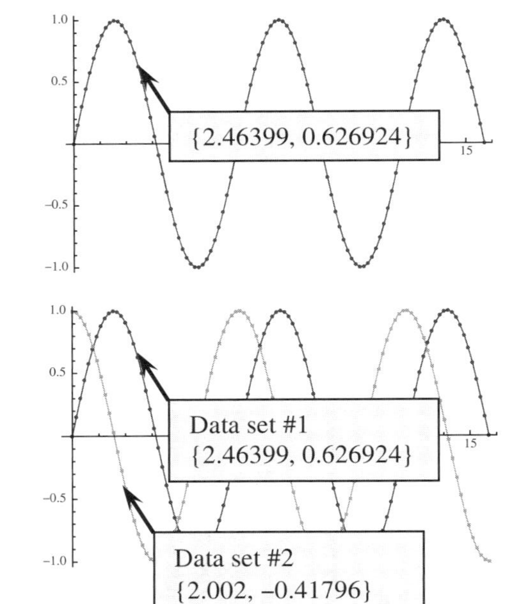

---

*Arrow indicates cursor position and text in box is produced by **Tooltip**. For graphs with two curves, both tooltips are shown; however, in actuality, only one can appear at a time.

†**Plotmarkers** is required otherwise **Tooltip** will not appear.

6.2.7 *Exporting Graphics*

Graphic entities can be saved to a directory of choice and in one of several formats by using **Export.** Its form is

```
Export["Path/FileName.ext",fig,"Graphics"]
```

where **Path** is the directory where the graphic is to reside, **FileName** is the file name that the user gives to the graphic, **.ext** is the extension to the file name that dictates the form in which the graphic is to be saved such as **.eps**, **.tif**, and **.jpg**, and "**Graphics**" is one of several descriptors that is used to define the type of graphic/image that is being exported. The quantity **fig** represents the graphic or image to be exported. It can be the graphic or image itself or a symbol defined by the figure.

6.3 3D Graphics

There are numerous three-dimensional plotting functions that are available in Mathematica. The ones that we shall discuss are listed in Tables 6.19. The plotting can be enhanced in the same manner that the 2D plotting functions were; that is, one may use the options appearing in Tables 6.5, 6.6, 6.7, 6.9, 6.12, and 6.18. However, there are three additional commands that are specific to 3D graphs: **Boxed**, **Mesh**, and **ViewPoint**. These are listed in Table 6.20. Also, for 3D graphics, one can rotate the display interactively by placing the cursor over the figure. A pair of curved lines, each with an arrowhead appearing head to tail, will become visible. Depressing the mouse button and moving this cursor will cause the figure to change orientation.

As was the case for 2D graphics, Mathematica provides commands that can create common 3D objects. These commands can be used to create separate figures or to enhance a figure. A list of these shapes is given in Table 6.21. They are created with the following command

```
Graphics3D[{s1,s2, ... c1,c2, ...},opt]
```

Table 6.19 3D plotting commands for plotting surfaces

Plotting type	Mathematica function
Basic	**Plot3D[{f,g, ...},{x,xs,xe},{y,ys,ye}]** **f** = $f(x,y)$, **g** = $g(x,y)$, ... are the z coordinates of the surfaces for **xs** $\le x \le$ **xe** and **ys** $\le y \le$ **ye**
Parametric	**ParametricPlot3D[{f,g,h},{u,us,ue},{v,vs,ve}]** **f** = $f(u,v)$, **g** = $g(u,v)$, and **h** = $h(u,v)$ are the (x,y,z) coordinates of a point on the surface
Surface of revolution	**RevolutionPlot3D[{fr,gz},{t,ts,te},{th,ths,the}]** **fr** = $f(t)$ is the radial coordinate value at t for **ts** $\le t \le$ **te** **gz** = $g(t)$ is the value of z at a value of t for **ts** $\le t \le$ **te** **th** = angle of rotation of **fr** and **gz** starting at **ths** and ending at **the**
List of values	**ListPlot3D[{{x1,y1,f1},{x2,y2,f2}, ...}]** **fN** = $f(x_n, y_n)$ = value of the z-coordinate corresponding to x_n = **xN** and y_n = **yN**

Table 6.20 Additional options for three-dimensional graphics: see Table 6.22 for typical usage

	Option	Instruction
	Boxed->*Instruction*	
Bounding box	Draw box (default)	**True**
	Omit box	**False**
	Mesh->*Instruction*	
Mesh	Draw mesh (default, using **n** = 15)	**True** (or **Mesh** omitted)
	Omit mesh	**False**
	Draw mesh using all regular points	**Full**
	Draw **n** equally spaced meshes	Positive integer
	ViewPoint->*Instruction**	
View point	Default	**{1.3,-2.4,2}**
	Top	**Top**
	Bottom	**Bottom**
	Front	**Front**
	Back	**Back**
	Left	**Left**
	Right	**Right**
	Combined	**{Top,Front}**,**{Top,Left}**, etc.

*There are numerous ways to invoke this option; see **ViewPoint** in *Documentation Center* for other choices

where **opt** are options such as **Axes**, **Mesh**, **Boxed**, **ViewPoint**, and **MaxRecursion** that can be used to modify the figure and

sK={col1K,EdgeForm[{col2K,thkK,dashK}],Opacity[n],shape}
cK={colK,thk,dash,pla}
colK, **col1K**, **col2K** = color from Table 6.5
thkK = **Thickness[...]**: line thickness from Table 6.6
dashK = **Dashing[...]**: dashed lines from Table 6.6
Opacity[n] = degree of opaqueness: n is a number from 0 to 1, where 1 is opaque and 0 is invisible. If omitted, the shape is opaque.
shape = **Cone**, **Cuboid**, **Cylinder**, **Sphere**, or **Tube**
pla = **Point**, **Line**, **Polygon**, or **Arrow**

In addition to the 3D shapes listed in Table 6.21, one can create a three-dimensional object of arbitrary complexity by using

```
GraphicsComplex[vec,prim]
```

Table 6.21 Three-dimensional shapes

Shape	Mathematica function
Connected lines	`Line[{{x1,y1,z1},{x2,y2,z2}, ...}]`
Connected lines with the last point containing an arrowhead	`Arrow[{{x1,y1,z1},{x2,y2,z2}, ...}]`
Points	`Point[{{x1,y1,z1},{x2,y2,z2}, ...}]`
Polygon	`Polygon[{{x1,y1,z1},{x2,y2,z2}, ...}]` `{xN,yN,zN}` = coordinates of vertices
Cone	`Cone[{{x1,y1,z1},{x2,y2,z2}},r]` `{x1,y1,z1}` = location of center of base `{x2,y2,z2}` = location of tip of cone `r` = radius of base
Cylinder	`Cylinder[{{x1,y1,z1},{x2,y2,z2}},r]` `{x1,y1,z1}` = location of center of one end of cylinder `{x2,y2,z2}` = location of center of the other end of cylinder `r` = radius of cylinder
Cuboid (rectangular parallelepiped)	`Cuboid[{xl,yl,zl},{xu,yu,zu}]` `{xl,yl,zl}` = coordinates of lower left-hand corner of cuboid `{xu,yu,zu}` = coordinates of upper right-hand corner of cuboid
Sphere	`Sphere[{x1,y1,z1},r]` `{x1,y1,z1}` = location of center of sphere `r` = radius of sphere
Tube (connected lines made into a cylindrical tube)	`Tube[{{x1,y1,z1},{x2,y2,z2}, ...},r]` `r` = radius of the tube

where **vec** is a list of *K* triplets of numbers and **prim** denotes the connectivity of the vertices of the object. Thus,

```
vec={{x1,y1,z1},{x2,y2,z2}, ... {xK,yK,zK}};
```

The argument of **prim** is a list of lists of integers, with each integer corresponding to a point in **vec**. The order of these integers is used by **ComplexGraphics** to create a surface with these vertices for each element of **prim**. For example,

```
prim[{{1,2,5,6}, ...}
```

indicates that the vertices of the first implementation of **prim** is composed of the points connected in the order given; that is,

```
{{x1,y1,z1},{x2,y2,z2},{x5,y5,z5},{x6,y6,z6}};
```

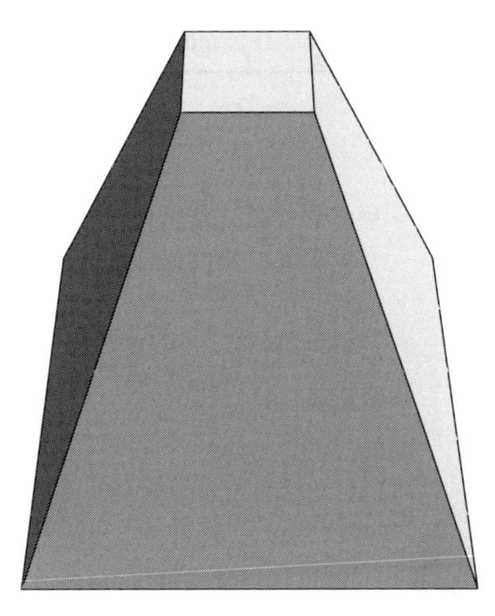

Figure 6.12 Frustum of a right pyramid created with **GraphicsComplex**

To illustrate the use of this command, we shall create a frustum of a right pyramid whose base is 1 by 1 units and whose top is 0.25 by 0.25 units and 2 units from the base. This object consists of six 4-sided polygons and its creation, which is shown in Figure 6.12, is obtained from

```
vec={{0,-1/4,-1/4},{0,1/4,-1/4},{0,1/4,1/4},{0,-1/4,1/4},
  {2,-1,-1},{2,1,-1},{2,1,1},{2,-1,1}};
n={{1,2,3,4},{5,6,7,8},{1,2,6,5},{3,4,8,7},{1,4,8,5},
  {2,3,7,6}};
Graphics3D[{Opacity[1],GraphicsComplex[vec,Polygon[n]]},
  Boxed->False,ViewPoint->{Bottom,Left}]
```

The detailed characteristics of a three-dimensional shape can be significantly altered by the options selected. To give an indication of how these options can alter a three-dimensional figure, we have chosen to show the effects of several of them on the parametric surface

$$x = a^v \cos v(1 + \cos u)$$
$$y = -a^v \sin v(1 + \cos u)$$
$$z = -ba^v(1 + \sin u)$$

for $0 \leq u \leq 2\pi$, $-15 \leq u \leq 6$, $a = 1.13$, and $b = 1.14$. The program that displays this shape is given in Table 6.22 as a function of several options.

We shall now illustrate the use of the three-dimensional plot commands appearing in Tables 6.19 to 6.21. **ParametricPlot3D** has been illustrated in Table 6.22. The following examples illustrate the use of **Graphics3D**, **Plot3D**, **ListPlot3D**, and **RevolutionPlot3D**.

Table 6.22 Illustration of the effects of several 3D graphics options

```
a=1.13;   b=1.14;
shell[u_,v_]:={a^v Cos[v] (1+Cos[u]),-a^v Sin[v] (1+Cos[u]),
  -b a^v (1+Sin[u])}
ParametricPlot3D[shell[u,v],{u,0,2 π},{v,-15,6},PlotRange->All,
  Axes->None,options]
```

options: none

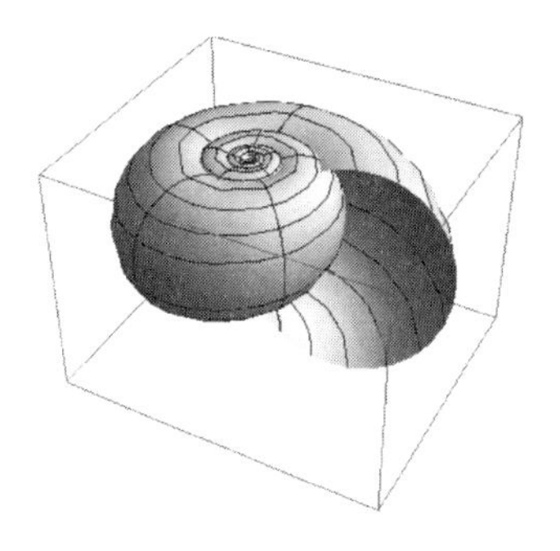

options: `Boxed->False`

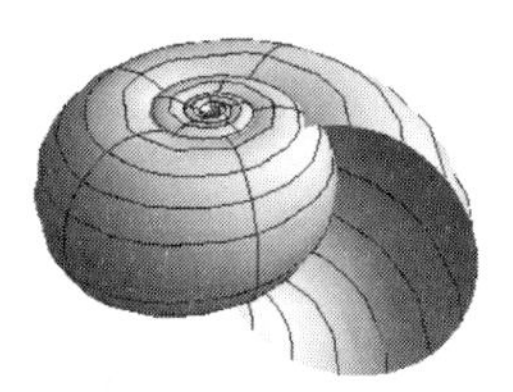

options: `Boxed->False,Mesh->False`

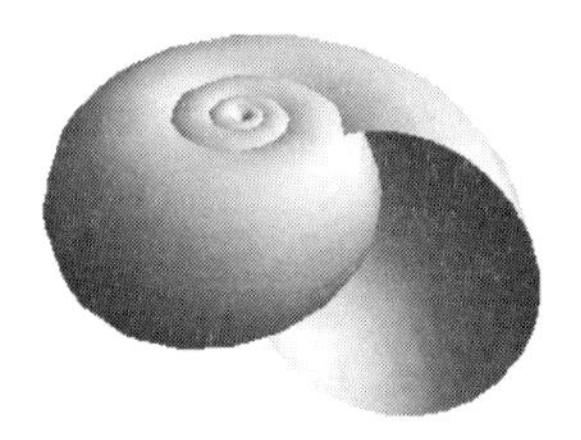

options: `Boxed->False, Mesh->False,MaxRecursion->4`

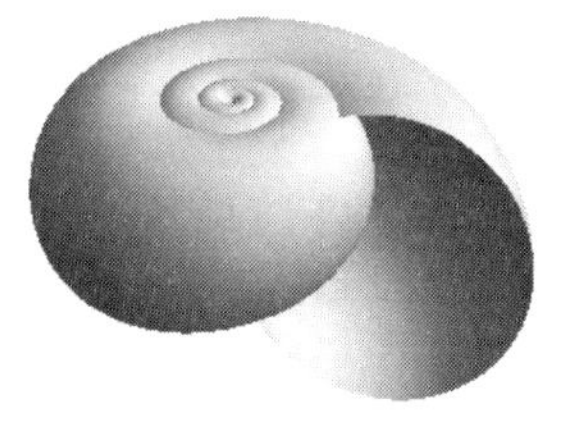

options: `Boxed->False,Mesh->False, MaxRecursion->4, PlotStyle->Opacity[0.6]`

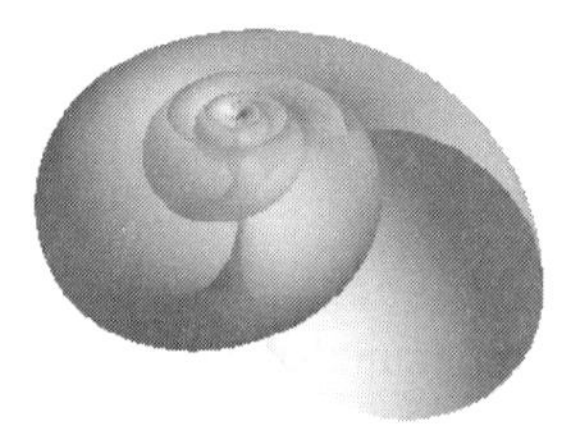

options: `Boxed->False, Mesh->False,MaxRecursion->4, PlotStyle->Opacity[0.6], ViewPoint->{Bottom,Right}`

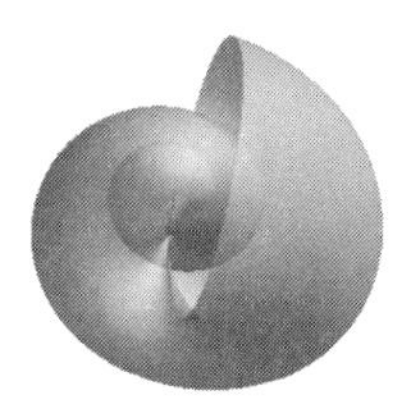

Example 6.7

Curves in Space

The parametric equations describing a sine wave on a cylindrical surface are given by

$$x = b\cos t$$
$$y = b\sin t$$
$$z = c\cos at$$

where $0 \leq t \leq 2\pi$ and we shall assume that $a = 0.4$, $b = 10$, and $c = 3$. In this program, we have included a cylindrical surface to better visualize the sine wave's spatial orientation and the axes to indicate the magnitude of the curve. **Graphics3D** is used to create the figure, the curve is created with **Line**, and the cylinder is created with **Cylinder**. The program is

```
a=10.;   b=10.;   c=3.;
tub=Table[{b Cos[t],b Sin[t],c Cos[a t]},{t,0,2 π,π/100}];
Graphics3D[{{Black,Thickness[Large],Line[tub]},
  {Opacity[0.6],Cylinder[{{0,0,-3},{0,0,3}},10]}},
    Axes->True]
```

and its output is shown in Figure 6.13.

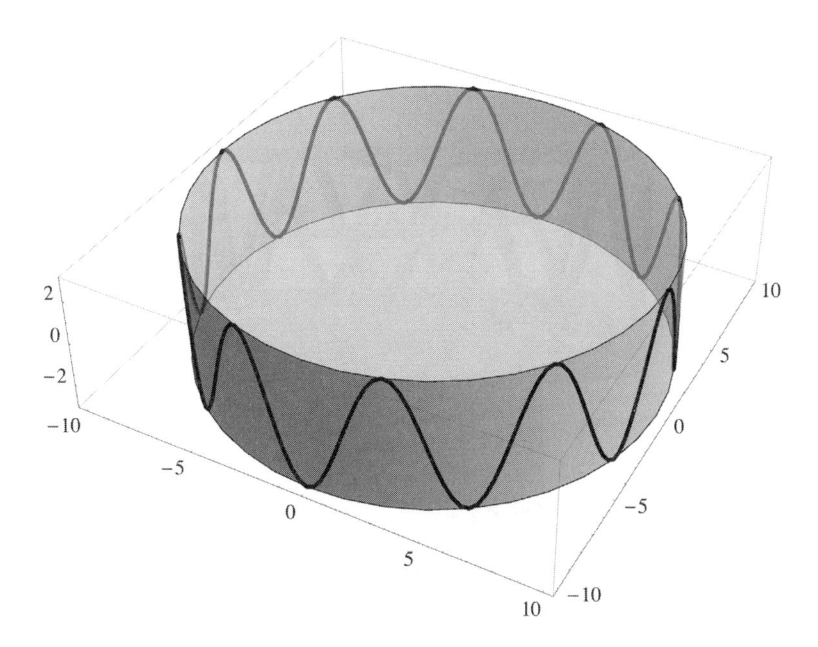

Figure 6.13 Sine wave on the surface of a cylinder

Example 6.8

A Collection of 3D Shapes

To illustrate the use of the common 3D shapes listed in Table 6.21, we shall create the image shown in Figure 6.14. The sine-wave-like object is created by using **tub** in Example 6.7 and **Tube**. The program is

```
a=10.;  b=10.;  c= 3.;
tub=Table[{b Cos[t],b Sin[t],c Cos[a t]},{t,0,2 π,π/100}];
pgon={{-5,-5,-0.5},{-5,5,0.5},{5,5,0.5},{5,-5,-0.5}};
Graphics3D[{{Red,Tube[tub,0.3]},
 {Orange,Cylinder[{{0,0,-3},{0,0,-6}},10]},
 {LightBlue,Cone[{{0,0,3},{0,0,12}},10]},
 {Yellow,Opacity[0.4],Sphere[{0,0,-2},3]},
 {Magenta,Opacity[0.7],Cuboid[{-5,-2.5,-12},{5,2.5,-6}]},
 {Magenta,Polygon[pgon]}},
 ViewPoint->{Front,Left},Boxed->False]
```

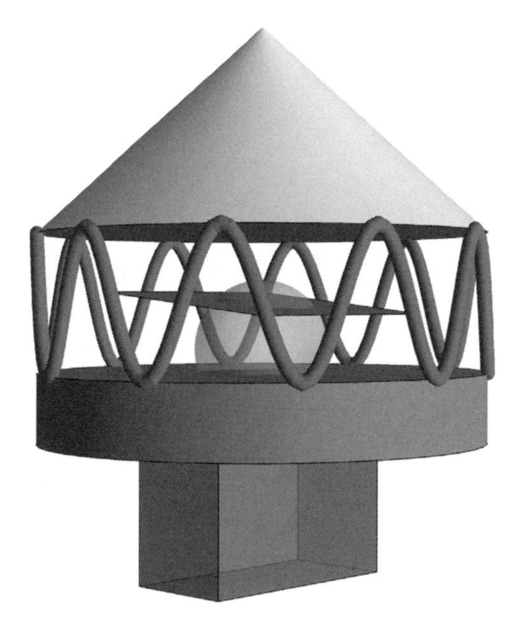

Figure 6.14 Arbitrary arrangement of built-in 3D shapes

Example 6.9

Intersecting Surfaces

`Plot3D` is used to draw the intersecting surfaces

$$f(x,y) = x^4 + 3x^2 + y^2 - 2x - 2y - 2x^2y + 6$$

$$g(x,y) = 100\cos(x^2 + y)$$

for $-3 \leq x \leq 3$ and $-3 \leq y \leq 13$. To emphasize the boundaries of the surface, we shall use **Filling** and select the color of the fill using **FillingStyle**. The execution of the following program results in Figure 6.15.

```
fz[x_,y_]:=x^4+3 x^2+y^2-2 y-2 x-2 x^2 y+6
g[x_,y_]:=100 Cos[x^2+y]
Plot3D[{fz[x,y],g[x,y]},{x,-3,3},{y,-3,13},PlotRange->All,
  Filling->Bottom,FillingStyle->Magenta,MaxRecursion->4,
  Axes->False,Boxed->False]
```

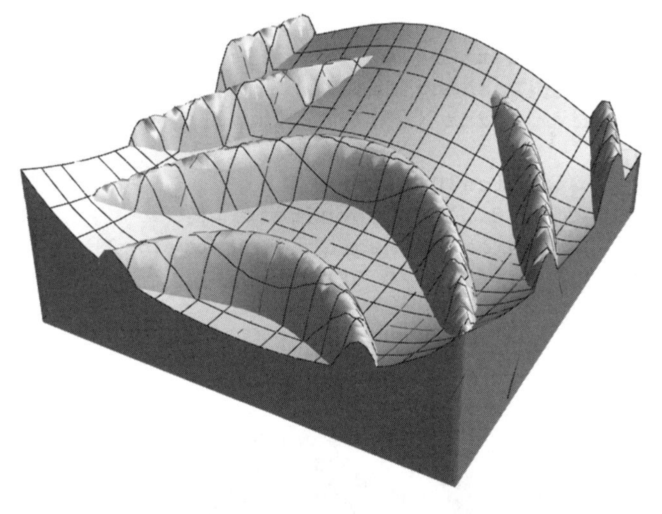

Figure 6.15 Intersection of two surfaces and the use of **Filling**

Example 6.10

Mode Shape of a Circular Membrane

The mode shape of a clamped solid circular membrane that has one nodal diameter and one nodal circle is given by

$$w_{11}(\eta,\theta) = J_1(7.0156\eta)\cos\theta$$

The nodal diameter is determined by that value of θ for which $w_{11} = 0$ and the nodal circle determined by that value of η for which $w_{11} = 0$. From the above equation, it is seen that the nodal diameter connects the polar coordinate points $(1,\pi/2)$ and $(1,3\pi/2)$. It can be shown that the nondimensional radius of the nodal circle is $\eta = 0.546$ and $\eta = 1$. Since the three-dimensional plotting function plots in Cartesian coordinates only, we have to convert from the (x,y) system to the (η,θ) system and use **ListPlot3D**. **Graphics3D** and **Line** will be used to create the nodal circle and nodal diameter. In addition, we shall show two forms of an additional enhancement to highlight the spatial depth of the mode shape: a plane drawn at $z = 0$ and a very thin cylinder whose center coincides with the center of the membrane. Each of these planes will be displayed with opacity of 0.3. The results will be displayed in three separate figures.

The program is

```
coor=Table[
 Table[{η Cos[θ],η Sin[θ],BesselJ[1,7.0156 η] Cos[θ]},
   {η,Range[0,1,0.05]}],{θ,Range[0,2 π,π/24]}];
rn=0.546;
rect={{-1,-1,0},{-1,1,0},{1,1,0},{1,-1,0}};
cir=Table[{rn Cos[θ],rn Sin[θ],0},
 {θ,Range[0,2. π,π/12.]}];
lp1=ListPlot3D[{coor},Boxed->False,Axes->False,
 Mesh->None,ViewPoint->{Top,Front}];
gp1=Graphics3D[{{Dashing[Small],Line[{cir}]},
 {Dashing[Medium],Line[{{0,1,0},{0,-1,0}}]}}];
Show[lp1,gp1]
Show[lp1,gp1,Graphics3D[
 {Magenta,Opacity[0.3],Polygon[rect]}]]
Show[lp1,gp1,Graphics3D[{Magenta,Opacity[0.3],
 Cylinder[{{0,0,0},{0,0,-0.0001}},1]}]]
```

The execution of this program yields Figure 6.16.

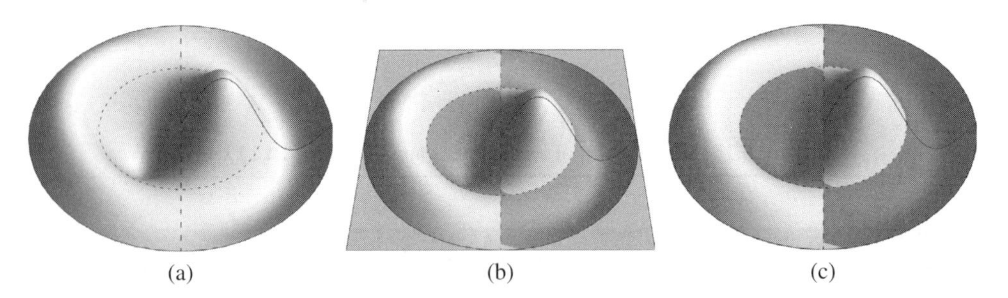

Figure 6.16 Mode shape of a clamped circular membrane where dashed lines indicate locations of nodal circle and nodal diameter: (a) no horizontal plane; (b) with horizontal plane; (c) with horizontal plane represented by a very thin cylinder

Example 6.11

Surface of Revolution

We shall create a surface of revolution from the following generatrix

$$r = 0.3 + e^{-z} \sin 3z$$

for $0 \leq z \leq 2.5$ and $0 \leq \theta \leq 3\pi/2$. The command is

```
RevolutionPlot3D[{0.3+Exp[-z] Sin[3 z],z},{z,0,2.5},
  {thet,0,3 π/2},Axes->False,Boxed->False,
  PlotStyle->FaceForm[Yellow,Cyan]]
```

where we have used **FaceForm** to select the colors of the inside surface and the outside surface, respectively. The execution of this statement produces Figure 6.17.

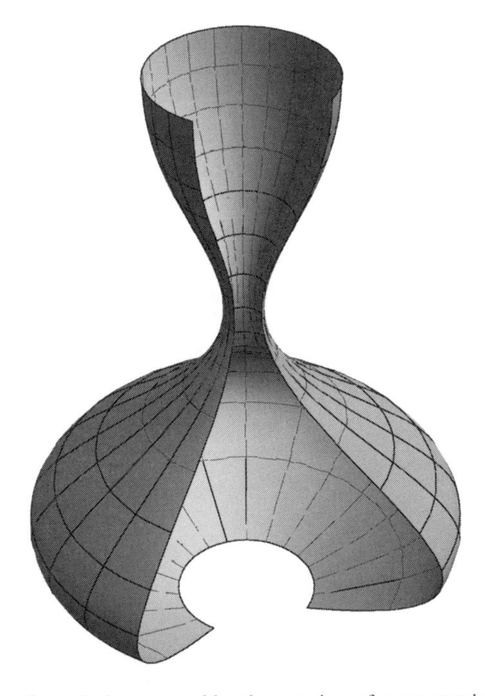

Figure 6.17 Surface of revolution created by the rotation of a generatrix through $3\pi/2$ radians

6.4 Summary of Functions Introduced in Chapter 6

The 2D plot commands are listed in Tables 6.1 to 6.4 and the built-in 2D geometric shapes are listed in Table 6.11. The options that are available to these graphics commands are listed in Tables 6.5, 6.6, 6.7, 6.9, 6.10, 6.12, 6.14, 6.15, and 6.18. The 3D plotting functions are given in Table 6.19 and the options specific to 3D graphics are listed in Table 6.20. The built-in 3D geometric shapes are listed in Table 6.21.

References

[1] P. K. Kundu and I. M. Cohen, *Fluid Mechanics*, 4th edn, Academic Press, Burlington, Massachusetts, 2008, p. 753.
[2] E. W. Weisstein, *CRC Concise Encyclopedia of Mathematics*, 2nd edn, Chapman Hall/CRC, Boca Raton, 2003, p. 2129–30.
[3] A. Leissa, Vibration of Shells, NASA Sp-288, 1973, p. 44.
[4] F. P. Incropera and D. P. Dewitt, *Introduction to Heat Transfer*, 4th edn, John Wiley & Sons, New York, 2002, p. 675.

Exercises

Section 6.2

6.1 The relationship between an oblique shock wave at an angle σ and the deflection angle δ as a function of the incident Mach number M_1 is [1]

$$\delta = \tan^{-1}\left[\frac{2\cot\sigma\left(M_1^2\sin^2\sigma - 1\right)}{M_1^2(\gamma + \cos(2\sigma)) + 2}\right] \tag{6.1}$$

where γ is the ratio of the specific heats of the gas. For a given M_1, there is a value of σ for which δ is a maximum. We denote these values as σ_{max} and δ_{max}. Then for a given M_1 those values of $\sigma > \sigma_{max}$ for which $\delta < \delta_{max}$ the downstream Mach number M_2 is less than 1 and the shock waves are called strong shock waves. For those values $\sigma < \sigma_{max}$ for which $\delta < \delta_{max}$, the downstream Mach number M_2 is greater than 1 and the shock waves are called weak shock waves. With this information, use Eq. (6.1) to obtain Figure 6.18. This figure was obtained with $\gamma = 1.4$.

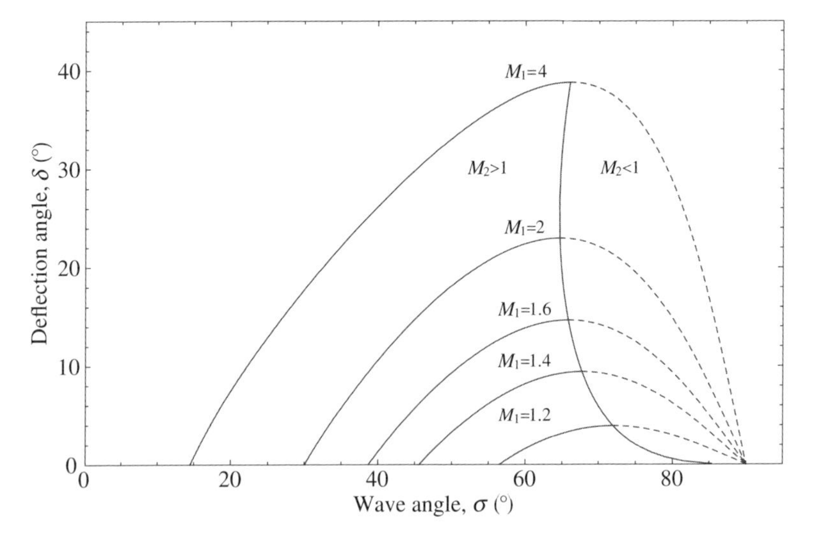

Figure 6.18 Solution to Exercise 6.1

6.2 For a given value of y, the value of p is determined from

$$(1 + k_3)h^3 + (2(1 + k_3) + k_1k_3 - (1 + k_3)y)h^2 + k_1k_3(1 + k_2 - y)h - yk_1k_2k_3 = 0$$

where $h = 10^{-p}$, $k_1 = 2.5 \times 10^{-4}$, $k_2 = 5.6 \times 10^{-11}$, and $k_3 = 1.7 \times 10^{-3}$. Plot the value of p when $0.00001 \le y \le 2.5$.

6.3 The Cartesian locations of bi-cylinder coordinates are given by

$$x = \frac{a \sinh \eta}{\cosh \eta - \cos \varphi}$$

$$y = \frac{a \sin \varphi}{\cosh \eta - \cos \varphi}$$

If $a = 1.5$, $-1 \le \eta \le 1$, and $0 \le \varphi \le 2\pi$, replicate Figure 6.19.

6.4 One can draw n circles, $n \ge 3$, that are tangent to a central circle of radius r_b and to each adjacent circle as shown in Figure 6.20 for $n = 10$. The radius of the outer

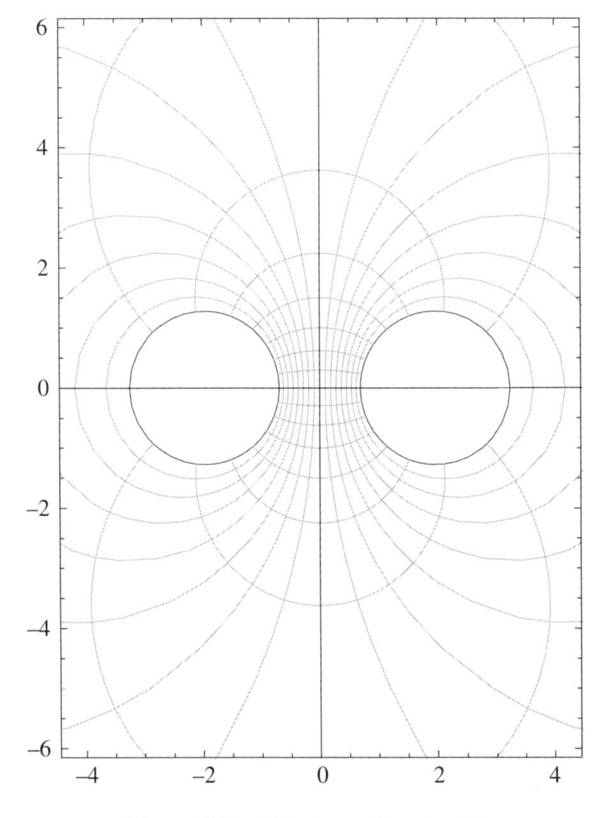

Figure 6.19 Solution to Exercise 6.3

Figure 6.20 Solution to Exercise 6.4

circles is

$$r_s = \frac{r_b \sin(\pi/n)}{1 - \sin(\pi/n)}$$

Write a program that replicates this figure and similar figures for arbitrary n. In this case, the outer circles (disks) are blue.

6.5 A Pappus chain is a series of n tangent circles inscribed in the area between three semicircles as shown in Figure 6.21. The radius of the outer circle is R_o and the radii of the other two circles are R_L and R_R, $R_L \geq R_R$. The location of the center of the nth circle is [2]

$$x_n = \frac{r(1+r)}{2D}$$
$$y_n = \frac{nr(1-r)}{D}$$
$$D = n^2(1-r)^2 + r$$

and its radius is

$$r_n = \frac{r(1-r)}{2D}$$

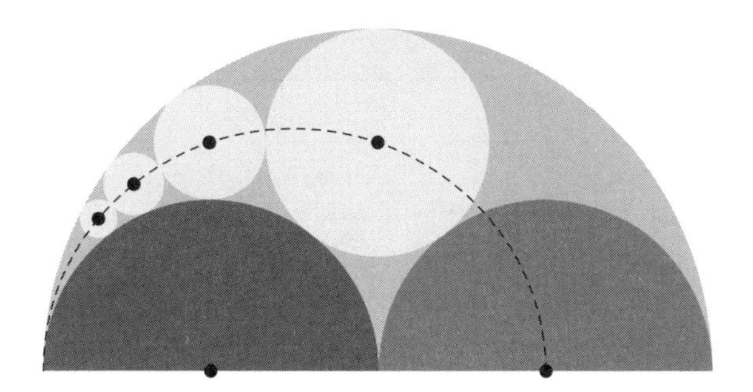

Figure 6.21 Solution to Exercise 6.5

where $r = 2R_L/(R_L + R_R)$. The ellipse that connects the centers of the circles is given by

$$\left[\frac{4x - (1+r)}{1+r}\right]^2 + \left[\frac{2y}{\sqrt{r}}\right]^2 = 1 \quad 0 \le x \le (1+r)/2$$

Replicate Figure 6.21, which has been obtained for $n = 4$, $R_o = 1/2$, $R_L = 1/4$, and $R_R = 1/4$. Color the four complete circles yellow and their background green. The right semicircle is colored magenta and the left one red.

6.6 A piece-wise linear map is defined by the points with the coordinates

$$\begin{aligned} x_{n+1} &= 1 - y_n + |x_n| \\ y_{n+1} &= x_n \qquad\qquad n = 1, 2, \ldots, N \end{aligned}$$

Obtain a plot for the coordinate points when $x_1 = 1$, $y_1 = 3.65$, and $N = 15{,}000$. The resulting figure is sometimes referred to as the gingerbread man.

6.7 The following relation is used to determine the value of a pipe's coefficient of friction λ for a given Reynolds' number Re, pipe diameter d, and surface roughness k

$$\frac{1}{\sqrt{\lambda}} + 2\log_{10}\left[\frac{2.51}{\mathrm{Re}\sqrt{\lambda}} + \frac{0.27}{d/k}\right] = 0 \quad 4 \times 10^3 \le \mathrm{Re} \le 10^7$$

Using this relation, replicate Figure 6.22.

6.8 The following equation will plot a Joukowsky airfoil over the range $-0.5 \le x \le 0.5$

$$y = \sqrt{\frac{1}{4} + \frac{1}{64h^2} - x^2} - \frac{1}{8h} \pm \frac{3}{8}t(1 - 2x)\sqrt{1 - (2x)^2}$$

where the plus sign describes the top of the airfoil and the negative sign its bottom. With $h = 0.08$ and $t = 0.13$, replicate the Figure 6.23.

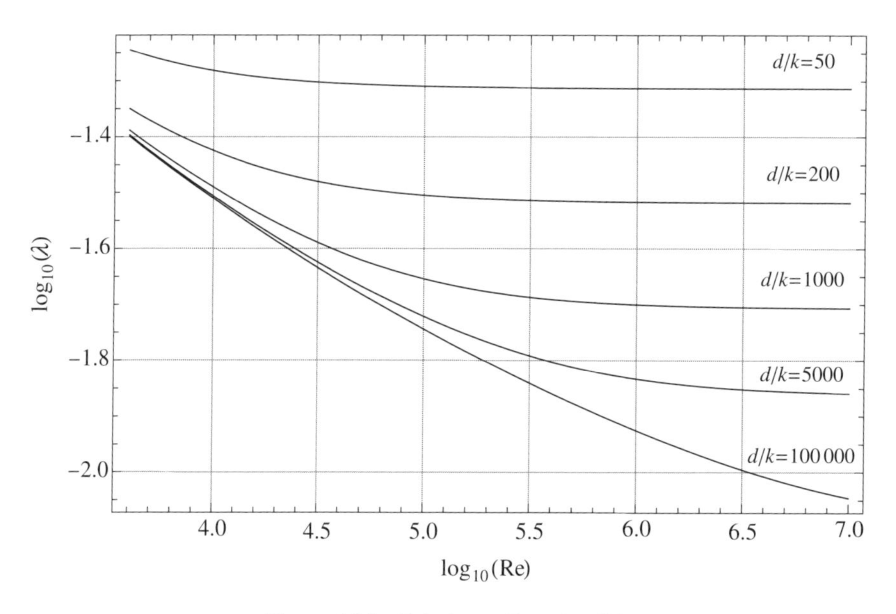

Figure 6.22 Solution to Exercise 6.7

6.9 A normalized structural damping factor describing the losses in a vibrating thin plate with a damping layer is given by

$$\eta = \frac{\alpha\xi(3 + 6\xi + 4\xi^2 + 2\alpha\xi^3 + \alpha^2\xi^4)}{(1 + \alpha\xi)(1 + \alpha^2\xi^4 + 2\alpha\xi(2 + 3\xi + 2\xi^2))}$$

where α is the ratio of the layers' moduli and ξ is the ratio of the layers' thickness. Replicate the logarithmic plot shown in Figure 6.24. The x-locations of the curves' labels and their individual orientations have been done iteratively.

6.10 Given the following polynomial [3]

$$\Omega^6 - k_2\Omega^4 + k_1\Omega^2 - k_0 - h\delta = 0 \tag{6.2}$$

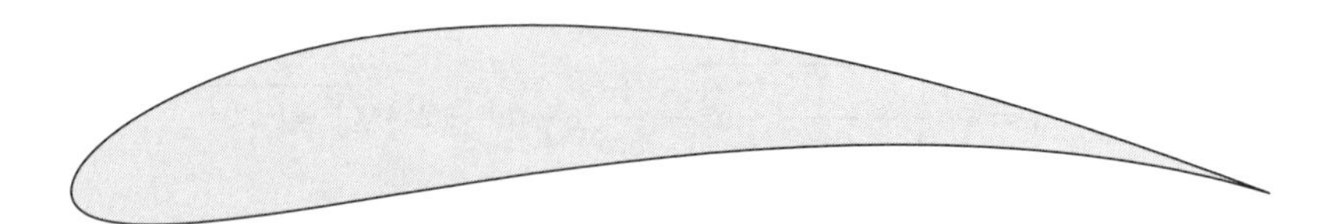

Figure 6.23 Solution to Exercise 6.8

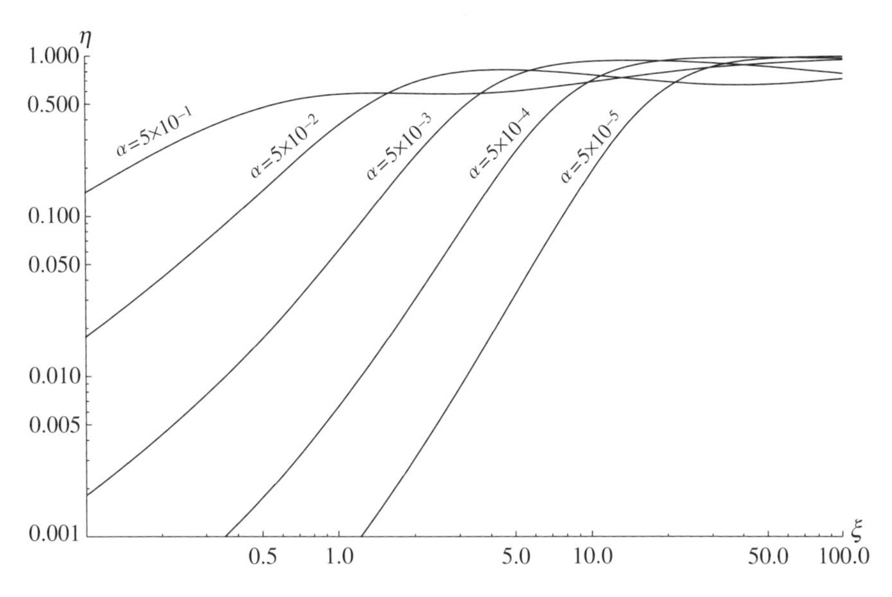

Figure 6.24 Solution to Exercise 6.9

where

$$k_2 = 1 + \frac{1}{2}(3 - \nu)(n^2 + \lambda^2) + h(n^2 + \lambda^2)^2$$

$$k_1 = \frac{1}{2}(1 - \nu)[(3 + 2\nu)\lambda^2 + n^2 + (n^2 + \lambda^2)^2 + \frac{3 - \nu}{1 - \nu}h(n^2 + \lambda^2)^3]$$

$$k_0 = \frac{1}{2}(1 - \nu)[(1 - \nu^2)\lambda^4 + h(n^2 + \lambda^2)^4]$$

$$\delta = \frac{1}{2}(1 - \nu)[2(2 - \nu)n^2\lambda^2 + n^4 - 2\nu\lambda^6 - 6n^2\lambda^4 - 2(4 - \nu)n^4\lambda^2 - 2n^6]$$

For $\nu = 0.3$, $h = 10^{-5}$, and $n = 1, 2, 3, 4, 5$, plot the smallest positive value of Ω that satisfies Eq. (6.2) for $0.5 \leq \pi/\lambda \leq 100$. The results should look like those shown in Figure 6.25. The placement of the labels on the curves should appear as shown even if the value of h changes.

6.11 The spectral emissive power is given by the Planck distribution [4]

$$E(\lambda, T) = \frac{c_1}{\lambda^5(e^{c_2/(\lambda T)} - 1)}$$

where c_1 =3.742 × 10^8 W μm^4·m^{-2} and c_2 = 1.4388 × 10^4 μm K. For T = {2000., 1000., 400., 100.} K, obtain a plot of E versus λ. The result should look like that shown in Figure 6.26. Show in tabular form that for the four temperatures given, $\lambda_{max}T_n$ = 2898 μm K.

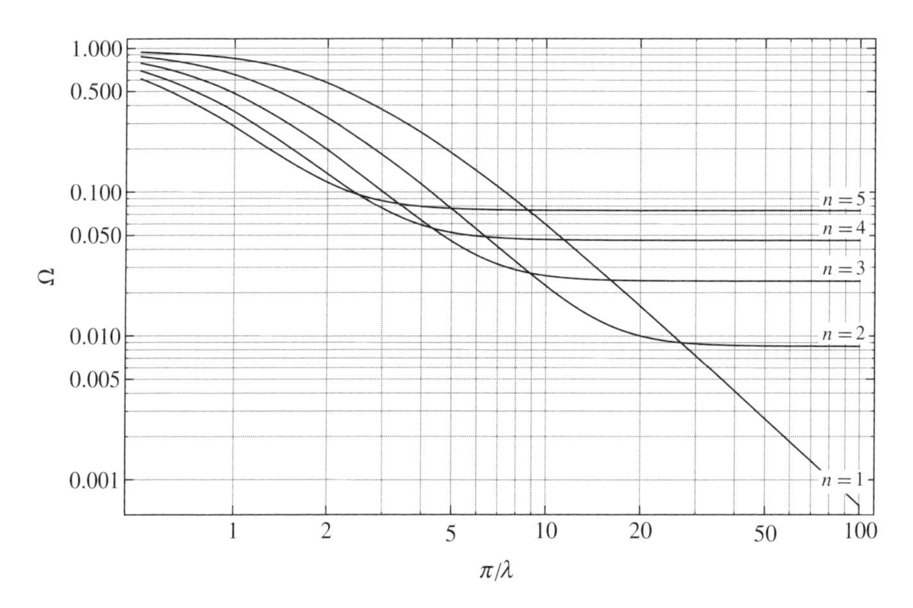

Figure 6.25 Solution to Exercise 6.10

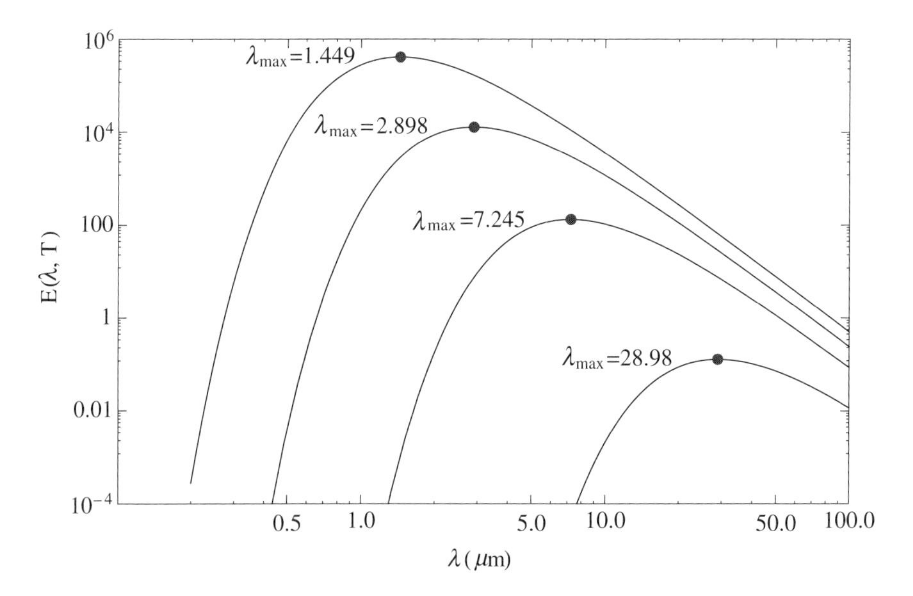

Figure 6.26 Solution to Exercise 6.11

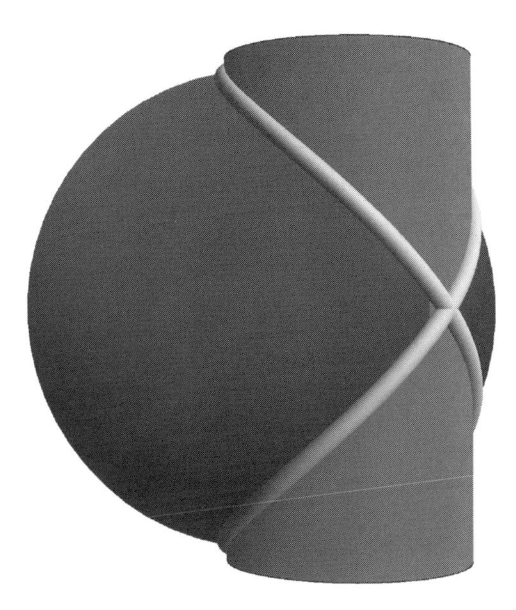

Figure 6.27 Solution to Exercise 6.14

Section 6.3

6.12 Plot the surface described by

$$u(\xi, \eta) = 8 \sum_{n=1,3,5}^{25} \frac{e^{-n\pi\xi}}{(n\pi)^3} \sin(n\pi\eta) \quad 0 \le \eta \le 1, \quad 0 \le \xi \le 0.7$$

over the region indicated. Label the axes.

6.13 Plot the surface described by

$$u(\eta, \tau) = \frac{2}{a\pi(1-a)} \sum_{n=1}^{50} \frac{\sin n\pi a}{n^2} \sin(n\pi\eta)\cos(n\pi\tau) \quad 0 \le \eta \le 1, \quad 0 \le \tau \le 2$$

over the region indicated when $a = 0.25$. Label the axes.

6.14 The curve that results from the intersection of a sphere of radius $2a$ centered at the origin and a cylinder of radius a centered at (a,0,0) is given by the parametric equations

$$x = a(1 + \cos\varphi)$$

$$y = a\sin\varphi$$

$$z = 2a\sin(\varphi/2)$$

where $0 \le \varphi \le 4\pi$. If $a = 1$ and the length of the cylinder is 2, then replicate Figure 6.27. Use a tube radius of 0.04. Color the sphere blue, the cylinder magenta, and the tube yellow.

6.15 Plot the following parametric equations over the regions indicated.

(a) Figure eight torus

$$x = \cos(u)(c + \sin(v)\cos(u) - \sin(2v)\sin(u)/2)$$
$$y = \sin(u)(c + \sin(v)\cos(u) - \sin(2v)\sin(u)/2)$$
$$z = \sin(u)\sin(v) + \cos(u)\sin(2 * v)/2 \qquad c = 1; \quad -\pi \le u, v \le \pi$$

(b) Seashell

$$x = 2(1 - e^{u/(6\pi)})\cos(u)\cos^2(0.5v)$$
$$y = 2(-1 + e^{u/(6\pi)})\sin(u)\cos^2(0.5v)$$
$$z = 1 - e^{u/(3\pi)} - \sin(v) + e^{u/(6\pi)}\sin(v) \quad 0 \le v \le 2\pi; \quad 0 \le u \le 6\pi$$

(c) Astroidal ellipsoid

$$x = (a\cos u \cos v)^3$$
$$y = (b\sin u \cos v)^3$$
$$x = (c\sin v)^3 \quad a = b = c = 1; \quad -\pi/2 \le u \le \pi/2; \quad -\pi \le v \le \pi$$

7

Interactive Graphics

7.1 Interactive Graphics: Manipulate[]

The **Manipulate** command is a straightforward way create an interactive environment for the manipulation of the parameters $u_1, u_2, \ldots$ of an expression $f(x, y, u_1, u_2, \ldots)$. The output of **Manipulate** can be numbers, symbolic expressions, and graphics. The **Manipulate** command has a very wide range of interactive capabilities. We shall discuss several of them in detail and follow their introduction with examples.

To illustrate the capabilities of **Manipulate**, the following general form is assumed

```
Manipulate[expr,
  text1,
  Item[object1,pos1]
  {spec1},
  Delimiter,
  text2,
  Item[object2,pos2]
  {spec2},
  Delimiter,
  ...,
  ControlPlacement->{pl1,pl2, ...},
  Initialization:>(proc),
  SaveDefinitions->True
  TrackedSymbols:>{u1,u2, ...}]
```

Each of these terms will be discussed in what follows.

The **Manipulate** command creates two types of output. In our case, the first type will be primarily a graphical display of the results, which can be composed of one or more fully annotated figures with each figure containing one or more curves. The procedure to create the graphic is denoted **expr**, which represents $f(x, y, u_1, u_2, \ldots)$. If more than one expression is used to create **expr**, then each expression is terminated with a semicolon, except for the last expression, which is terminated with a comma. The second type of output is a set of control

An Engineer's Guide to Mathematica®, First Edition. Edward B. Magrab.

Companion Website: www.wiley.com/go/magrab

devices that provide the interactivity; that is, the ability to change the parameters $u_1, u_2, \ldots,$ (denoted **uN**) that appear in $f(x, y, u_1, u_2, \ldots)$. The control devices that we shall consider are the slider/animator, slider, 2D slider, radio buttons, setter buttons, popup menu, locator, angular gauge, and horizontal gauge. Each of these control devices requires a different set of specifications (denoted **specN**, and that is a function of **uN**) and can appear in any order after **expr**. The specifications for each of these control devices are presented in Table 7.1 along

Table 7.1 Several **Manipulate** control devices, their usage, and their representative output

```
Manipulate[expr,
   Style["Text",Attributes],
   Item[object,Alignment->pos],
   {spec,Enabled->tf,ContinuousAction->ft}]
```

where **spec** is a control-device-dependent sequence of instructions, **Attributes** are those given in Figure 6.2, and

tf = **True** (default) or **False**; used to enable (true) or disable (false) control device

ft = **True** (default) or **False**; when **True** changes in variables are continuously acted on, whereas when **False** only final value is acted on

---

Slider with Animator (Used to change a variable's value with the option to bypass slider to enter a value manually or by automatically changing the variable throughout its range; that is, animating the display. If **ControlType** is omitted and the form of **spec** shown below is used, then this is the default control element.)

Control specification

spec is replaced with

```
{var,var_init,label},lower,upper,incr,Appearance->lst
```

where

var = variable name that is used in **expr**

var = **var_init** when **Manipulate** is first executed

label = optional label to left of slider: general form given in Figure 6.2

lst = **"Labeled"** displays value of **var** to the right of slider; if omitted, no display

= **"Open"** display buttons below slider as shown in Output below

= **"Closed"** buttons below slider not displayed (default)

= **{"Labeled","Open"}** displays **var** and displays buttons below slider

lower ≤ **var** ≤ **upper** in increments of **incr**; if **incr** omitted **var** varies "continuously"

Example

```
Manipulate[None,Style["Animator/Slider",Bold],
   {{xi,0.11,"Damping coefficient"},0,0.7,0.01,
   Appearance->{"Labeled","Open"}}]
```

Output

Animator/Slider

Damping coefficient 0.11

0.11

Table 7.1 (*Continued*)

Slider (Used to change a variable's value.)

Control specification

`spec` is replaced with

```
{var,var_init,label},lower,upper,incr,
   ControlType->Slider,Appearance->"Labeled"
```

For definitions, see **Slider with Animator** above

Example

```
Manipulate[None,Style["Slider (no animator)",Bold],
   {{xi,0.11,"Damping coefficient"},0,0.7,0.01,
   Appearance->"Labeled",ControlType->Slider}]
```

Output

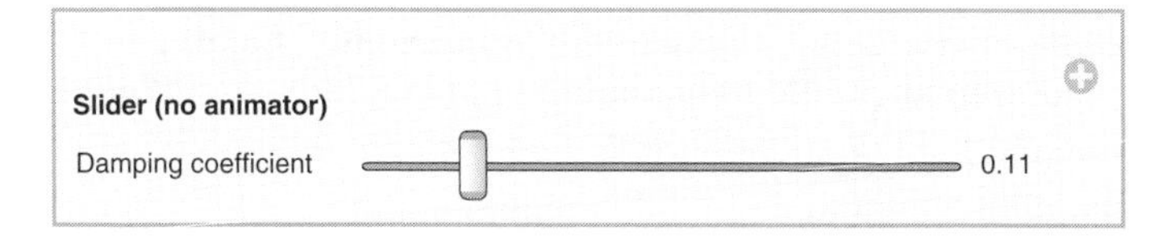

Slider2D (Used to change the values of two variables simultaneously.)

Control specification

`spec` is replaced with

```
var,{xmin,ymin},{xmax,ymax},ControlType->Slider2D
```

where

`var` = variable name that is used in `expr`; it is a two-element list

`{xmin,ymin}` and `{xmax,ymax}`, respectively, define the lower and upper limits of `var`

Example

```
Manipulate[None,Style["Slider2D",Bold],
   {vc,{1,3},{3,6},ControlType->Slider2D}]
```

Output

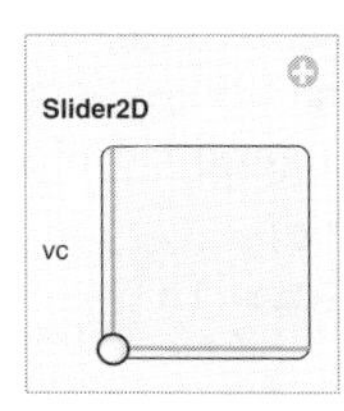

Input Field: Blank Editable Field (Means to enter manually a number, expression, or string.)

Control specification

`spec` is replaced with

```
w,ControlType->InputField
```

where

`w` = entry followed by *Enter*; typically a number, expression, or string

Example

```
Manipulate[None,Style["Input Field",Bold],
   {w,ControlType->InputField}]
```

Output

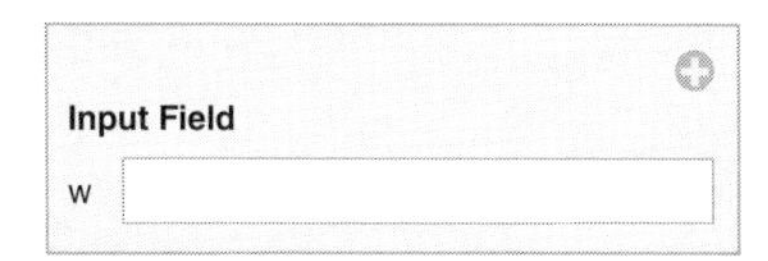

(*continued*)

Table 7.1 (*Continued*)

Radio Buttons (Used to select one of two or more mutually exclusive states for a variable.)

Control specification

spec is replaced with

```
{var,var_init,label},{but1->label1,but2->label2, ...},
    ControlType->RadioButtonBar
```

where

- **var** = **butK**
- **var_init = butL**; specifies the button that is selected on initialization
- **label** = optional label to left of first button: general form given in Figure 6.2
- **labeln** = label to the right of each button: general form given in Figure 6.2

Example

```
Manipulate[None,Style["Radio Buttons",Bold],
  {{inspan,1,"Attachments?"},{1->"No",2->"Yes"},
  ControlType->RadioButtonBar}]
```

Output

Radio Buttons

Attachments? No Yes

Setter Bar (Used to select one of two or more mutually exclusive states for a variable.)

Control specification

spec is replaced with

```
  {var,var_init,label},{but1->label1,but2->label2, ...},
    ControlType->SetterBar
```

where

- **var** = **butK**
- **var_init = butL**; specifies the button that is selected on initialization
- **label** = optional label to left of first button: general form given in Figure 6.2
- **labeln** = label appearing in each button: general form given in Figure 6.2

Example

```
Manipulate[None,Style["Setter Bar",Bold],
  {{inspan,2,"Select"},{1-> Graphics[Circle[],
  ImageSize->30],2->"Choice 2",
  3->TraditionalForm[Exp[a] Sin[Cos[x]]]},
  ControlType->SetterBar}]
```

Output

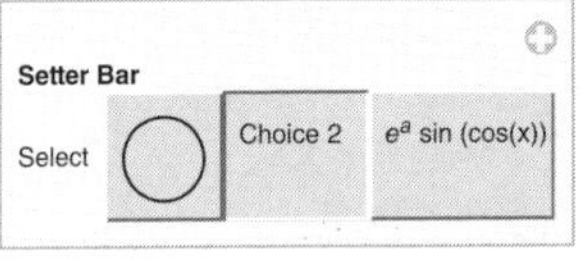

Table 7.1 (*Continued*)

Popup Menu: Form 1 (Used to select one of two or more mutually exclusive states for a variable.)

Control specification

`spec` is replaced with

```
{var,sel_init,label},{sel1->label1,sel2->label2, ...},
   ControlType->PopupMenu
```

where

`var` = `seln`
`var` = `sel_init`; specifies the entry that is selected on initialization
`label` = optional label to left of menu tab: general form given in Figure 6.2
`labeln` = label of each entry in the menu: general form given in Figure 6.2

Example

```
Manipulate[None,Style["Popup Menu: Form 1",Bold],
   {{bc,2,"Boundary conditions"},{1->"Clamped-Clamped",
      2->"Hinged-Hinged", 3->"Clamped-Free"},
      ControlType->PopupMenu}]
```

Output

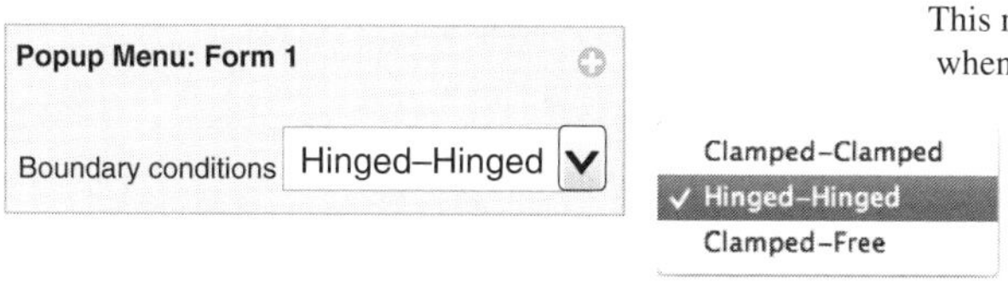

Popup Menu: Form 1
Boundary conditions Hinged–Hinged

Clamped–Clamped
✓ Hinged–Hinged
Clamped–Free

This menu appears when ∨ is clicked

Popup Menu: Form 2 (Selects one of two or more mutually exclusive states for a variable.)

Control specification

`spec` is replaced with

```
{expr,fcn_init,label},{fcn1->label1},fcn2->
  label2, ...}, ControlType->PopupMenu
```

where

`expr` = `fcnK`
`expr` = `fcn_init`; specifies the expression that is selected on initialization
`label` = optional label to left of menu tab: general form given in Figure 6.2
`labelK` =`TraditionalForm[fcnK]`

Example

```
Manipulate[None,Style["Popup Menu: Form 2",Bold],
   {{expr,Cos[Cos[x]],"Functions"},
      {Sin[x]->TraditionalForm[Sin[x]],
      x^2/Sqrt[1+x^3]->TraditionalForm[x^2/Sqrt[1+x^3]],
      Cos[Cos[x]]->TraditionalForm[Cos[Cos[x]]]},
      ControlType->PopupMenu}]
```

Output

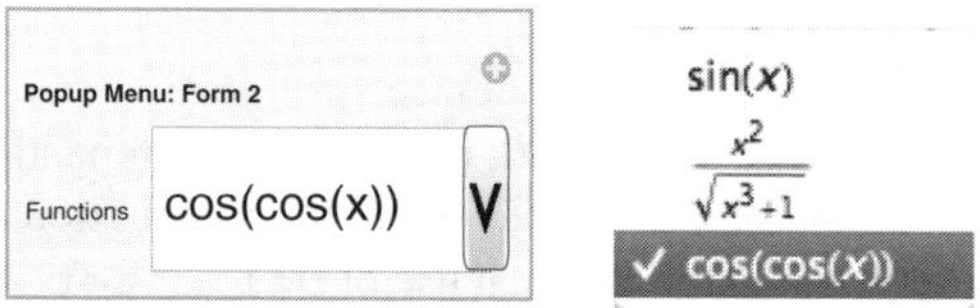

Popup Menu: Form 2
Functions cos(cos(x))

$\sin(x)$
$\frac{x^2}{\sqrt{x^3+1}}$
✓ $\cos(\cos(x))$

This menu appears when ∨ is clicked

(*continued*)

Table 7.1 (*Continued*)

Locator (Used to determine the *x*-*y* coordinate values of a location on a graph.)

Control specification	**spec** is replaced with `var,{xinit,yinit},ControlType->Locator` where **var** = variable name that is used in **expr** **{xinit,yinit}** are the coordinates of the initial location of the locator
Example	`Manipulate[Graphics[Circle[{0,0},1]],` `{bb,{0.5,0.75},ControlType->Locator}]`
Output	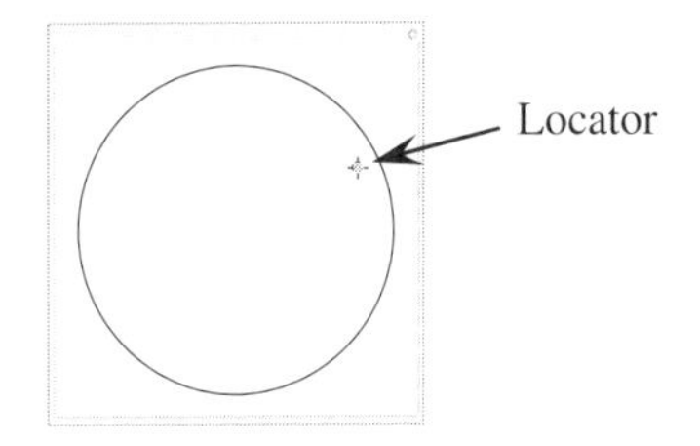

Item (Used to place a Mathematica object or an imported image.)

Control specification	`Item[object,Alignment->loc] (* See text *)`

Example	`Manipulate[None,`
	`Style["Item centered",Bold],`
	`Item[beam,Alignment->Center],Style["Slider",Bold],`
	`{{r,0,"label"},-2,2,1,Appearance->"Labeled"},`
	`Initialization:>(beam=Show[{Graphics[Line[{{0,0},{1,0}}],`
	`ImageSize->Tiny,PlotLabel->"Concentrated"]},`
	`Graphics[{Arrowheads[0.1],`
	`Arrow[{{0.5,0.25},{0.5,0}}]}]])]`

Output	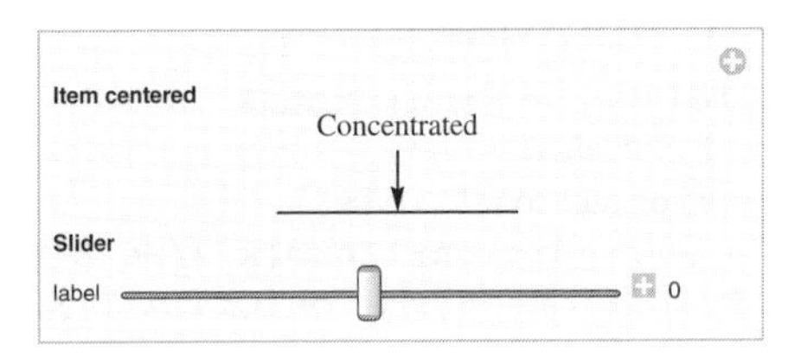

with examples of their respective usage. Each control device specification can be intermixed with: optional text display, which is denoted by **textN**; the optional command **Item**, which is used to place a Mathematica-generated object such as a graphic or equation or to place an imported image from an external source; and the optional use of **Delimiter**, which places a horizontal line to visually delineate text or a control device or a group of control devices. One can employ any number of the control devices as are meaningful and these control devices need not be different. It will be seen that the slider tends to be used most often. In the general form

Table 7.1 (*Continued*)

Angular Gauge (Used to change a variable's value by moving the needle of the gauge or by clicking on a value appearing on the gauge's scale.)[†]

Control specification	**{spec,Enabled->tf,ContinuousAction->ft}** is replaced with **{{var,var_init,label},lower,upper,AngularGauge[##, GaugeLabels->{label2,"Value"},ImageSize->Size, ScaleDivisions->{major,minor}] &}** where **var** = variable name **var** = **var_init** when **Manipulate** is first executed **label** = optional label to left of angular gauge: general form given in Figure 6.2 **lable2**= label to appear near the center of the gauge **"Value"** specifies that the numerical value corresponding to the position of the needle is to appear as a digital readout **major** = approximate number of major-scale divisions to appear on gauge face **minor** = approximate number of minor-scale divisions to appear between major scale divisions **Size** = size of gauge displayed image: typically **Tiny**, **Small**, **Medium**, and **Large** **lower** ≤ **var** ≤ **upper** specifies range of the numerical values of the gauge
Example	**Manipulate[None,{{alp,30,Style["α",14]},0,90, AngularGauge[##,GaugeLabels->{"α (°)","Value"}, ImageSize->Small,ScaleDivisions->{20,5}] &}]**
Output	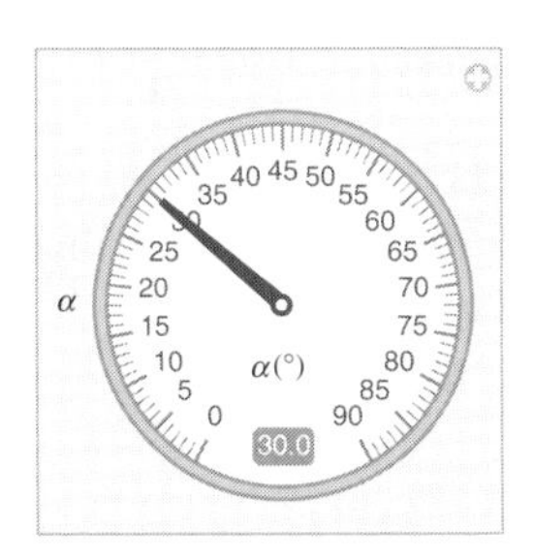

(*continued*)

depicted above, **textN** is usually given by the general form shown in Figure 6.2; that is, by **Style["Message",** ***color***, ***font style***, ***font attribute*****]**; the form **"Message"** without using **Style** can also be used.

The command **Item** is given by

```
Item[object,Alignment->pos]
```

where **object** is the object to be placed and **pos** is the position of the object: **Left** (default), **Right**, or **Center**. The object can be a graphic generated with Mathematica commands as illustrated in the subheading **Item** in Table 7.1, or it can be a picture in one of many formats.

Table 7.1 (*Continued*)

Horizontal Gauge (Used to change a variable's value by moving the gauge indicator or by clicking on a value appearing on the gauge's scale.)†

Control specification

`{spec,Enabled->tf,ContinuousAction->ft}` is replaced with

```
{{var,var_init,label},lower,upper,HorizontalGauge[##,
    GaugeLabels->{label2,"Value"},ImageSize->Size,
    ScaleDivisions->{major,minor},
    GaugeMarkers->{"Name"}] &}
```

where the variable names have the same meaning as those given for the angular gauge and **"Name"** indicates the built-in style of the gauge. The available styles can be found by entering

```
ChartElementData["HorizontalGauge","Markers"]
```

Example

```
Manipulate[None,
  {{r,2.6,Style["r",14]},2,3,HorizontalGauge[##,
    GaugeLabels->{"Value"},ImageSize->Small,
    ScaleDivisions->{5,2},
    GaugeMarkers->{"GlassRectangle"}]&}]
```

Output

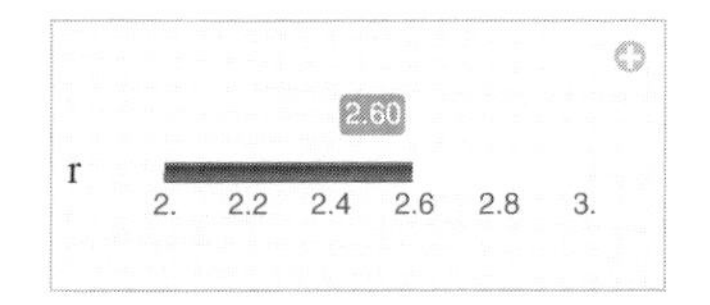

† See **AngularGauge** and **HorizontalGauge** in the *Documentation Center* for a description of the various options associated with these commands.

After the control elements have been specified, there are numerous options that are available. Four that are frequently employed are now introduced. The first is

```
ControlPlacement->{pl1,pl2, ...}
```

which is used to position each control device and each message (**textN**) with respect to the output area in the order that they appear. If omitted, the control devices are placed at the top. The value of each **plN** can be **Left**, **Right**, **Top**, or **Bottom**. If only one position is selected, then all control devices are placed at that location.

The second option is

```
Initialization:>(proc)
```

where **proc** is a procedure that, with respect to **expr**, can contain any constants that need to be given values, any solutions to one or more equations that need to be obtained, any data sets that need to be initialized, and any functions that need to be created. In addition, it can contain any figures that must be created or inserted prior to using **Item**. Each of the types of operations would be contained in **proc** and would be of the form **(c1=operation1; c2=operation2; ...)**, where **operationN** is a specified operation needed to evaluate or define or assign a value to **cN**. The parentheses are required if there is more than just **c1**. The usage of **Initialization** is illustrated in Table 7.1 under the subheading **Item**.

The third option is

```
TrackedSymbols:>{u1,u2, ...}]
```

which is used to specify which parameters appearing in the control device definitions should trigger updates when changed. It is a list of the parameter names **u1**, **u2**, ... It is good practice to always use **TrackedSymbols**.

The fourth option is

```
SaveDefinitions->True
```

which is needed when converting a notebook that contains **Manipulate** to a CDF file. It ensures that all data sets and function definitions are embedded in **Manipulate** prior to its conversion. Although this function is typically performed by **Initialization**, in some cases the amount of data and/or the number and complexity of the function definitions may make their inclusion in **Initialization** awkward or inconvenient. In this case, one does two things. The first is to place the data and function definitions prior to **Manipulate** but in the same cell. The second is to employ the option statement shown above.

We shall discuss in a little more detail the animator/slider presented in Table 7.1. This control device is the most capable of all the control devices. Referring to the example output of this device shown in Table 7.1, clicking on the minus sign to the right of the slider bar removes the display of the symbols under the slider. The minus sign now becomes a plus sign; clicking on it will again display these symbols. The box that contains the number 0.11 can be used to enter a numerical value manually. If the value 0.25 is typed and *Enter* is pressed, the new value for **var** will be 0.25 and 0.25 will appear at the right of the slider. If the number entered is greater than **upper** or less than **lower**, the number will be used in the evaluation of **expr**, but the appropriate end of the slider will turn red, indicating that a limit has been exceeded.

The six buttons under the slider will now be discussed. The "–" button will decrement the current value of **var** by **incr**, whereas the "+" button will increment the current value of **var** by **incr**. Clicking on the button with the solid triangle will cause the slider to assume all values of **var** between **lower** and **upper** sequentially at a given rate, thereby creating an animation of the image with respect to that variable. Each new value of **var** is obtained by adding/subtracting **incr** to it. Clicking on this button, which now displays two parallel bars, will stop this calculation. Whether the continuous calculation is adding or subtracting **incr** depends on the direction of the arrow of the right-most button. When the arrow is pointing to the right (→) it will be adding **incr**, when it is pointing to the left (←) it will be subtracting it, and when the arrow is pointing in both directions (↔) it will add it until it reaches the maximum value and then subtract it until it reaches its minimum value and continue in this fashion until stopped. Lastly, the rate at which the continuous calculations occur can be increased or decreased, respectively, by clicking one or more times on the button with the two upward pointing arrowheads and can be decreased by clicking one or more times on the button with the downward pointing double arrowheads.

The use of **Manipulate** is now illustrated with several examples.

Example 7.1

Basic Elements of an Interactive Graph

We shall use three control devices by considering the evaluation of any of the following functions

$$f_1(x) = \tanh(x)$$
$$f_2(x) = \frac{x^2}{\sqrt{1+x^3}} \qquad 0.1 \le x \le 10$$
$$f_3(x) = \cos(\cos x)$$

two ways: as given or by using them as the power of e; that is,

$$e^{f_n(x)} \qquad n = 1, 2, 3$$

We shall represent the selection of $f_n(x)$ as a popup menu that will be placed on the top of the area displaying the numerical result, the evaluation option with a pair of radio buttons that will be placed on its right, and the selection of x with a slider that will be placed on the bottom of the display area. The output area will be a box in which the numerical result appears. For all three functions, the value of x will vary over the range $0.1 \le x \le 10$ in increments of 0.1. The program is

```
Manipulate[If[c==1,expr/.x->y,Exp[expr/.x->y]],
  Style["Select a function",Bold],
                    (* Create popup menu *)
{{expr,x^2/Sqrt[1+x^3],"Function"},
   {Tanh[x]->TraditionalForm[Tanh[x]],
   x^2/Sqrt[1+x^3]->TraditionalForm[x^2/Sqrt[1+x^3]],
   Cos[Cos[x]]->TraditionalForm[Cos[Cos[x]]]},
   ControlType->PopupMenu},
Delimiter,
                    (* Create radio buttons *)
Style["Evaluate as",Bold],
{{c,1," "},{1->"Function",2->"e^Function"},
  ControlType->RadioButtonBar},
Delimiter,
Style["Select a value for x",Bold],
                    (* Create slider *)
{{y,4.,"x"},0.1,10,0.1,Appearance->"Labeled",
  ControlType->Slider},
        (* Specify placement of labels and control devices *)
ControlPlacement:>{Top,Top,Right,Right,Bottom,Bottom},
TrackedSymbols:>{expr,c,y}]
```

which results in the initial display shown in Figure 7.1.

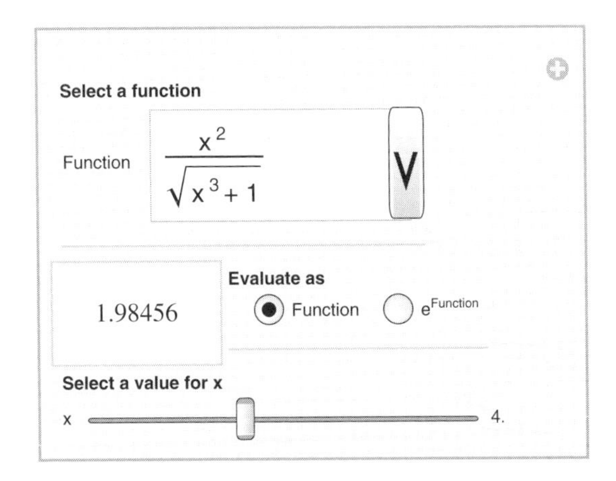

Figure 7.1 Initial interactive display created by **`Manipulate`** for Example 7.1

Example 7.2

Sum of Two Sinusoidal Signals

We shall use **`Manipulate`** to graphically illustrate the adding of two sine waves, each of which has an amplitude of A_n, a frequency of ω_n, and phase angle θ_n. Thus,

$$f_n(t) = A_n \sin(\omega_n t + \theta_n)$$
$$f_{sum}(t) = f_1(t) + f_2(t)$$

The sum $f_{sum}(t)$ will be displayed in one figure, and the magnitude of each A_n as a function of frequency ω_n will be displayed in another figure that is placed below the first figure. The ranges for each of these parameters are: $0 \le A_n \le 1$, $0 \le \omega_n \le 10$, and $0 \le \theta_n \le 180°$. To be able to see both the details of $f_{sum}(t)$ and its envelope over a wide frequency range, we shall also add the ability to change the extent of the display's time axis from 0 to t_{end}. All control devices will be located at the left of the display area. The two signals will be color-coded.

The program is as follows.

```
Manipulate[
  GraphicsColumn[{
    Plot[a1 Sin[ω1 t+θ1 Degree]+a2 Sin[ω1 t+θ2 Degree],
      {t,0,tend},PlotRange->{{0,tend},{-2,2}},
      AxesLabel->{Style["t",Italic],"Amplitude"},
      LabelStyle->{12},PlotLabel->labp],
    ListLinePlot[{{{ω1,0},{ω1,a1}},{{ω2,0},{ω2,a2}}},
      PlotRange->{{0,10.1},{0,1}},PlotStyle->{Blue,Red},
      AxesLabel->{"ωk","Ak"},LabelStyle->{12}]}],
  Style["Sine Wave 1",Bold,11],
```

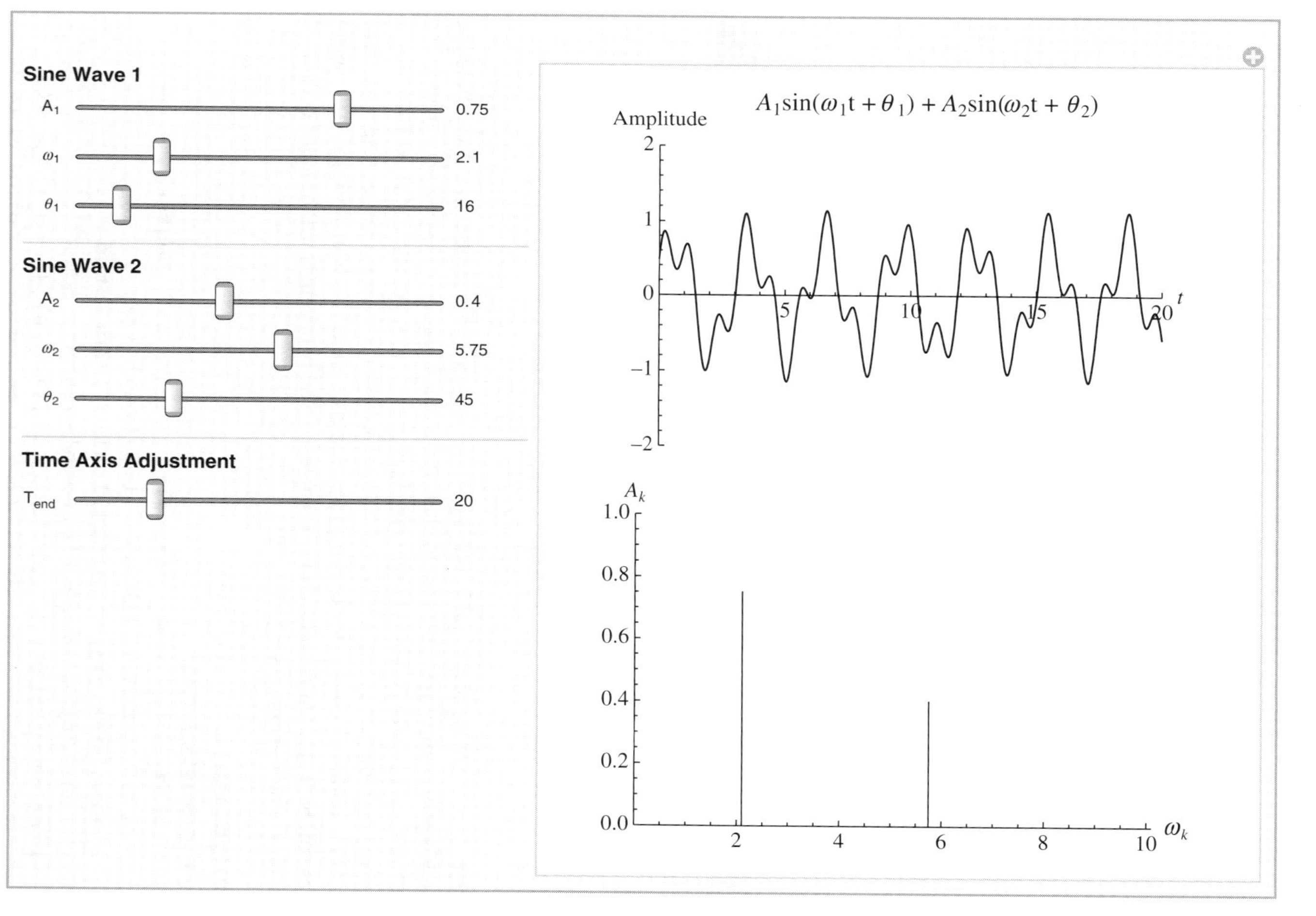

Figure 7.2 Initial interactive display created by `Manipulate` for Example 7.2

```
(* Create sliders for first signal *)
{{a1,0.75,"A1"},0,1,0.05,Appearance->"Labeled",
   ControlType->Slider},
{{ω1,2.1,"ω1"},0,10,0.05,Appearance->"Labeled",
   ControlType->Slider},
{{θ1,16,"θ1"},0,180,1,Appearance->"Labeled",
   ControlType->Slider},
 Delimiter,
 Style["Sine Wave 2",Bold,11],
(* Create sliders for second signal *)
{{a2,0.4,"A2"},0,1,0.05,Appearance->"Labeled",
   ControlType->Slider},
{{ω2,5.75,"ω2"},0,10,0.05,Appearance->"Labeled",
   ControlType->Slider},
{{θ2,45,"θ2"},0,180,1,Appearance->"Labeled",
   ControlType->Slider},
 Delimiter,
 Style["Time Axis Adjustment",Bold,11],
(* Create slider for time axis *)
{{tend,10,"tend"},1,100,0.5,Appearance->"Labeled",
   ControlType->Slider},
TrackedSymbols:>{a1,a2,ω1,ω2,θ1,θ2,tend},
ControlPlacement->Left,
Initialization:>
   (labp=Row[{Style["A1sin(ω1t+θ1)",14,Blue],
      Style["+A2sin(ω2t+θ2)",14,Red]}]))]
```

The execution of this program results in the initial display shown in Figure 7.2.

Example 7.3

Steerable Sonar/Radar Array

We shall use **Manipulate** and **PolarPlot** to demonstrate the effects that several parameters have on a steerable sonar/radar array composed of N radiators with identical amplitude excitation. If we consider the radiators to be oriented vertically and spaced a distance d apart, then the normalized angular output of the array at a large radial distance from the center of the array and oriented at an angle θ down from the vertical is proportional to

$$AF_n = \left| \frac{\sin\left[kdN\left(\cos\theta - \cos\theta_0\right)/2\right]}{N\sin\left[kd\left(\cos\theta - \cos\theta_0\right)/2\right]} \right|$$

where AF_n is called the normalized array factor, $k = 2\pi/\lambda$ is the wave number and λ is the wavelength of the wave propagating at a frequency ω, and θ_0 is the array's steering angle.

We shall plot AF_n as a function of θ and the parameters N, kd, and θ_0. The expression AF_n is represented by a function that is created in **Initialization**. The program is as follows.

```
Manipulate[PolarPlot[afn[θ],{θ,0,2 π},PlotRange->All],
  {{kd,3,"kd"},0.1,10,0.1,ControlType->Slider,
    Appearance->"Labeled"},
  {{nN,8,"N"},2,25,1,ControlType->Slider,
    Appearance->"Labeled"},
```

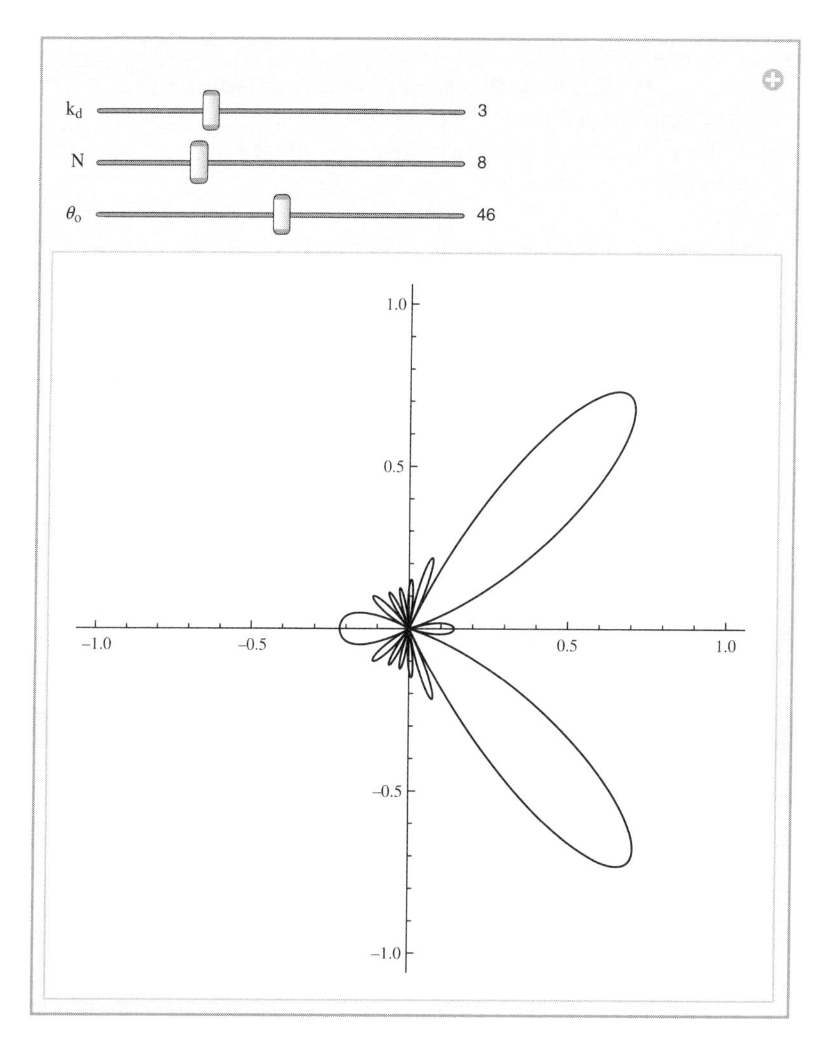

Figure 7.3 Initial interactive display created by **Manipulate** for the normalized array factor of a steerable array in Example 7.3

```
{{θ0,46,"θ0"},0,90,1,ControlType->Slider,
  Appearance->"Labeled"},
Initialization:>(afn[θ_]:=
  Abs[Sin[kd nN (Cos[θ]-Cos[θ0 Degree])/2]/
  Sin[kd (Cos[θ]-Cos[θ0 Degree])/2]/nN]),
TrackedSymbols:>{kd,nN,θ0}]
```

The result is shown in Figure 7.3.

Example 7.4

von Mises Stress in a Stretched Plate with a Hole

We shall use **Manipulate**, **ListContourPlot**, and a locator to display a contour map of the von Mises equivalent tensile stress in a rectangular plate that has a hole of nondimensional radius 1 and to display the value of the von Mises stress at an arbitrary position. The force is applied in the x-direction. **Manipulate** will be used to provide a locator that determines a position on the plate. Using the coordinates of this position, the van Mises stress is computed and displayed. The von Mises stress in the nondimensional polar coordinates $(\eta,\ \theta)$ is given by

$$\bar{\sigma}(\eta,\theta) = \sqrt{\frac{1}{2}\left[(\bar{\sigma}_{rr}-\bar{\sigma}_{\theta\theta})^2+\bar{\sigma}_{rr}^2+\bar{\sigma}_{\theta\theta}^2\right]+3\bar{\tau}_{r\theta}^2} \qquad \eta \geq 1, \quad 0 \leq \theta \leq 2\pi$$

where

$$\bar{\sigma}_{rr} = 1-\frac{1}{\eta^2}+\left(1+\frac{3}{\eta^4}-\frac{4}{\eta^2}\right)\cos(2\theta)$$

$$\bar{\sigma}_{\theta\theta} = 1+\frac{1}{\eta^2}-\left(1+\frac{3}{\eta^4}\right)\cos(2\theta)$$

$$\bar{\tau}_{r\theta} = -\left(1-\frac{3}{\eta^4}+\frac{2}{\eta^2}\right)\sin(2\theta)$$

The plotting function **ListContourPlot** requires that the coordinate triplet be in Cartesian coordinates. Therefore, we shall use **Table** to create the coordinates $(\eta\cos\theta, \eta\sin\theta, \bar{\sigma}(\eta,\theta))$ for $1 \leq \eta \leq 3$ and $0 \leq \theta \leq 2\pi$. This array will be created with a function placed in **Initialization**. The output of **Locator** will be in Cartesian coordinates in the form $u = \{x_l, y_l\}$; therefore, the von Mises stress will be determined from

$$\bar{\sigma}_{vm}\left(\sqrt{x_l^2+y_l^2}, \tan^{-1}(y_l/x_l)\right)$$

The program is as follows.

```
Manipulate[η=Sqrt[u[[1]]^2+u[[2]]^2];
  θ=ArcTan[u[[1]],u[[2]]];
  lab=Row[{"η = ",NumberForm[η,3]," θ = ",
    NumberForm[θ/Degree,3],"° σ̄vm = ",
    NumberForm[vonmis[η,θ],3]}];
  Show[ListContourPlot[Flatten[coord1,1],
    ContourLabels->All,
    RegionFunction->Function[{x,y},x^2+y^2>1]],
    Graphics[{{Black,Circle[{0,0},1]},
    Arrow[{{2.2,0},{2.8,0}}],Arrow[{{-2.2,0},{-2.8,0}}]}],
    PlotRange->All,AspectRatio->Automatic,Frame->False,
    PlotLabel->lab],
  {{u,{1.5,1.5}},ControlType->Locator,
    ContinuousAction->False},
```

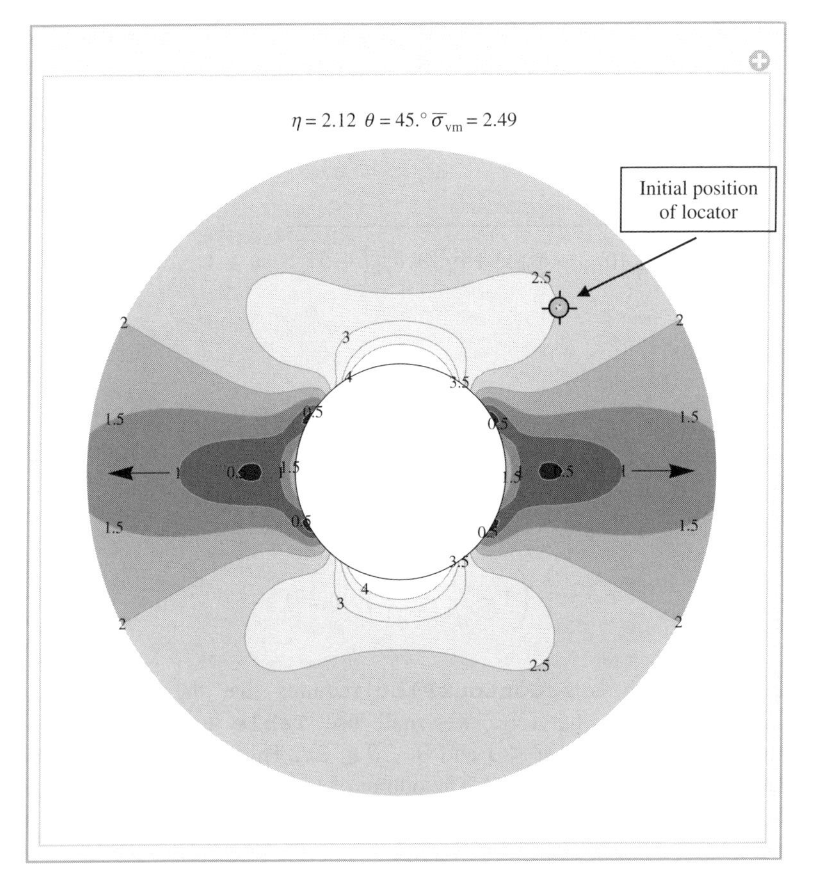

Figure 7.4 Initial interactive display created by `Manipulate` for obtaining the value of the von Mises stress at any location (η, θ) exterior to the hole, as described in Example 7.4

```
TrackedSymbols:>{u},
Initialization:>(vonmis[η_,θ_]:=
  (sr=1-1/η^2+(1+3/η^4-4/η^2) Cos[2 θ];
  st=1+1/η^2-(1+3/η^4) Cos[2 θ];
  srt=-(1-3/η^4+2/η^2) Sin[2 θ];
  Sqrt[0.5 ((sr-st)^2+sr^2+st^2)+3 srt^2]);
  coord1=Table[Table[{η Cos[θ],η Sin[θ],
    vonmis[η,θ]},{η,1,3,0.05}],{θ,0,2 π,π/64}];)]
```

To improve the response time of the locator, we have set **ContinuousAction** to **False**. Also, we have chosen to label the contour lines by using the option **ContourLabels-> All**. The output of **Table** contains in an extra set of braces, which is removed by using **Flatten** on only the first level. The degree symbol (°) was obtained from the *Special Characters* template after clicking on the *Symbols* tab and then the ∞ tab. Lastly, the hole is emphasized in two ways: by using a circle and by limiting the plot to only display the region $x^2 + y^2 > 1$. This latter requirement is met by using the option **RegionFunction**. The results are shown in Figure 7.4. Using a mouse to move the locator will alter the numbers appearing at the top of the contour plot. In addition, passing the cursor over any of the contour lines will result is the appearance of its value. This is redundant information with that shown in Figure 7.4; however, for use in a different venue (e.g., a publication) the labels should be used.

Example 7.5

Analysis of Beams

In Example 4.20, **DSolve** was used to obtain the symbolic solution to a thin beam hinged at both ends and subjected to a concentrated load at an arbitrary location. We shall extend those results to consider three sets of boundary conditions – clamped at both ends, hinged at both ends, and clamped at one end and free at the other (cantilever) – and two types of loading – concentrated and uniformly distributed over a portion of the beam. The objective is to create an interactive figure that displays, for a given loading and set of boundary conditions, the displacement y, rotation of a cross section y', the moment y'', and the shear force y''' as a function of the position of the load.

The nondimensional form of the governing equation for the beam is written as

$$\frac{d^4y}{dx^4} = g(x) \quad 0 \le x \le 1$$

where, for a concentrated load

$$g(x) = \delta(x - x_0) \quad 0 \le x_0 \le 1$$

and $\delta(x)$ is the delta function and for a uniform load that spans $0 \le a \le x \le b \le 1$

$$g(x) = u(x - a) - u(x - b) \quad a < b$$

where $u(x)$ is the unit step function.

The three sets of boundary conditions are

Hinged at both ends	Clamped at both ends	Cantilever
$y(0) = y''(0) = 0$	$y(0) = y'(0) = 0$	$y(0) = y'(0) = 0$
$y(1) = y''(1) = 0$	$y(1) = y'(1) = 0$	$y''(1) = y'''(1) = 0$

The program that follows produces the initial display of the interactive figure shown in Figure 7.5. The images accompanying the loading selection are generated in **Initialization**. All the controls are placed on the left. Depending on the load selected, one set of sliders is disabled. Lastly, for the uniform load, the limits of the selection of a and b have to be coupled so that $a < b$. The program is as follows.

```
Manipulate[
                    (* Select loading *)
 If[loading==2,load=DiracDelta[x-etao],
   load=UnitStep[x-a]-UnitStep[x-b]];
(* Select boundary conditions and obtain symbolic solution to equation*)
 Which[bc==1,
   beam=y[x]/.Quiet[DSolve[{y'''' [x]==load,y[0]==0,
     y'' [0]==0,y'' [1]==0,y[1]==0},y[x],x]],
  bc==2,
  beam=y[x]/.Quiet[DSolve[{y'''' [x]==load,y[0]==0,
     y' [0]==0,y' [1]==0,y[1]==0},y[x],x]],
   bc==3,
   beam=y[x]/.Quiet[DSolve[{y'''' [x]==load,y[0]==0,
     y' [0]==0,y'' [1]==0,y''' [1]==0},y[x],x]]];
        (* Plot solution and its derivatives on a 2×2 grid *)
 GraphicsGrid[{{Plot[Evaluate[beam],{x,0,1},
     PlotRange->All,PlotLabel->"Displacement"],
   Plot[Evaluate[D[beam,x]],{x,0,1},PlotRange->All,
     PlotLabel->"Rotation"]},
   {Plot[Evaluate[D[beam,x,x]],{x,0,1},PlotRange->All,
     PlotLabel->"Moment"],
   Plot[Evaluate[D[beam,x,x,x]],{x,0,1},
     PlotRange->All,PlotLabel->"Shear force"]}}],
              (* Create radio buttons and sliders *)
 Style["Loading",Bold,11],
 {{loading,1," "},{1->f1,2->f2},ControlType->RadioButton},
 Delimiter,
 Style["Boundary conditions",Bold,11],
 {{bc,1," "},{1->"Hinged both ends",
   2->"Clamped both ends",3->"Cantilever"},
   ControlType-> RadioButton},
```

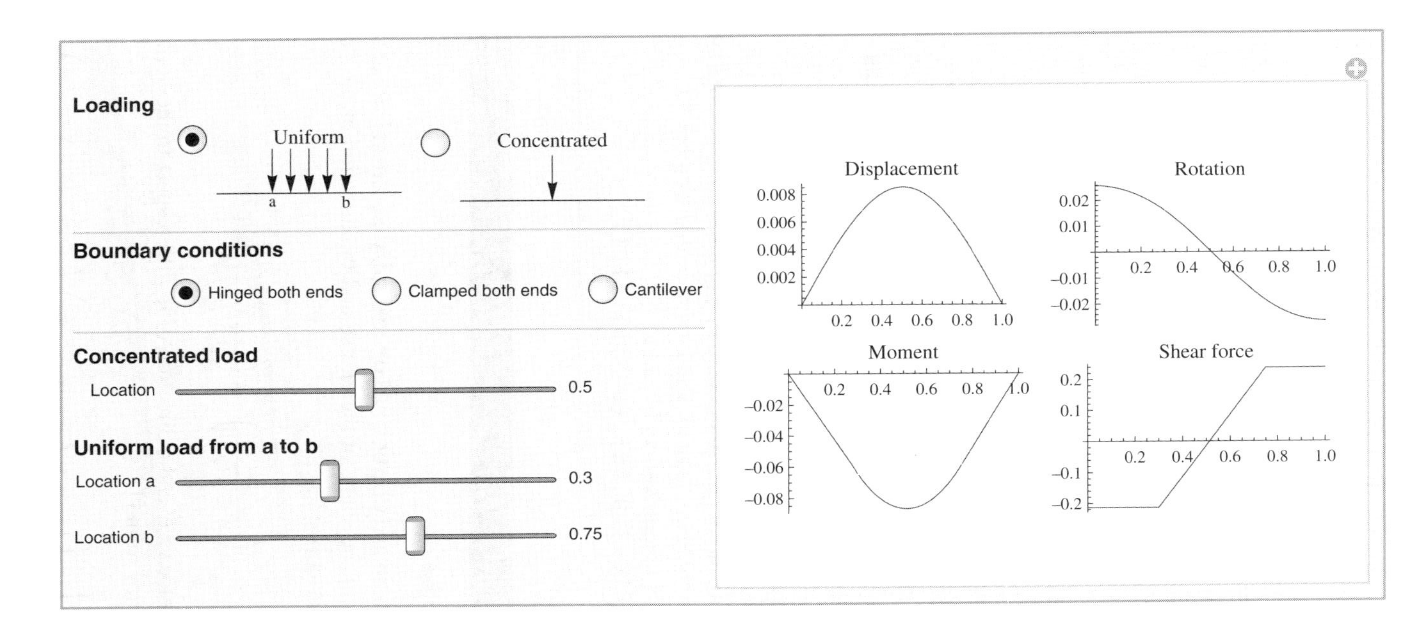

Figure 7.5 Initial interactive display created by `Manipulate` for obtaining the values that are proportional to the displacement, rotation, moment, and shear force in a beam for two types of loading and their respective locations, and its boundary conditions

```
Delimiter,
Style["Concentrated load",Bold,11],
{{etao,0.5,"Location"},0.01,0.99,0.01,
  Appearance->"Labeled",ControlType->Slider,
  Enabled->If[loading==2,True,False]},
Style["Uniform load from a to b",Bold,11],
{{a,0.3,"Location a"},0.01,b-0.01,0.01,
  ControlType->Slider,Appearance->"Labeled",
  Enabled->If[loading==1,True,False]},
{{b,0.75,"Location b"},a+0.01,0.99,0.01,
  ControlType->Slider,Appearance->"Labeled",
  Enabled->If[loading==1,True,False]},
ControlPlacement->Left,
TrackedSymbols:>{etao,loading,bc,a,b},
            (* Create images for the two types of loading *)
Initialization:>(
  f1=Show[Graphics[Line[{{0,0},{1,0}}],
    PlotRange->{-0.1,0.25},Axes->False,ImageSize->Tiny,
    PlotLabel->"Uniform",Epilog->{Text["a",{0.3,-0.05}],
    Text["b",{0.7,-0.05}]}],
    Table[Graphics[{Arrowheads[0.1],
    Arrow[{{xx,0.25},{xx,0}}]}],{xx,0.3,0.7,0.1}]];
  f2=Show[Graphics[Line[{{0,0},{1,0}}],Axes->False,
    ImageSize->Tiny,PlotLabel->"Concentrated"],
    Graphics[{Arrowheads[0.1],
      Arrow[{{0.5,0.25},{0.5,0}}]}]])]
```

Example 7.6

Flow Around an Ellipse

We shall use **Manipulate** and **ContourPlot** to display the streamlines of the flow around an ellipse as a function of the shape of the ellipse and the direction of the flow. The ellipse under consideration is described by

$$\left(\frac{x}{a-b^2/a}\right)^2+\left(\frac{y}{a+b^2/a}\right)^2=1$$

Then, for a flow with velocity U and a direction of α with respect to the horizontal axis, the streamlines are determined from [1]

$$\psi(x,y)=\operatorname{Im}\left[U\left\{ze^{-j\alpha}-\left(\frac{a^2}{b^2}e^{j\alpha}+e^{-j\alpha}\right)\left(\frac{z}{2}\pm\sqrt{\left(\frac{z}{2}\right)^2+b^2}\right)\right\}\right]$$

where Im means the imaginary part of its argument and $z=x+jy$.

In the program that follows, we take $U = 1$ and $a = 1$ and note that when $b \rightarrow 1$, the ellipse approaches a flat plate. In addition, the plus sign before the radical gives the streamlines in the right-hand plane and the negative sign those in the left-hand plane. Hence, some logic is required to correctly choose the sign, which depends on the sign of x.

```
Manipulate[ContourPlot[
    If[x<0,Im[fzell[x,y,a,b,α Degree,-1]],
      Im[fzell[x,y,a,b,α Degree,1]]],
    {x,-10,10},{y,-10,10}, Contours->40,
    ContourShading->False,FrameLabel->{"x","y"},
    Epilog->{Red,EdgeForm[{Thick,Black}],
      Disk[{0,0},{Abs[a-b^2/a],a+b^2/a}]}],
```

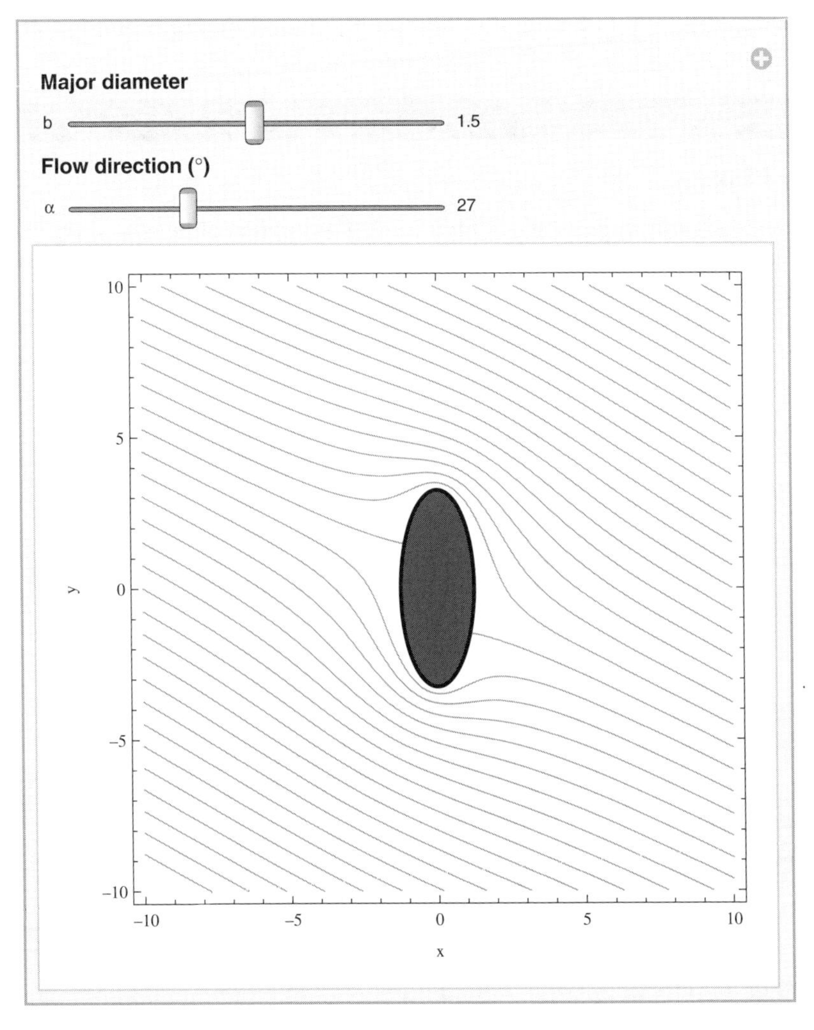

Figure 7.6 Initial interactive display created by **Manipulate** for obtaining the streamlines of flow around an ellipse for various elliptical shapes and flow directions

```
Style["Major diameter",Bold,11],
{{b,1.5,"b"},1,2,0.01,Appearance->"Labeled",
  ControlType->Slider},
Style[Row[{"Flow direction (",Degree,")"}],Bold,11],
{{α,27,"α"},0,90,1,Appearance->"Labeled",
  ControlType->Slider},
TrackedSymbols:>{b,α},
Initialization:>(fzell[x_,y_,a_,b_,α_,s_]:=(z=x+I y;
  z Exp[-I α]-(a^2/b^2 Exp[I α]+Exp[-I α])*
  (z/2+s Sqrt[(z/2)^2+b^2])); a=1.)]
```

The execution of this program produces the initial display of the interactive figure shown in Figure 7.6.

Example 7.7

Four-Bar Linkage

Consider the four-bar linkage shown in Figure 7.7. The location of point $P = (x_P, y_P)$ on the rigid triangular plate defined by the vertices Q, S, and P as a function of the rotation angle θ of the link of length a is given by

$$x_P = a\cos\theta + r\cos(\alpha + \delta)$$
$$y_P = a\sin\theta + r\sin(\alpha + \delta)$$

where

$$\delta = \tan^{-1}\frac{b\sin[\psi(\theta)] - a\sin\theta}{g + b\cos[\psi(\theta)] - a\cos\theta}$$
$$\psi(\theta) = \tan^{-1}\frac{B(\theta)}{A(\theta)} + \cos^{-1}\frac{-C(\theta)}{\sqrt{A^2(\theta) + B^2(\theta)}}$$

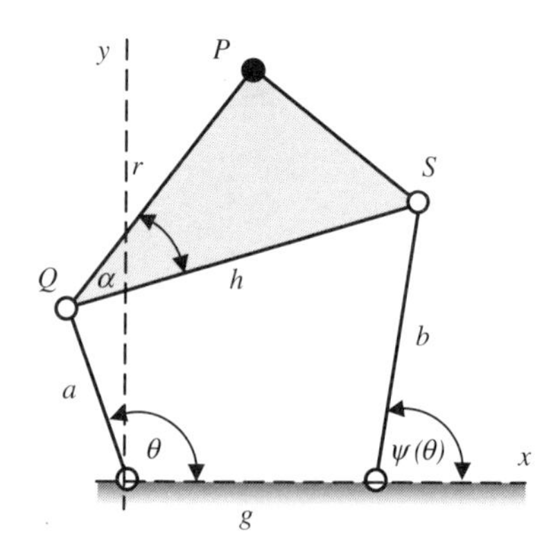

Figure 7.7 Four-bar linkage

and

$$A(\theta) = 2bg - 2ab\cos\theta$$
$$B(\theta) = -2ab\sin\theta$$
$$C(\theta) = a^2 + b^2 + g^2 - h^2 - 2ag\cos\theta$$

For an appropriate selection of values for the link lengths, the link of length a will rotate through 360°. During this complete rotation, $\psi_{min} \leq \psi \leq \psi_{max}$, where ψ_{min} and ψ_{max} have to be determined numerically. The following values for the link lengths are chosen: $a = 1$, $b = 1.7$, $h = 2$, and $g = 1.5$. For these values, the arccosine in the definition of $\psi(\theta)$ will remain real.

We shall create an animation of the path of P as θ varies from 0 to 360°, display a shaded sector delineating the range specified by ψ_{min} and ψ_{max}, and display the circular path traversed by point Q. The following program displays the results shown in Figure 7.8. Five functions

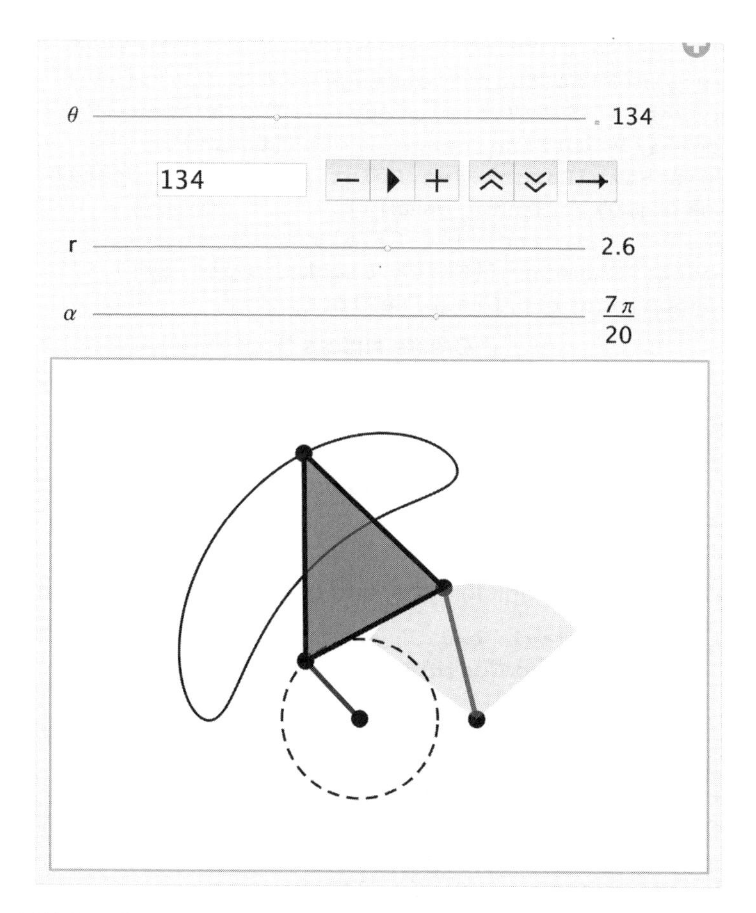

Figure 7.8 Initial configuration of the interactive display created by **`Manipulate`** for animating the motion of a four-bar linkage

representing the definitions given above are created in **Initialization**. The animation is started by clicking on the solid triangle under the slider bar that is labeled θ.

```
Manipulate[
        (* Determine four-bar linkage orientation for a given θ *)
 ptQ={a Cos[θ Degree],a Sin[θ Degree]};
 ptS={g+b Cos[psi[θ Degree]],b Sin[psi[θ Degree]]};
 fbar={{0,0},ptQ,ptS,{g,0}};
 p4={a Cos[θ Degree]+r Cos[α+del[θ]],a Sin[θ Degree]+
   r Sin[α+del[θ]]};
 pgn={ptQ,p4,ptS};
                (* Determine coordinates of P *)
 pt4=Table[{a Cos[θ Degree]+r Cos[α+del[θ]],
   a Sin[θ Degree]+r Sin[α+del[θ]]},{θ,0,360,1}];
      (* Plot four bar linkage, path of P, sector for ψ, and path of Q *)
 Show[ListLinePlot[fbar,PlotRange->{{-3.5,4},{-1.5,4}},
   AspectRatio->5.5/7.5,PlotStyle->Thick,Axes->False,
   Epilog->{{PointSize[Large],Point[fbar]},
     {PointSize[Large],Red,Point[p4]},{Cyan,Opacity[0.4],
     Disk[{g,0},b,{psmn,psmx}]},
     {Dashed,Red,Circle[{0,0},a]},
   { Magenta,EdgeForm[{Thick,Black}],Opacity[0.6],
     Polygon[pgn]}}],ListLinePlot[pt4]],
                     (* Create sliders *)
 {{θ,134,"θ"},0,360,1.,Appearance->{"Labeled","Open"}},
 {{r,2.6,"r"},h,3,0.1,Appearance->"Labeled",
   ControlType->Slider},
 {{α,7 π/20,"α"},0,π/2,π/20,Appearance->"Labeled",
   ControlType->Slider},
 TrackedSymbols:>{θ,r,α},
 (* Create functions, set link lengths, and determine max and min of ψ *)
 Initialization:>(a=1; b=1.7; h=2.; g=1.5;
   aA[θ_]:=2 b g-2 a b Cos[θ];
   bB[θ_]:=-2 a b Sin[θ];
   cC[θ_]:=a^2+b^2+g^2-h^2-2 a g Cos[θ];
   psi[θ_]:=ArcTan[aA[θ],bB[θ]]+ArcCos[-cC[θ]/
     Sqrt[aA[θ]^2+bB[θ]^2]];
   del[θ_]:=ArcTan[g+b Cos[psi[θ Degree]]-
     a Cos[θ Degree],b Sin[psi[θ Degree]]-a Sin[θ Degree]];
  psmn=FindMinValue[psi[θ],{θ,π}];
  psmx=FindMaxValue[psi[θ],{θ,π/10}];)]
```

References

[1] G. Currie, *Fundamentals of Fluid Mechanics*, 2nd edn, McGraw-Hill, Inc., New York, 1993, p. 98.
[2] B. Balachandran and E. B. Magrab, *Vibrations*, 2nd edn, Cengage Learning, Toronto Ontario, 2009, pp. 165–7.

Exercises

7.1 One model of machine tool chatter gives that the region of instability can be determined from the following equation [2]

$$\frac{1}{Q} + \frac{K}{k\Omega} + \frac{\mu k_1}{k}\frac{\sin(\omega/\Omega)}{\omega} = 0$$

where Q is the quality factor, k_1 is the cutting stiffness, K is a penetration rate coefficient, k is the work-piece stiffness, μ is the overlap factor ($0 \le \mu \le 1$), Ω is the nondimensional work piece rotation speed, and ω is the chatter frequency that is the root of

$$\omega^2 = 1 + \frac{k_1}{k}(1 - \mu\cos(\omega/\Omega))$$

Thus, for a given k_1/k, K/k, and μ, a plot of Q versus Ω will display the regions where chatter occurs. The locus of the region where the system is chatter-free is given by

$$Q \to Q_m = \frac{1}{B - K/(k\Omega)}$$

where

$$B = \sqrt{2}\sqrt{1 + \frac{k_1}{k} - \sqrt{\left(1 + \frac{k_1}{k}\right)^2 - \left(\frac{\mu k_1}{k}\right)^2}}$$

Create an interactive graphic that displays the regions of chatter that looks like that shown in Figure 7.9. The ranges for the various parameters are: $0 \le \mu \le 1$, $0.07 \le k_1/k \le 0.08$, and $0.0015 \le K/k \le 0.0035$. Use the **`RegionFunction`** option to limit Q to the region $0 < Q < 50$.

7.2 The nondimensional magnitude a_o of the amplitude response of the mass of a single degree-of-freedom system with a spring that is proportional to the cube of the displacement of the mass and undergoing harmonic oscillations can be obtained from

$$\Omega^2 = 1 + a_o^2 - 2\zeta^2 \pm \sqrt{\frac{S_o^2}{a_o^2} - 4\zeta^2\left(1 + a_o^2 - \zeta^2\right)} \quad \text{(a)}$$

where the plus sign displays the right portion of the curve and the minus sign the left portion. The quantity Ω is the nondimensional frequency ratio, S_o is related to the magnitude of the harmonic input, and ζ is the damping factor. The curve that is midway between the two curves given by Eq. (a) is called the spine and is determined from

$$\Omega^2 = 1 + a_o^2 - 2\zeta^2 \quad \text{(b)}$$

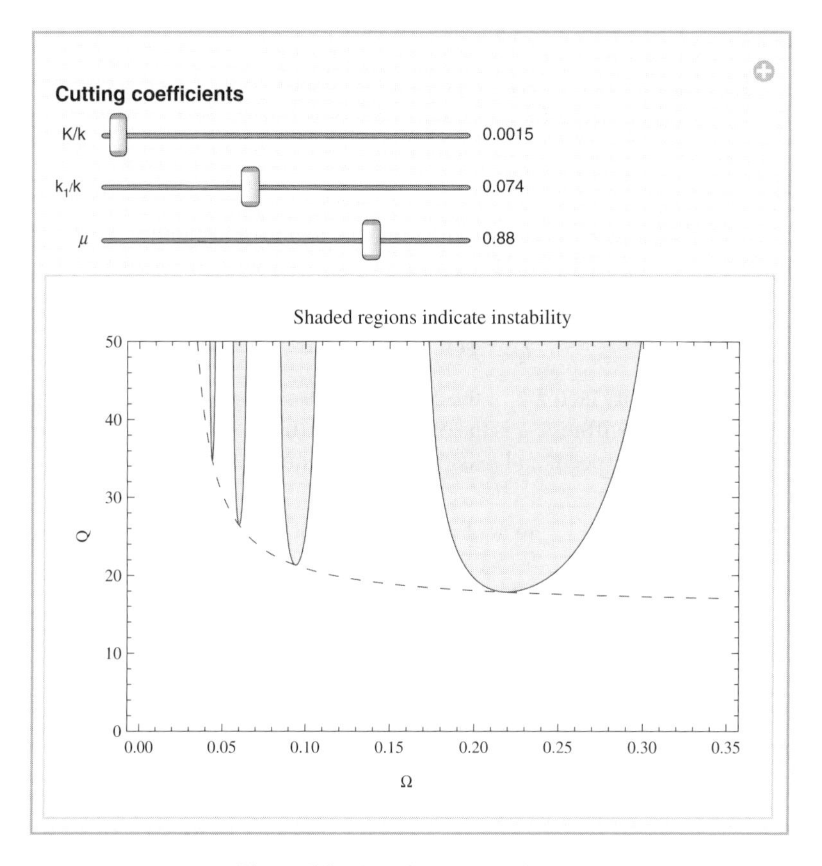

Figure 7.9 Solution to Exercise 7.1

The maximum value of a_o is obtained from

$$a_{\max} = \sqrt{-\frac{1}{2} + \frac{1}{2}\sqrt{(1-\zeta^2)^2 + \frac{S_o^2}{\zeta^2}}}$$

and the frequency at which it occurs is

$$\Omega_{\max}^2 = 1 + a_{\max}^2 - 2\zeta^2$$

In the plotting of the curve corresponding to the plus branch of the curve; that is, the right portion, there is an inflection point. This inflection point is determined numerically by finding the value of a_o that causes Ω to be a minimum.

The procedure to plot these curves and display their values is as follows. One selects a value of a_o and uses Eq. (a) to determine the corresponding values of Ω for the left and right curves. In a similar fashion, the spines of the curves are determined using

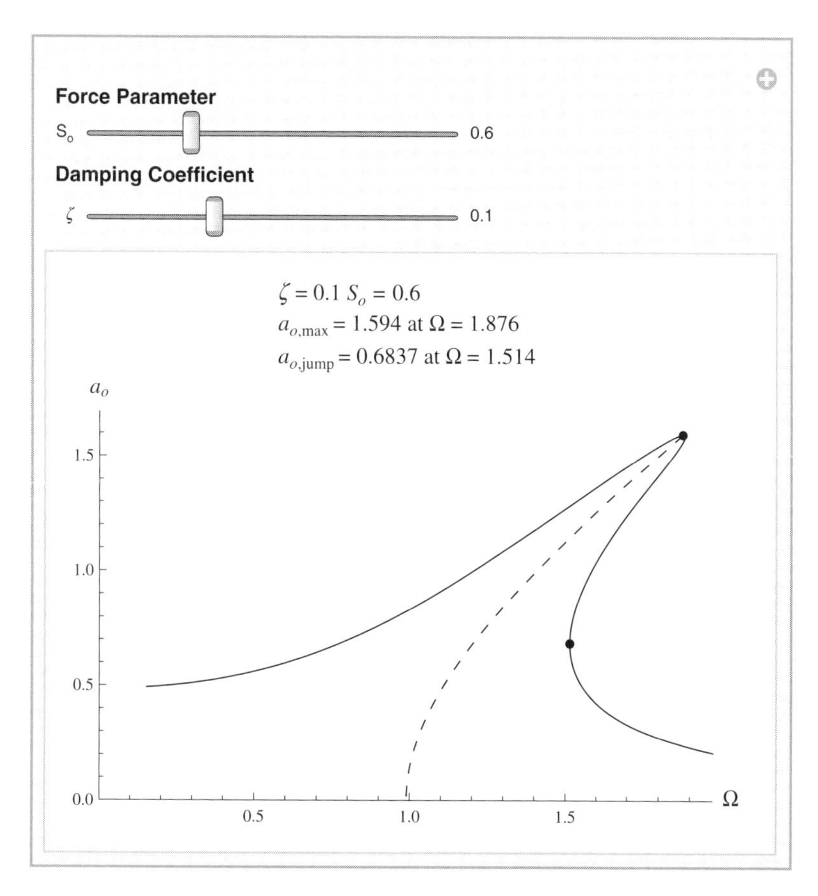

Figure 7.10 Solution to Exercise 7.2

Eq. (b). Then the location of the maximum value of $a_o = a_{\max}$ is computed and, lastly, the inflection point is determined from the right side of Eq. (a).

Create an interactive graphic that does these computations and displays the results as shown in Figure 7.10. Let $0.1 \le S_o \le 2$ and $0.02 \le a_o \le a_{\max}$.

Part II

Engineering Applications

8

Vibrations of Spring–Mass Systems and Thin Beams

8.1 Introduction

The time-varying response and the frequency response of single and two degrees-of-freedom systems for various loadings, initial conditions, and nonlinearities are examined by creating suitable interactive environments to explore their characteristics as a function of the parameters that govern the respective systems. Then, the determination of the natural frequencies and mode shapes for thin beams subject to various boundary conditions and loadings are examined. The following applications will be considered:

Single degree-of-freedom systems –

- effects of spectral content of periodic forces on the time-varying response;
- effects of squeeze film damping and viscous fluid damping on the amplitude response and phase response;
- stability in the presence of an electrostatic force;
- parameters that maximize the average power of a piezoelectric energy harvester.

Two degrees-of-freedom systems –

- effects of systems parameters on the amplitude response functions;
- system parameters that produce an enhanced piezoelectric energy harvester.

Thin beams –

- effects of in-span attachments on the natural frequencies and modes shapes of thin cantilever beams;
- effects of an electrostatic force on the lowest natural frequency and the stability of beams;
- effects of in-span attachments on the response of a cantilever beam to an impulse force.

For each of these topics, we shall use **`Manipulate`** to create an interactive graphic to explore the system response to the variation of its parameters.

An Engineer's Guide to Mathematica®, First Edition. Edward B. Magrab.

Companion Website: www.wiley.com/go/magrab

8.2 Single Degree-of-Freedom Systems

8.2.1 Periodic Force on a Single Degree-of-Freedom System

Consider a single degree-of-freedom system with a mass m (kg), a spring of stiffness k (N·m^{-1}), and a viscous damper with damping coefficient c (N·s·m^{-1}). The mass is subjected to periodic force of magnitude F_o (N) and period $T = 2\pi/\omega_o$ (s), where ω_o (rad·s^{-1}) is the fundamental frequency of the forcing. If the periodic force is expressed as a Fourier series, then the displacement response of the mass is [1, p. 259–264]

$$x(\tau) = c_0 + \sum_{l=1}^{\infty} c_l H(\Omega_l) \sin(\Omega_l \tau - \theta(\Omega_l) - \psi_l) \tag{8.1}$$

where $\Omega_0 = \omega_o/\omega_n$, $\Omega_l = l\Omega_0$,

$$\begin{aligned} H(\Omega_l) &= \frac{1}{\sqrt{\left(1-\Omega_l^2\right)^2 + (2\zeta\Omega_l)^2}}, \qquad \theta(\Omega_l) = \tan^{-1}\frac{2\zeta\Omega_l}{1-\Omega_l^2} \\ c_l &= \sqrt{a_l^2 + b_l^2}, \qquad c_0 = \frac{a_0}{2}, \qquad \psi_l = \tan^{-1}\frac{a_l}{b_l} \end{aligned} \tag{8.2}$$

and

$$\omega_n = \sqrt{\frac{k}{m}} \text{rad} \cdot \text{s}^{-1}, \quad \tau = \omega_n t, \quad 2\zeta = \frac{c}{m\omega_n}, \quad y = \frac{x}{F_o/k} \tag{8.3}$$

Two types of periodic pulses are considered; a single pulse and a double pulse, which are shown in Figure 8.1. For the single pulse of duration τ_d, the coefficients are

$$\frac{a_0}{2} = \alpha, \quad a_l = \frac{1}{l\pi}\sin(2\pi\alpha l), \quad b_l = \frac{1}{l\pi}[1 - \cos(2\pi\alpha l)] \quad l = 1, 2, \ldots, \tag{8.4}$$

and, therefore,

$$c_l = 2\alpha\left|\frac{\sin(\pi\alpha l)}{\pi\alpha l}\right| \quad \text{and} \quad \psi_l = \tan^{-1}\frac{\sin(2\pi\alpha l)}{1-\cos(2\pi\alpha l)} \tag{8.5}$$

where $\alpha = \Omega_0\tau_d/2\pi$. For the double pulse that is positive for $0 \le \tau \le \tau_d$ and negative for $\tau_d \le \tau \le 2\tau_d$, the coefficients are

$$\begin{aligned} a_0 &= 0 \\ a_l &= \frac{1}{l\pi}\{2\sin(2\pi\alpha l) - \sin(2\pi\alpha l)\} \qquad l = 1, 2, \ldots \\ b_l &= \frac{1}{l\pi}\{1 - 2\cos(2\pi\alpha l) + \cos(4\pi\alpha l)\} \end{aligned} \tag{8.6}$$

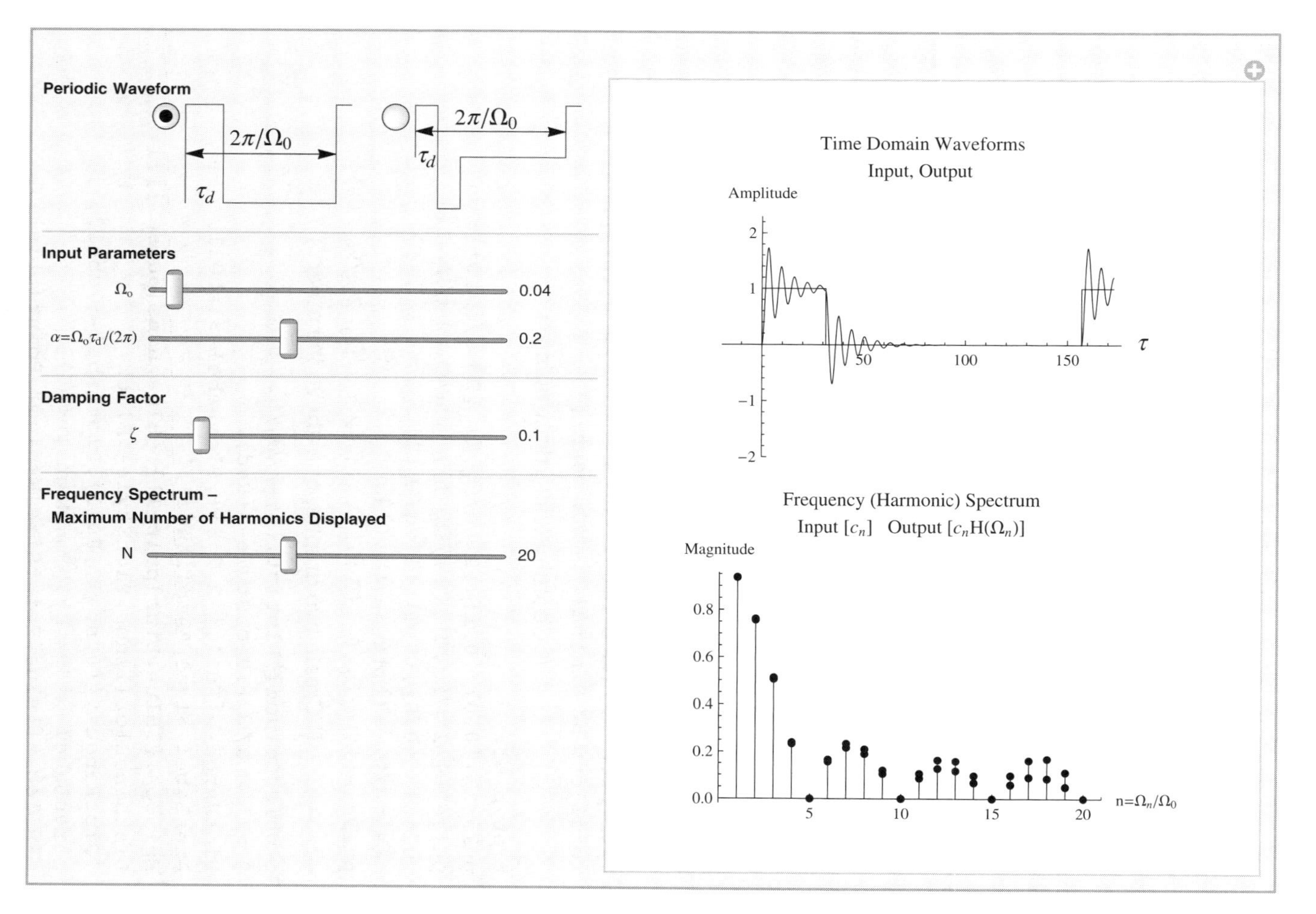

Figure 8.1 Initial configuration of the interactive graph to explore the response of a single degree-of-freedom system to two different periodic waveforms

An interactive figure shown in Figure 8.1 shall be created so that one can explore how the spectral content of the periodic force affects the time domain waveform as a function of α, Ω_o, and ζ. The time-domain signal is given by Eq. (8.1), where the spectral content before being applied to the system is given by c_l and that of the mass is given by $c_l H(\Omega_l)$. We shall use 150 terms in the Fourier series expansion. The program that creates Figure 8.1 is as follows.

```
Manipulate[
                    (* Evaluate Eq. (8.2) *)
 hh=1/Sqrt[(1-(r Ω0)^2)^2+(2 ζ r Ω0)^2];
 thh=ArcTan[1-(r Ω0)^2,2 ζ r Ω0];
     (* Obtain coefficients from either Eq. (8.5) or Eq. (8.6) *)
 cnn=If[ptyp==1,Abs[Sin[r π α]/(r π α)],
  a1=(2 Sin[2 π α r]-Sin[4 π α r])/(π r);
  b1=(1-2 Cos[2 π α r]+Cos[4 π α r])/(π r);
  Sqrt[a1^2+b1^2]];
 psnn=If[ptyp==1,ArcTan[1-Cos[2 r π α],Sin[2 r π α]],
  ArcTan[b1,a1]];
 ptin=Table[{n,cnn[[n]]},{n,1,nn}];
 ptout=Table[{n,hh[[n]] cnn[[n]]},{n,1,nn}];
 lines=Table[{{n,0},{n,If[cnn[[n]]<hh[[n]] cnn[[n]],
  hh[[n]] cnn[[n]],cnn[[n]]]}},{n,1,nn}];
                    (* Sum series: Eq. (8.1) *)
 xt=If[ptyp==1,α+2 α Total[cnn hh Sin[r Ω0 t-thh+psnn]],
  Total[cnn hh Sin[r Ω0 t-thh+psnn]]];
                (* Coordinates to draw pulse *)
 If[ptyp==1,
  pulse1={{0,0},{0,1},{2 π α/Ω0,1},{2 π α/Ω0,0}};
  pulse2={{2 π/Ω0,0},{2 π/Ω0,1}, {π α/Ω0+2 π/Ω0,1}},
  pulse1={{0,0},{0,1},{2 π α/Ω0,1},{2 π α/Ω0,-1},
   {4 π α/Ω0,-1},{4 π α/Ω0,0}};
  pulse2={{2 π/Ω0,0},{2 π/Ω0,1}, {2 π α/Ω0+2 π/Ω0,1}}];
          (* Create two graphs, one above the other *)
 GraphicsColumn[{Plot[xt,{t,-π/Ω0/5.,
  2 π/Ω0+π α/Ω0},PlotStyle->{Red},
  PlotRange->{Full,{-2,2.3}},PlotLabel->label1,
  AxesLabel->{"τ","Amplitude"},
  Epilog->{{Blue,Line[pulse1]},{Blue,Line[pulse2]}}],
  ListLinePlot[lines,PlotStyle->Black,
   PlotRange->{{0,nn+1},Full},PlotLabel->label2,
   AxesLabel->{"n=Ωn/Ω0","Magnitude"},
   Epilog->{{Blue,PointSize[Medium],Point[ptin]},
   {Red,PointSize[Medium],Point[ptout]}}]}],
```

```
                  (* Create sliders and radio buttons *)
Style["Periodic Waveform",Bold],
{{ptyp,1," "},{1->labs,2->labd}, ControlType->RadioButton},
Delimiter,
Style["Input Parameters",Bold],
{{Ω0,0.04,"Ωo"},0.01,1,0.01,Appearance->"Labeled",
 ControlType->Slider},
{{α,0.2,la},0.02,0.49,0.01,Appearance->"Labeled",
 ControlType->Slider},
Delimiter,
Style["Damping Factor",Bold],
{{ζ,0.1,"ζ"},0.02,0.7,0.01,Appearance->"Labeled",
 ControlType->Slider},
Delimiter,
Style["Frequency Spectrum -",Bold],
Style["  Maximum Number of Harmonics Displayed",Bold],
{{nn,20,"N"},1,50,1,Appearance->"Labeled",
 ControlType->Slider},
ControlPlacement->Left,
Initialization:>(puls={{0,0},{0,1},{0.25,1},{0.25,0},
 {1,0},{1,1},{1.1,1}};
                        (* Radio button images *)
labs=ListLinePlot[puls,PlotRange->{{0,1.2},{-0.1,1}},
 Axes->False,ImageSize->Tiny,Epilog->{Arrowheads[0.1],
  Arrow[{{0,0.5},{1,0.5}}],Arrow[{{1,0.5},{0,0.5}}],
  Inset[Style["2π/Ω0",14],{0.5,0.65}],
  Inset[Style["τd",14],{0.125,0.1}]}];
puld={{0,0},{0,1},{0.15,1},{0.15,-1},{0.3,-1},{0.3,0},
 {1,0},{1,1},{1.1,1}};
labd=ListLinePlot[puld,PlotRange->{{0,1.2},{-1.1,1}},
 Axes->False,ImageSize->Tiny,
 Epilog->{Arrowheads[{-0.1,0.1}],
  Arrow[{{0,0.5},{1,0.5}}],
  Inset[Style["2π/Ω0",14],{0.5,0.75}],
  Inset[Style["τd",14],{0.075,0}]}];
                           (* Figure titles *)
label1=Column[{"Time Domain Waveforms",
 Row[{Style["Input, ",Blue],
 Style["Output ",Red]}]},Center];
label2=Column[{"Frequency (Harmonic) Spectrum",
 Row[{Style[Row[{"Input cn"}],Blue],
 Style[Row[{"  Output [cnH(Ωn)]"}],Red]}]},Center];
```

```
(* Slider label *)
la="α=Ω₀τd/(2π)"; r=Range[1,150];),
TrackedSymbols:>{Ω₀,α,ζ,nn,ptyp}]
```

8.2.2 Squeeze Film Damping and Viscous Fluid Damping

We shall examine the amplitude response and phase response of a single degree-of-freedom system undergoing harmonic oscillations at frequency ω when the system is subjected to squeeze film damping and viscous fluid damping. The single degree-of-freedom system has a mass m (kg) and a spring of stiffness k ($N{\cdot}m^{-1}$).

Squeeze Film Damping

Squeeze film air damping is caused by the entrapment of air between two parallel surfaces that are moving relative to each other in the normal direction. For this damping, the amplitude response and phase response, respectively, for a base-excited spring–mass system when the surface of the mass that is in contact with an air film has a rectangular shape are [2, Section 2.3]

$$\begin{aligned} H_{sq}(\Omega) &= \left\{[1 + r_k S_k(\beta, \sigma_n\Omega) - \Omega^2]^2 + [r_k S_d(\beta, \sigma_n\Omega)]^2\right\}^{-1/2} \\ \theta_{sq}(\Omega) &= \tan^{-1}\frac{r_k S_d(\beta, \sigma_n\Omega)}{1 + r_k S_k(\beta, \sigma_n\Omega) - \Omega^2}. \end{aligned} \tag{8.7}$$

where

$$\begin{aligned} S_k(\beta, \sigma_n\Omega) &= \frac{64(\sigma_n\Omega)^2}{\pi^8}\sum_{l=1,3,5}^{\infty}\sum_{m=1,3,5}^{\infty}\frac{1}{m^2l^2\{(m^2+(l/\beta)^2)^2+(\sigma_n\Omega)^2/\pi^4\}} \\ S_d(\beta, \sigma_n\Omega) &= \frac{64\sigma_n\Omega}{\pi^6}\sum_{l=1,3,5}^{\infty}\sum_{m=1,3,5}^{\infty}\frac{m^2+(l/\beta)^2}{m^2l^2\{(m^2+(l/\beta)^2)^2+(\sigma_n\Omega)^2/\pi^4\}} \end{aligned} \tag{8.8}$$

and

$$r_k = \frac{P_a A}{k h_o}, \qquad \sigma_n = \frac{12\mu\omega_n L^2}{P_a h_o^2}, \qquad \Omega = \frac{\omega}{\omega_n}$$

The quantity σ_n is the squeeze number at ω_n, where ω_n is defined in Eq. (8.3). In Eq. (8.8), $\beta = a/L$ is the aspect ratio of the rectangular shape such that for a narrow strip, $\beta \to \infty$ and for a square surface $\beta = 1$. The film depth is h_o, μ is the dynamic viscosity ($N{\cdot}s{\cdot}m^{-2}$) of the gas between the two surfaces, A is the area of the surfaces, and P_a is the atmospheric pressure in the gap h_o.

Viscous Fluid Damping

Approximations of viscous fluid damping are obtained by considering the mass to be a long rigid circular cylinder of length L, density ρ_m, and diameter b that is oscillating harmonically at a frequency ω in a fluid of infinite extent and density ρ_f. The response functions for a mass-excited system with viscous fluid damping are [2, Section 2.4]

$$\begin{aligned} H_{vf}(\Omega) &= \left[(1-\Omega^2 m_e(\Omega))^2 + (\Omega c_e(\Omega))^2\right]^{-1/2} \\ \psi_{vf}(\Omega) &= \tan^{-1}\frac{\Omega c_e(\Omega)}{1-\Omega^2 m_e(\Omega)}. \end{aligned} \tag{8.9}$$

where

$$\begin{aligned} m_e(\Omega) &= 1 + \frac{\rho_f}{\rho_m}\text{Real}(\Gamma(\text{Re}_n\Omega)) \\ c_e(\Omega) &= -\frac{\rho_f}{\rho_m}\Omega\text{Imag}(\Gamma(\text{Re}_n\Omega)) \\ \Gamma(\text{Re}_n\Omega) &= 1 + \frac{4K_1(\sqrt{j\text{Re}_n\Omega})}{\sqrt{j\text{Re}_n\Omega}K_0(\sqrt{j\text{Re}_n\Omega})}. \end{aligned} \tag{8.10}$$

and

$$\text{Re}_n = \frac{\rho_f \omega_n b^2}{4\mu_f} \tag{8.11}$$

In Eq. (8.10), $K_n(x)$ is the modified Bessel function of the second kind of order n, Re_n is the Reynolds number at frequency ω_n, and μ_f is the dynamic viscosity (N·s·m^{-2}). The quantity Γ is called the hydrodynamic function, and it is noted that Imag[Γ] is negative so that c_e is positive.

We shall create an interactive figure shown in Figure 8.2 from which one can explore the effects that the two types of damping have on the amplitude response and phase response of a single degree-of-freedom system and for each type of damping show how the corresponding damping parameters affect these responses. In Eq. (8.8), the maximum values of l and m are 51. The program that creates Figure 8.2 is as follows.

```
Manipulate[
 (* Select appropriate amplitude response function: Eq. (8.7) or Eq. (8.9) *)
  ampl=Which[funk==1,1/Sqrt[(1+rk srk[σ,Ω,β]-Ω^2)^2+
    (rk srd[σ,Ω,β])^2],
    funk==2,gc=1+4 BesselK[1,Sqrt[I ren Ω]]/
    (Sqrt[I ren Ω] BesselK[0,Sqrt[I ren Ω]]);
    1/Sqrt[(1-(1+rfrm Re[gc]) Ω^2)^2+(-rfrm Ω^2 Im[gc])^2]];
  (* Select appropriate phase response function: Eq. (8.7) or Eq. (8.9) *)
phase=Which[funk==1,
  ArcTan[1+rk srk[σ,Ω,β]-Ω^2,rk srd[σ,Ω,β]]/Degree,
    funk==2,
```

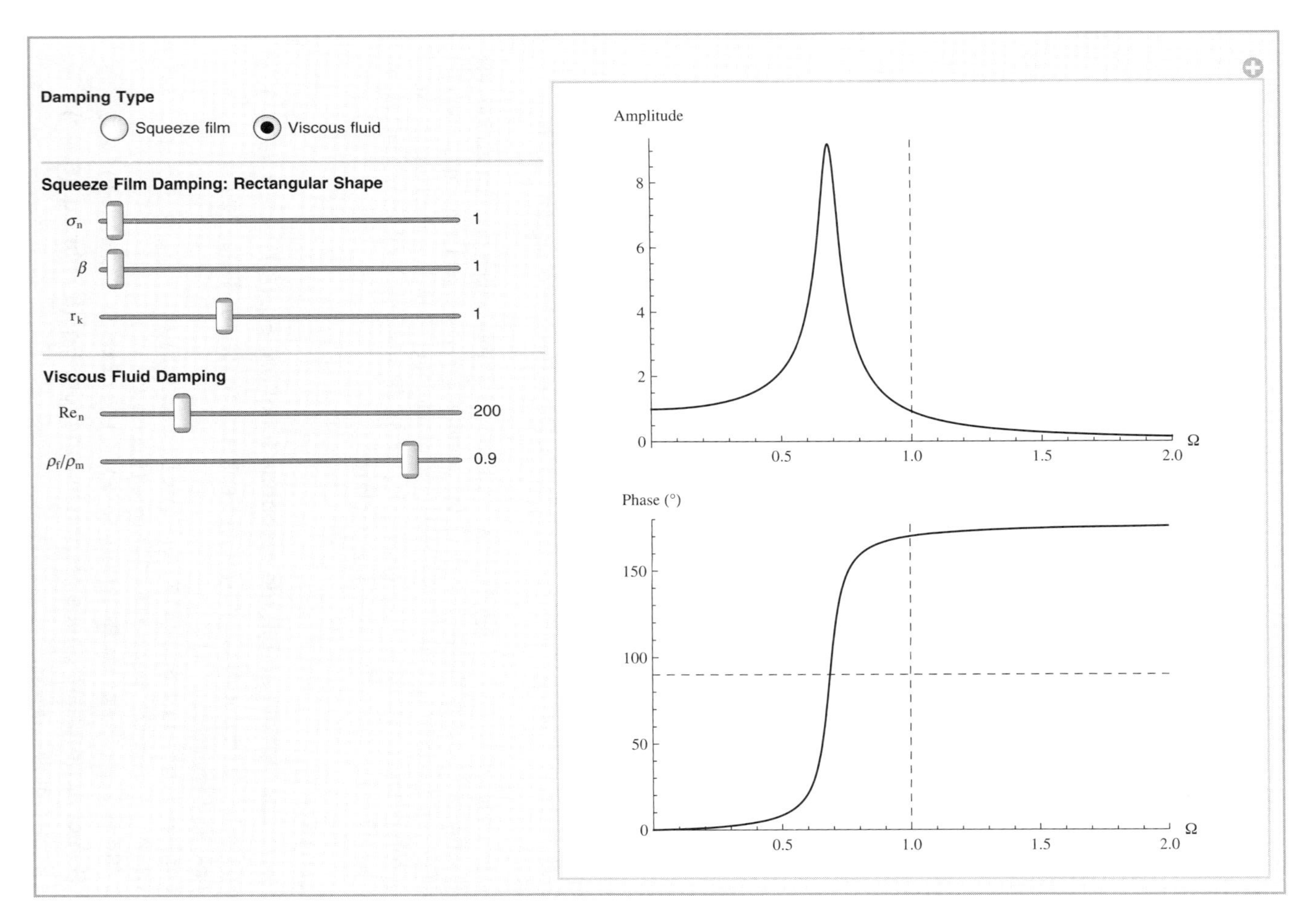

Figure 8.2 Initial configuration of the interactive graph to determine the amplitude response and phase response of a single degree-of-freedom system for two types of damping

```
  gc=1+4 BesselK[1,Sqrt[I ren Ω]]/(Sqrt[I ren Ω]*
    BesselK[0,Sqrt[I ren Ω]]);
  ArcTan[1-(1+rfrm Re[gc]) Ω^2,-rfrm Ω^2 Im[gc]]/Degree];
```

(* Create two graphs, one above the other *)

```
GraphicsColumn[{Plot[ampl,{Ω,0,2},
  PlotRange->{{0,2},All},AxesLabel->{"Ω","Amplitude"},
  Epilog->{Red,Dashed,Line[{{1,0},{1,500}}]}],
Plot[phase,{Ω,0,2},PlotRange->{{0,2},{0,180}},
  AxesLabel->{"Ω","Phase (°)])"},Epilog->{Red,Dashed,
    Line[{{{1,0},{1,180}},{{0,90},{2,90}}}]}]}],
```

(* Create sliders and popup menu *)

```
Style["Damping Type",Bold],
{{funk,2,""},{1->"Squeeze film",2->"Viscous fluid"},
  ControlType->RadioButton},
Delimiter,
Style["Squeeze Film Damping: Rectangular Shape",Bold],
{{σ,1,"σn"},0.5,200,0.5,Appearance->"Labeled",
  Enabled->If[funk==1,True,False],ControlType->Slider,
  ContinuousAction->False},
{{β,1,"β"},1,10,1,Appearance->"Labeled",
  ControlType->Slider,Enabled->If[funk==1,True,False],
  ContinuousAction->False},
{{rk,1,Style["rk ",Italic]},0.5,2,0.05,
  Appearance->"Labeled",ContinuousAction->False,
  Enabled->If[funk==1,True,False],ControlType->Slider},
Delimiter,
Style["Viscous Fluid Damping",Bold],
{{ren,200,"Ren"},0.1,1000,Appearance->"Labeled",
  Enabled->If[funk==2,True,False],ControlType->Slider,
  ContinuousAction->False},
{{rfrm,0.9,"ρf/ρm"},0.05,1,0.05,Appearance->"Labeled",
  ControlType->Slider,ContinuousAction->False,
  Enabled->If[funk==2,True,False]},
TrackedSymbols:>{σ,rk,ren,rfrm,funk,β},
```

(* Function representing Eq. (8.8) *)

```
Initialization:>(
  srk[σ_,Ω_,β_]:=64. (Ω σ)^2/π^8 Total[Table[Total[Table[
    1/(m^2 n^2 (m^2+(n/β)^2)^2+(σ Ω)^2/π^4),
    {m,1,51,2}]],{n,1,51,2}]];
  srd[σ_,Ω_,β_]:=64. (σ Ω)/π^6*
    Total[Table[Total[Table[(m^2+(n/β)^2)/
      (m^2 n^2 ((m^2+(n/β)^2)^2+(σ Ω)^2/π^4)),
      {m,1,51,2}]],{n,1,51,2}]])]
```

8.2.3 Electrostatic Attraction

Consider a single degree-of-freedom system whose mass m (kg) is suspended by a spring with spring constant k (N·m^{-1}) and a viscous damper with damping coefficient c (N·s·m^{-1}). The bottom surface of the mass is rectangular with area A and is flat and parallel to a stationary flat surface. The distance between the two surfaces is d_o and there is a time varying voltage $V_o v(\tau)$ that creates an electrostatic force, where $v(\tau)$ is the time-varying shape of the voltage of magnitude V_o (V). If the displacement of the mass is x and fringe correction factors are neglected, then the nondimensional form of the governing equation for this single degree-of-freedom system is given by [2, Section 2.5.2]

$$\frac{d^2w}{d\tau^2} + 2\zeta\frac{dw}{d\tau} + w = \frac{e_1^2 V_o^2 v^2(\tau)}{(1-w)^2} \tag{8.12}$$

where τ and ζ are given in Eq. (8.3), $w = x/d_o$, and

$$e_1^2 = \frac{\varepsilon_o A}{2kd_o^3}\text{V}^{-2}$$

The quantity ε_o is the permittivity of free space, which for air is 8.854×10^{-12} F·m^{-1}. Equation (8.12) must be solved numerically.

For a given $v(\tau)$ and ζ, the solution to Eq. (8.12) is only stable when $e_1^2 V_o^2 < (e_1^2 V_o^2)_{\text{max}}$. At this value of $(e_1^2 V_o^2)_{\text{max}}$, w is denoted w_{max}. This quantity is determined interactively from the numerical evaluation of Eq. (8.12). The static displacement of the spring–mass system due to the electrostatic force is determined from

$$w_{\text{static}}^3 - 2w_{\text{static}}^2 + w_{\text{static}} - e_1^2 V_o^2 = 0 \tag{8.13}$$

We shall create the interactive environment shown in Figure 8.3 from which one can obtain w_{max} from the displacement response as a function of ζ and $e_1^2 V_o^2$ when $v(\tau) = u(\tau)$, where $u(\tau)$ is the unit step function. The interactive determination of the values of w_{max} and $(e_1^2 V_o^2)_{\text{max}}$, is facilitated by incrementing $e_1^2 V_o^2$ in steps of 0.0001. Hence, the slider is displayed with the optional controls visible, thus enabling one to have more precise control over this parameter's values. In addition, because of the sensitivity of the solution method in the vicinity of $(e_1^2 V_o^2)_{\text{max}}$, the working precision of the numerical solution function is increased to 25. Also, the placement of the labels for w_{max} and w_{static} is a function of t_{end}. The program that creates Figure 8.3 is as follows.

```
Manipulate[
                    (* Determine wstatic from Eq. (8.13) *)
  fg=Min[x/.NSolve[x^3-2 x^2+x-eV==0,x]];
                    (* Solve Eq. (8.12) in terms of (eoVo)2 *)
  sol=Quiet[ParametricNDSolveValue[{w''[t]+
   2 ζ1 w'[t]+w[t]==eVV/(1-w[t])^2,w[0]==w'[0]==0},
   w,{t,0,tend},{eVV,ζ1},WorkingPrecision->25]];
```

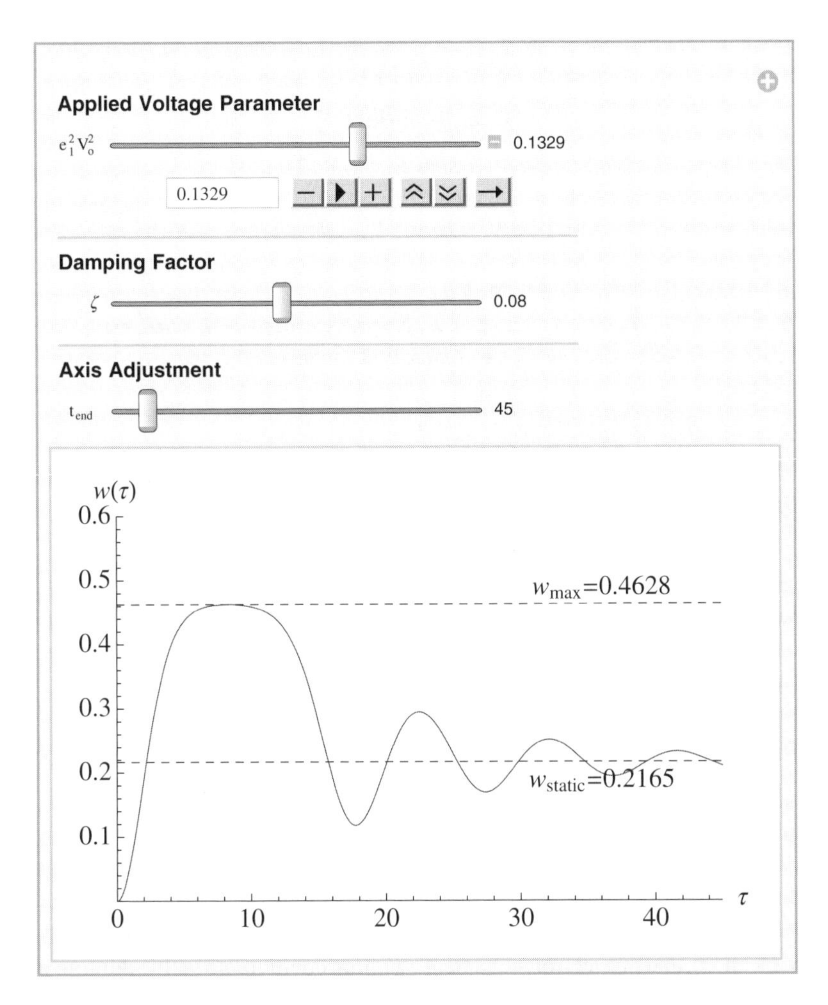

Figure 8.3 Initial configuration of the interactive graph to determine the displacement response of a single degree-of-freedom system to a suddenly applied electrostatic force

```
        (* Determine maximum amplitude of solution to Eq. (8.12) *)
bb=Quiet[NMaxValue[{sol[eV,ζ][t],2<t<18},t,
 WorkingPrecision->25]];
                          (* Plot results *)
Plot[sol[eV,ζ][t],{t,0,tend},LabelStyle->{14},
 PlotRange->{{0,tend},{0,0.6}},
 AxesLabel->{τ,TraditionalForm[w[τ]]},
 Epilog->{{Red,Dashed,Line[{{0,fg},{tend,fg}}]},
  {Red,Dashed,Line[{{0,bb},{tend,bb}}]},
  Inset[Style[Row[{"wmax=",NumberForm[bb,4]}],
   Red,14],{0.8 tend,bb+0.025}],
```

```
    Inset[Style[Row[{"wstatic=",NumberForm[fg,4]}],
     Red,14],{0.8 tend,fg-0.03}]}],
                    (* Create sliders *)
Style["Applied Voltage Parameter",Bold],
{{eV,0.1329,"e1^2Vo^2"},0.12,0.1387,0.0001,
  Appearance->{"Labeled","Open"}, ContinuousAction->False},
Delimiter,
Style["Damping Factor",Bold],
{{ζ,0.08,"ζ"},0.02,0.15,0.01,Appearance->"Labeled",
  ContinuousAction->False,ControlType->Slider},
Delimiter,
Style["Axis Adjustment",Bold],
{{tend, 45,"τend"},35,200,5, Appearance->"Labeled",
  ContinuousAction->False, ControlType->Slider},
TrackedSymbols:>{eV,ζ,tend}]
```

Referring to Figure 8.3, the initial parameters were chosen such that if the + button on the animator/slider is depressed once; that is, $e_1^2 V_o^2$ is increased to 0.1330, the system becomes unstable.

8.2.4 Single Degree-of-Freedom System Energy Harvester

Consider a single degree-of-freedom system with a mass m (kg), a spring of stiffness k (N·m^{-1}), and a viscous damper with damping coefficient c (N·s·m^{-1}) whose base is excited harmonically at a frequency ω and magnitude Y_o. We replace the spring with a piezoelectric element of area A and height h that is operating in its 33 mode and has the following properties: a dielectric constant ε_{33}^S measured at constant strain (F·m^{-1}), an elastic stiffness c_{33}^S measured at constant electric field (N·m^{-2}), and an electromechanical coupling coefficient k_{33}. In addition, the output of the piezoelectric element is connected to a load resistor R_L. For harmonic oscillations of frequency ω, the nondimensional average power P_{avg} in the resistive load can be expressed as [2, Sections 2.6.2 and 2.6.3]

$$P_{avg}(r_L, \Omega_E) = \frac{\Omega_E^6 r_L k_e^2}{2\left(A_R^2(r_L, \Omega_E) + B_R^2(r_L, \Omega_E)\right)} \tag{8.14}$$

where

$$A_R(r_L, \Omega_E) = 1 - \Omega_E^2(1 + 2\zeta_E r_L)$$

$$B_R(r_L, \Omega_E) = \Omega_E\left[2\zeta_E + r_L\left(1 + k_e^2\right) - \Omega_E^2 r_L\right]$$

$$k_e^2 = \frac{k_{33}^2}{1 - k_{33}^2}$$

and

$$\Omega_E = \frac{\omega}{\omega_n^E}, \quad 2\zeta_E = \frac{c}{m\omega_n^E}, \quad r_L = \omega_n^E C_{pe}^S R_L$$

$$\omega_n^E = \sqrt{\frac{K_{pe}^E}{m}} \text{rad} \cdot \text{s}^{-1}, \quad C_{pe}^S = A\varepsilon_{33}^S/h\text{F}$$

$$K_{pe}^E = Ac_{33}^E/h\text{N} \cdot \text{m}^{-1}$$

The value of r_L that maximizes P_{avg} is

$$r_{\text{opt}}(\Omega_E) = \frac{1}{\Omega_E}\sqrt{\frac{\Omega_E^4 + \left(4\zeta_E^2 - 2\right)\Omega_E^2 + 1}{\Omega_E^4 + \left(4\zeta_E^2 - 2\left(1 + k_e^2\right)\right)\Omega_E^2 + \left(1 + k_e^2\right)^2}} \tag{8.15}$$

which is a function of the frequency coefficient Ω_E. Thus, at a given value of Ω_E, any value of r_L that is different from r_{opt} will result in less average power. However, this value of r_{opt} does not necessarily give the largest maximum average power. That has to be determined from the value of $\Omega_E = \Omega_{E,\max}$ that maximizes $P_{avg}(r_{\text{opt}}(\Omega_{E,\max}), \Omega_{E,\max})$. The interactive environment to conduct these investigations in shown in Figure 8.4 and the program that created this figure is as follows.

```
Manipulate[
 (* Determine maximum power with Eq. (8.15) substituted in Eq. (8.14) *)
 pmax=NMaximize[{pavg[x,ζ,k33,ropt[x,ζ,k33]],
  2.5>=x>=1.06},x];
                    (* Plot results *)
 Plot[{pavg[Ω,ζ,k33,ropt[Ω,ζ,k33]],pavg[Ω,ζ,k33,rL]},
  {Ω,0,2.5},PlotRange->{{0,2.5},All},
  PlotLabel->Column[{lab1,Row[{"Pavg,max=",
  NumberForm[pmax[[1]],4]," at ΩE,max=",
  NumberForm[Ω/.pmax[[2]],4]," and ropt=",
  NumberForm[ropt[Ω/.pmax[[2]],z,k33],4]}]},Center],
  AxesLabel->{Ω,"Pavg"},LabelStyle->{14},
  PlotStyle->{Black,Blue}],
                    (* Create sliders *)
 Style["Damping Factor",Bold],
 {{ζ,0.05,"ζE"},0.01,0.25,0.01,Appearance->"Labeled"
  ControlType->Slider},
 Style["Coupling Coeficient",Bold],
 {{k33,0.74,"k33"},0.5,0.9,0.01,Appearance->"Labeled",
  ControlType->Slider},
 Style["Load Resistance",Bold],
 {{rL,0.69,"rL"},0.01,20,0.01,Appearance->"Labeled",
```

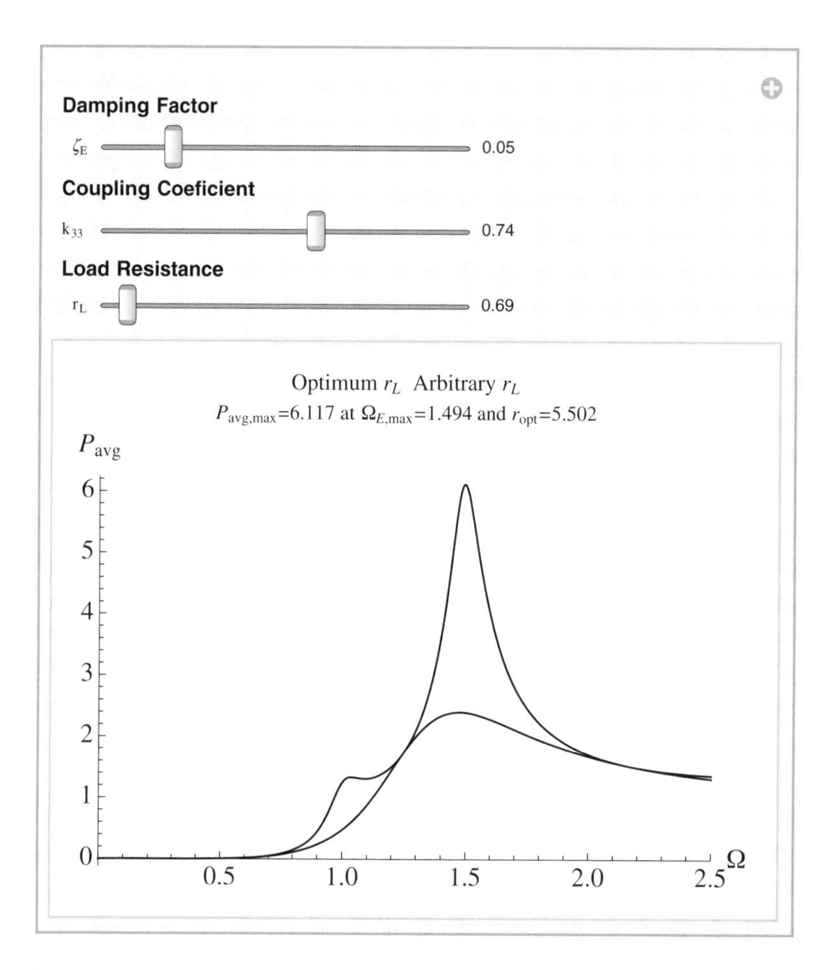

Figure 8.4 Initial configuration of the interactive graph to determine the normalized average power from a single degree-of-freedom piezoelectric energy harvester

```
  ControlType->Slider},
 Delimiter,
                    (* Create portions of a plot label *)
 Initialization:>(labl=Row[{Style["Optimum rL",Black,14],
  Style[" Arbitrary rL",Blue,14]}];
                   (* Function representing Eq. (8.15) *)
  ropt[Ω_,ζ_,k33_]:=(ke2=k33^2/(1-k33^2);
   1/Ω Sqrt[(Ω^4+(4 ζ^2-2) Ω^2+1)/
   (Ω^4+(4 ζ^2-2 (1+ke2)) Ω^2+(1+ke2)^2)]);
                   (* Function representing Eq. (8.14) *)
  pavg[Ω_,ζ_,k33_,rL_]:=(ke2=k33^2/(1-k33^2);
   0.5 Ω^6 rL ke2/((1-Ω^2 (1+2 ζ rL))^2+
```

```
    (Ω (2 ζ+rL (1+ke2)-rL Ω^2))^2));),
  TrackedSymbols:>{ζ,k33,rL}]
```

8.3 Two Degrees-of-Freedom Systems

8.3.1 Governing Equations

Consider a single degree-of-freedom system composed of a mass m_1, a viscous damper with damping coefficient c_1, and a spring with stiffness k_1. Another single degree-of-freedom system composed of a mass m_2, a viscous damper with damping coefficient c_2, and a spring with stiffness k_2 is attached to m_1. The nondimensional governing equations of motion of a two degrees-of-freedom systems are given in Example 4.30 and, for convenience, are repeated here. Thus,

$$\begin{aligned}\frac{d^2x_1}{d\tau^2} + (2\zeta_1 + 2\zeta_2 m_r\omega_r)\frac{dx_1}{d\tau} + \left(1 + m_r\omega_r^2\right)x_1 - 2\zeta_2 m_r\omega_r\frac{dx_2}{d\tau} - m_r\omega_r^2 x_2 &= \frac{f_1(\tau)}{k_1}\\ \frac{d^2x_2}{d\tau^2} + 2\zeta_2\omega_r\frac{dx_2}{d\tau} + \omega_r^2 x_2 - 2\zeta_2\omega_r\frac{dx_1}{d\tau} - \omega_r^2 x_1 &= \frac{f_2(\tau)}{k_1 m_r}\end{aligned} \tag{8.16}$$

where

$$\omega_r = \frac{\omega_{n2}}{\omega_{n1}} = \frac{1}{\sqrt{m_r}}\sqrt{\frac{k_2}{k_1}}, \quad m_r = \frac{m_2}{m_1}, \quad \tau = \omega_{n1}t$$

$$2\zeta_j = \frac{c_j}{m_j\omega_{nj}}, \quad \omega_{nj} = \sqrt{\frac{k_j}{m_j}}\text{rad}\cdot\text{s}^{-1} \quad j = 1, 2$$

and $f_l(\tau)$, $l = 1, 2$ is the force applied to mass m_l.

8.3.2 Response to Harmonic Excitation: Amplitude Response Functions

It is assumed that the applied force is harmonic and of the form $f_l(\tau) = F_l e^{j\Omega\tau}$, $l = 1, 2$. Then, if it is assumed that a solution to Eq. (8.16) is of the form $x_l(\tau) = X_l e^{j\Omega\tau}$, it is found that

$$\begin{aligned}X_1(j\Omega) &= \frac{1}{k_1 D(j\Omega)}[F_1 E(j\Omega) + F_2 B(j\Omega)/m_r]\\ X_2(j\Omega) &= \frac{1}{k_1 D(j\Omega)}[F_1 C(j\Omega) + F_2 A(j\Omega)/m_r]\end{aligned} \tag{8.17}$$

where $\Omega = \omega/\omega_{n1}$, and

$$\begin{aligned}
A(j\Omega) &= -\Omega^2 + 2(\zeta_1 + \zeta_2 m_r \omega_r) j\Omega + 1 + m_r \omega_r^2 \\
B(j\Omega) &= 2\zeta_2 m_r \omega_r j\Omega + m_r \omega_r^2 \\
C(j\Omega) &= B(j\Omega)/m_r \\
E(j\Omega) &= -\Omega^2 + 2\zeta_2 \omega_r j\Omega + \omega_r^2 \\
D(j\Omega) &= \Omega^4 - j[2\zeta_1 + 2\zeta_2 \omega_r m_r + 2\zeta_2 \omega_r]\Omega^3 \\
&\quad - [1 + m_r \omega_r^2 + \omega_r^2 + 4\zeta_1 \zeta_2 \omega_r]\Omega^2 + j[2\zeta_2 \omega_r + 2\zeta_1 \omega_r^2]\Omega + \omega_r^2
\end{aligned} \tag{8.18}$$

For Eq. (8.17), two scenarios are considered: (i) $F_1 \neq 0$ and $F_2 = 0$ and (ii) $F_2 \neq 0$ and $F_1 = 0$. For these assumptions, it is found that the frequency-response functions are

$$\begin{aligned}
H_{11}(j\Omega) &= \frac{E(j\Omega)}{D(j\Omega)} \\
H_{21}(j\Omega) &= \frac{C(j\Omega)}{D(j\Omega)} \\
H_{12}(j\Omega) &= \frac{B(j\Omega)}{m_r D(j\Omega)} = H_{21}(j\Omega) \\
H_{22}(j\Omega) &= \frac{A(j\Omega)}{m_r D(j\Omega)}
\end{aligned} \tag{8.19}$$

where, in H_{ij}, the subscript i refers to the response of mass m_i and the subscript j indicates that the force is applied to m_j. The magnitudes of the frequency response functions are given by $|H_{ij}|$.

The natural frequencies of the systems are determined by finding the values of Ω that satisfy $D(\Omega) = 0$ when $\zeta_1 = \zeta_2 = 0$. These operations yield

$$\Omega_{1,2} = \sqrt{\frac{1}{2}\left[1 + \omega_r^2(1 + m_r) \mp \sqrt{\left(1 + \omega_r^2(1 + m_r)\right)^2 - 4\omega_r^2}\right]} \tag{8.20}$$

The magnitudes of the frequency response functions $|H_{ij}|$ and the natural frequencies $\Omega_{1,2}$ are placed in an interactive environment shown in Figure 8.5, where the effects that the parameters ω_r, m_r, ζ_1, and ζ_1 can have on these quantities are displayed. The option of filling the region under each curve has also been provided.

The program that creates this interactive environment is as follows.

```
Manipulate[mr=mr ; ωr=wr; a1=1+wr^2 (1+mr);
                              (* Eq. (8.20) *)
  Ω1=Sqrt[0.5 (a1-Sqrt[a1^2-4 ωr^2])];
  Ω2=Sqrt[0.5 (a1+Sqrt[a1^2-4 ωr^2])];
                     (* Red dashed lines for display *)
  ws={{{Red,Dashed,Line[{{Ω1,0},{Ω1,100}}]}},
    {{Red,Dashed,Line[{{Ω2,0},{Ω2,300}}]}}};
```

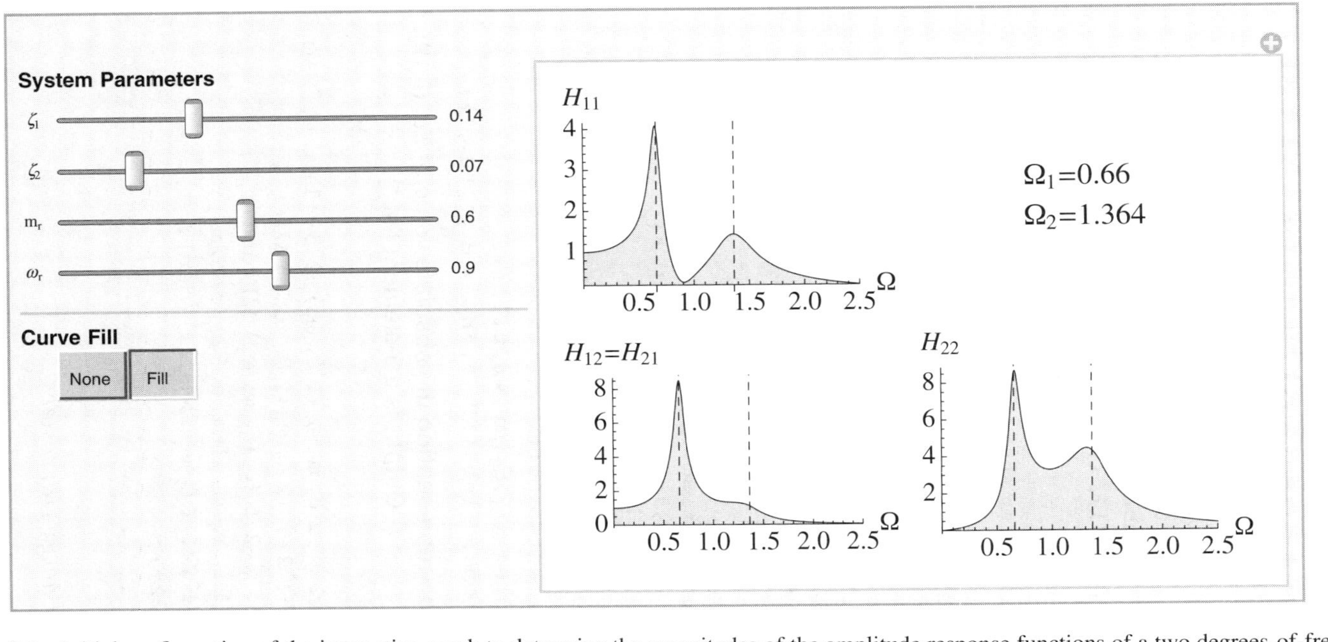

Figure 8.5 Initial configuration of the interactive graph to determine the magnitudes of the amplitude response functions of a two degrees-of-freedom system. Also displayed are the system's natural frequencies

```
               (* Results displayed in a 2-by-2 graphics grid *)
plt[n_]:=Plot[Hij[Ω][[n]],{Ω,0.0,2.5},
 AxesLabel->{"Ω",lab[[n]]},PlotRange->{{0,2.5},All},
 PlotStyle->Blue,LabelStyle->{14},
 Filling->If[fill==2,Axis,False],Epilog->ws];
GraphicsGrid[{{plt[1],Style[Column[
  {Row[{"Ω1=", NumberForm[Ω1,3]}],
  Row[{"Ω2=",NumberForm[Ω2,4]}]}],16,Red]},
   {plt[2],plt[3]}}],
                        (* Create sliders *)
Style["System Parameters", Bold],
{{ζ1,0.14,"ζ1"},0,0.4,0.01,Appearance->"Labeled",
 ControlType->Slider},
{{ζ2,0.07,"ζ2"},0,0.4,0.01,Appearance->"Labeled",
 ControlType->Slider},
{{mr,0.6,"mr"},0.01,1.2,0.01,Appearance->"Labeled",
 ControlType->Slider},
{{wr,0.9,"ωr"},0.01,1.5,0.01,Appearance->"Labeled",
 ControlType->Slider},
Delimiter,
Style["Curve Fill",Bold],
{{fill,2," "},{1->"None",2->" Fill  "},
 ControlType->SetterBar},
                     (* Place controls at left *)
ControlPlacement-> Left,
TrackedSymbols:>{ζ1,ζ2,mr,wr,fill},
Initialization:>( lab={"H11","H12 = H21","H22"};
       (* Function representing the absolute value of Eq. (8.19). *)
                    (* Output is a 3-element list *)
Hij[Ω_]:=(
 den=Ω^4-I (2 ζ1+2 ζ2 mr ωr+2 ζ2 ωr) Ω^3-(1+mr ωr^2+ωr^2+
  4 ζ1 ζ2 ωr) Ω^2+I (2 ζ2 ωr+2 ζ1 ωr^2) Ω+ωr^2;
 h22=Abs[(-Ω^2+2 I (ζ1+ζ2 mr ωr) Ω)/den/mr];
 h12=Abs[(2 I ζ2 mr ωr Ω+mr ωr^2)/den/mr];
 h11=Abs[(-Ω^2+2 I ζ2 ωr Ω+ωr^2)/den];
 {h11,h12,h22});
```

8.3.3 Enhanced Energy Harvester

We shall create an enhanced energy harvester by replacing the spring k_2 in the two degrees-of-freedom system given by Eq. (8.16) with a piezoelectric element as defined in Section 8.2.4. The output of this element is connected to a load resistor R_L. In addition, the base that is supporting k_1 and c_1 is now assumed to move an amount x_3. Then, in Eq. (8.16), the base excitation can be accounted if $f_2 = 0$ and f_1 is replaced with

$$f_1 = x_3 + 2\zeta_1 \frac{dx_3}{d\tau}$$

If harmonic excitation of the form $x_l(\tau) = X_l e^{j\Omega\tau}$, $l = 1, 2, 3$, is assumed, it can be shown that the normalized average power $P'_{avg}(\Omega)$ into the load resistor can be written as [2, Section 2.7.3]

$$P'_{avg}(\Omega) = \frac{r'_L c_o(\Omega)}{|f_1 r'_L + f_2 + j(g_1 r'_L + g_2)|^2} \tag{8.21}$$

where $r'_L = C^S_{pe} R_L \omega_{n1}$,

$$\begin{aligned}
f_1 &= \Omega\{(B_i - A_i)\alpha_{33}/m_r + B_i C_r + B_r C_i - A_i E_r - A_r E_i + (C_i - E_i)\alpha_{33}\} \\
f_2 &= B_i C_i - B_r C_r - A_i E_i + A_r E_r \\
g_1 &= \Omega\{(A_r - B_r)\alpha_{33}/m_r + B_i C_i - B_r C_r - A_i E_i + A_r E_r + (E_r - C_r)\alpha_{33}\} \\
g_2 &= A_i E_r - B_r C_i - B_i C_r + A_r E_i \\
\alpha_{33} &= k_e^2 \omega_r^2 m_r
\end{aligned} \tag{8.22}$$

and

$$\begin{aligned}
c_o(\Omega) &= \frac{k_e^2}{2\omega_r} |j\Omega(1 + 2j\zeta_1\Omega)(C_r - E_r + j(C_i - E_i))|^2 \\
A_r + jA_i &= -\Omega^2 + 1 + m_r\omega_r^2 + 2j(\zeta_1 + \zeta_2 m_r \omega_r)\Omega \\
B_r + jB_i &= m_r\omega_r^2 + 2j\zeta_2 m_r \omega_r \Omega \\
C_r + jC_i &= \omega_r^2 + 2j\zeta_2\omega_r\Omega \\
E_r + jE_i &= -\Omega^2 + \omega_r^2 + 2j\zeta_2\omega_r\Omega \\
k_e^2 &= \frac{k_{33}^2}{1 - k_{33}^2}
\end{aligned} \tag{8.23}$$

It can be shown that the maximum average power at a given frequency can be obtained when $r'_L = r'_{L,\text{opt}}$, which is determined from

$$r'_{L,\text{opt}}(\Omega) = \sqrt{\frac{f_2^2 + g_2^2}{f_1^2 + g_1^2}} \tag{8.24}$$

Thus, at a given value of Ω, any value of r'_L that is different from $r'_{L,\text{opt}}$ will result in less average power. However, this value of $r'_{L,\text{opt}}$ does not necessarily give the largest maximum average power. That has to be determined numerically.

We shall create an interactive environment shown in Figure 8.6 that plots $P'_{avg}(\Omega)$ for the optimum value $r'_{L,\text{opt}}$ at each value of Ω and allows one to explore this curve as a function

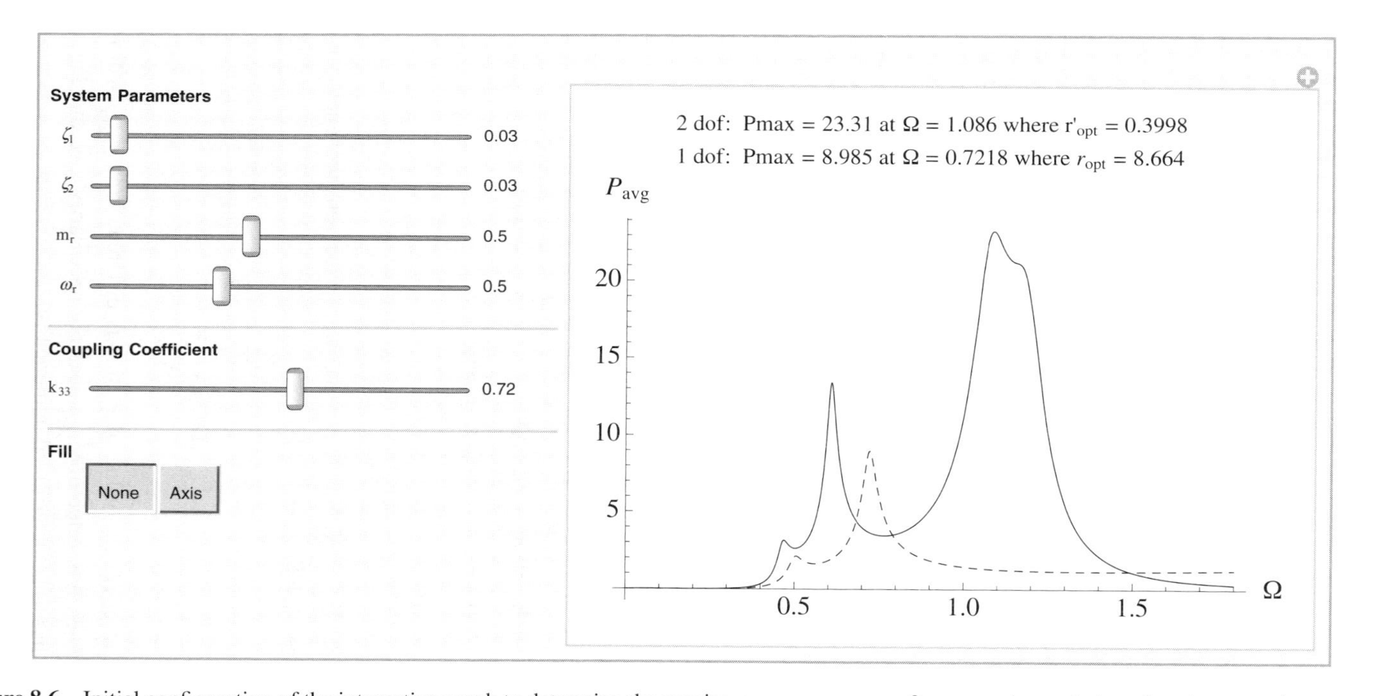

Figure 8.6 Initial configuration of the interactive graph to determine the maximum average power from an enhanced piezoelectric energy harvester and from a single degree-of-freedom energy harvester

of ω_r, m_r, ζ_1, ζ_2, and k_{33}. For each set of these parameters, the maximum average power, the frequency at which it occurs, and the corresponding value of r'_L are determined and displayed. In addition, we have included for comparison purposes the equivalent curve for the single degree-of-freedom system, which are given by Eqs. (8.14) and (8.15). To compare the notation of the two degrees-of-freedom energy harvester with that of the single degree-of-freedom energy harvester, we note the following: $\omega_n = \omega_{2n}$, $Y_o = X_3$, $\zeta_E = \zeta_2$, and $\Omega_E = \Omega/\omega_r$. The program that creates this interactive environment is as follows.

```
Manipulate[ωr=wr; mr=mr; ke2=k33^2/(1-k33^2);
  pt=pavg2[rvv,rLopt[rvv]];
  lb=First[Ordering[pt,-1]];
   (* Maximum average power from two degrees-of-freedom system *)
  rx=Quiet[FindMaximum[pavg2[Om2,rLopt[Om2]],
   {Om2,rvv[[lb]]}]];
  Omax=Om2/.Last[rx]; ropt2=rLopt[Omax];
  pt1=pavg1[rvv/ωr,rLopt1[rvv/ωr]];
  lb1=First[Ordering[pt1,-1]];
   (* Maximum average power from single degree-of-freedom system *)
  rx1=Quiet[FindMaximum[{pavg1[Om1/ωr,rLopt1[Om1/ωr]],
   {0<Om1<1.8}},{Om1,rvv[[lb1]]}]];
  Omax1=Om1/.Last[rx1];
  ropt11=rLopt1[Omax1/wr];
                        (* Plot results *)
  Plot[{pavg2[x,rLopt[x]],pavg1[x/ωr,rLopt1[x/ωr]]},
   {x,0,1.8},Filling->fill,PlotRange->All,
   AxesLabel->{Ω,"Pavg"},LabelStyle->14,
   PlotStyle->{Blue,{Red,Dashed}},
   PlotLabel->Column[{Style[Row[
    {lab2,NumberForm[First[rx],4]," at Ω = ",
    NumberForm[Omax, 4],lab4,NumberForm[ropt2,4]}],
    Blue,12],Style[Row[{lab3,NumberForm[First[rx1],4],
   " at Ω = ",NumberForm[Omax1,4],lab5,
    NumberForm[ropt11,4]}], Red,12]}]],
                  (* Create sliders and setter bar *)
  Style["System Parameters",Bold],
  {{ζ1,0.03,"ζ1"},0.02,0.35,0.01,Appearance->"Labeled",
   ControlType->Slider},
  {{ζ2,0.03,"ζ2"},0.02,0.35,0.01,Appearance->"Labeled",
   ControlType->Slider},
  {{mr,0.5,"mr"},0,1.2,0.01,Appearance->"Labeled",
   ControlType->Slider},
  {{wr,0.5,"ωr"},0,1.5,0.01,Appearance->"Labeled",
   ControlType->Slider},
  Delimiter,
```

```
Style["Coupling Coefficient",Bold],
{{k33,0.72,"k33"},0.5,0.9,0.01,Appearance->"Labeled",
 ControlType->Slider},
Delimiter,
Style["Fill",Bold],
{{fill,None," "},{None,Axis},ControlType->SetterBar},
TrackedSymbols:>(ζ1,ζ2,mr,wr,fill,k33},
```

(* Create labels, a list of numbers, and several functions *)

```
Initialization:>(lab2="2 dof:  Pmax = ";
 lab3="1 dof:  Pmax = ";   lab4=" where r′opt = ";
 lab5=" where ropt = ";   rvv=Range[0.01,1.8,0.01];
```

(* Function representing Eqs. (8.22) and (8.23). *)
(* Output is a 5-element list. *)

```
fgc[Ω_]:=(a33=ke2 ωr^2 mr; cC=ωr^2+2 I ζ2 ωr Ω;
 aA=-Ω^2+1+mr ωr^2+2 I (ζ1+ζ2 mr ωr) Ω;
 bB=mr ωr^2+2 I ζ2 mr ωr Ω; eE=-Ω^2+ωr^2+2 I ζ2 ωr Ω;
 co=0.5 ke2/ωr Abs[I Ω (1+2 I ζ1 Ω) (Re[cC]-Re[eE]+
  I (Im[cC]-Im[eE]))]^2;
 f1=Ω ((Im[bB]-Im[aA]) a33/mr+Im[bB] Re[cC]+
  Re[bB] Im[cC]-Im[aA] Re[eE]-Re[aA] Im[eE]+
  (Im[cC]-Im[eE]) a33);
 f2=Im[bB] Im[cC]-Re[bB] Re[cC]-
  Im[aA] Im[eE]+Re[aA] Re[eE];
 g1=Ω ((Re[aA]-Re[bB]) a33/mr+Im[bB] Im[cC]-
  Re[bB] Re[cC]-Im[aA] Im[eE]+Re[aA] Re[eE]+
  (Re[eE]-Re[cC]) a33);
 g2=Im[aA] Re[eE]-Re[bB] Im[cC]-Im[bB] Re[cC]+
  Re[aA] Im[eE]; {f1,f2,g1,g2,co});
```

(* Function representing Eq. (8.21) *)

```
pavg2[Ω_,rL_]:=(q=fgc[Ω];
 rL q[[5]]/Abs[q[[1]] rL+q[[2]]+I (q[[3]] rL+q[[4]])]^2);
```

(* Function representing Eq. (8.24) *)

```
rLopt[Ω_]:=(q=fgc[Ω];
 Sqrt[(q[[2]]^2+q[[4]]^2)/(q[[1]]^2+q[[3]]^2)]);
```

(* Function representing Eq. (8.15) *)

```
rLopt1[Ω_]:=(1/Ω Sqrt[(Ω^4+(4 ζ2^2-2) Ω^2+1)/
 (Ω^4+(4 ζ2^2-2 (1+ke2)) Ω^2+(1+ke2)^2)]);
```

(* Function representing Eq. (8.14) *)

```
pavg1[Ω_,rL_]:=(aR=1-Ω^2 (1+2 ζ2 rL);
 bR=Ω (2 ζ2+rL (1+ke2)-rL Ω^2);
 0.5 Ω^6 rL ke2/(aR^2+bR^2));)]
```

8.4 Thin Beams

8.4.1 Natural Frequencies and Mode Shapes of a Cantilever Beam with In-Span Attachments

A cantilever beam of constant cross section has a length L (m), cross-sectional area A (m^2), area moment of inertia I (m^4), Young's modulus E (N·m^{-2}), and density ρ (kg·m^{-3}). There are two in-span attachments: one is a mass M_i (kg) that is attached at $x = L_m$, $0 < L_m \le L$ and the other is a translation spring of stiffness k_i (N·m^{-1}) that is attached at $x = L_s$, $0 < L_s \le L$. It is mentioned that, in general, L_m and L_s are independent; however, only the case for $L_m = L_s$ will be considered here.

The natural frequency coefficients Ω_n for the case where $\eta_s = \eta_m$ are determined from [2, Section 3.3.3]

$$\Omega_n^3 D^{(3)}(\Omega_n) + \left(m_i\Omega_n^4 - K_i\right) H_n^{(3)}(\Omega_n, \eta_s, \eta_s) = 0 \tag{8.25}$$

where

$$\begin{aligned} D^{(3)}(\Omega_n) &= R(\Omega_n)T(\Omega_n) - Q^2(\Omega_n) \\ H^{(3)}(\Omega_n, x, y) &= T(\Omega_n x)[T(\Omega_n)R(\Omega_n[1-y]) - Q(\Omega_n)Q(\Omega_n[1-y])] \\ &\quad + S(\Omega_n x)[R(\Omega_n)Q(\Omega_n[1-y]) - Q(\Omega_n)R(\Omega_n[1-y])] \end{aligned} \tag{8.26}$$

and

$$\begin{aligned} Q(\Omega x) &= \frac{1}{2}[\cos(\Omega x) + \cosh(\Omega x)] \\ R(\Omega x) &= \frac{1}{2}[\sin(\Omega x) + \sinh(\Omega x)] \\ S(\Omega x) &= \frac{1}{2}[-\cos(\Omega x) + \cosh(\Omega x)] \\ T(\Omega x) &= \frac{1}{2}[-\sin(\Omega x) + \sinh(\Omega x)] \end{aligned} \tag{8.27}$$

In addition,

$$\begin{aligned} \eta &= \frac{x}{L}, \quad \eta_m = \frac{L_m}{L}, \quad \eta_s = \frac{L_s}{L}, \quad t_o = L^2\sqrt{\frac{A\rho}{EI}} \text{ s} \\ m_i &= \frac{M_i}{m_b}, \quad m_b = \rho AL \text{ kg}, \quad K_i = \frac{k_i L^3}{EI} \end{aligned} \tag{8.28}$$

It is noted that when $K_i \to \infty$, we have the case of a beam with an intermediate rigid support for which the displacement is zero but the rotation is unrestrained.

The mode shapes are given by

$$Y_n(\eta) = \frac{H_n^{(3)}(\Omega_n, \eta, \eta_s)}{D^{(3)}(\Omega_n)} - T(\Omega_n[\eta - \eta_s])u(\eta - \eta_s) \tag{8.29}$$

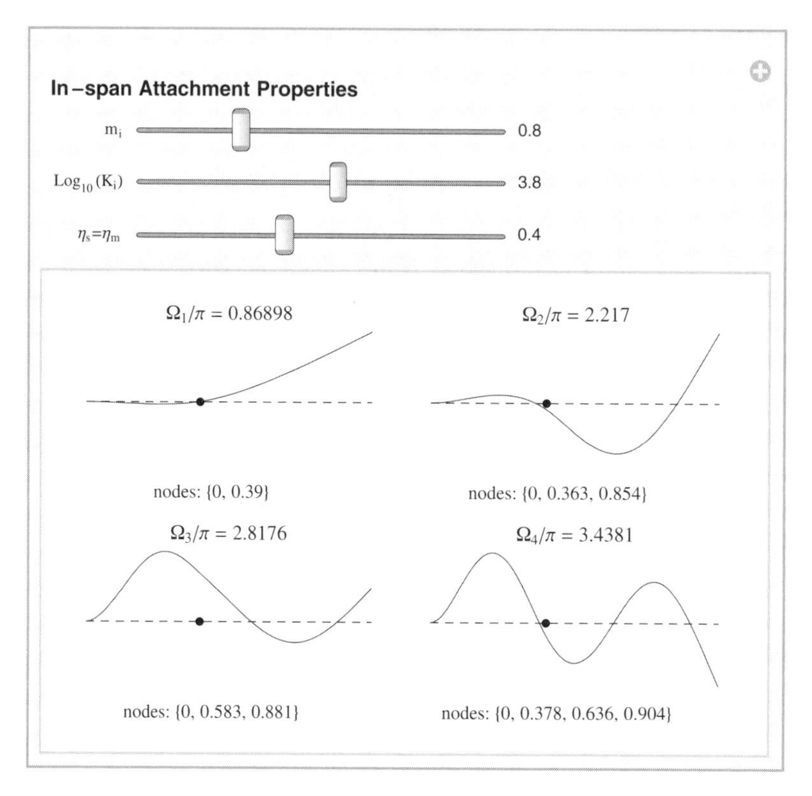

Figure 8.7 Initial configuration of the interactive graph to obtain the natural frequencies, modes shapes, and node points of a cantilever beam for various combinations of in-span attachments

where $u(\eta)$ is the unit step function. The normalized mode shape is given by $Y_n(\eta)/\max|Y_n(\eta)|$. The node points are those values of η for which

$$Y_n(\eta) = 0 \tag{8.30}$$

To improve the numerical evaluation process, the magnitude of K_i appearing in Eq, (8.25) is monitored. Whenever the value exceeds 500, the equation is first divided by K_i and then numerically evaluated.

An interactive environment shown in Figure 8.7 is created to explore the effects of attachments on the lowest four natural frequencies and their associated mode shapes and node points. The program that creates Figure 8.7 is as follows.

```
Manipulate[Ki=10^kKi; w=ce8[Ωe]; nf={};
                    (* Determine Ωn from Eq. (8.25) *)
 Do[If[w[[n]] w[[n+1]]<0,AppendTo[nf,
   (x/.Quiet[FindRoot[ce8[x],{x,Ωe[[n]], Ωe[[n+1]]}]])/π];
   If[Length[nf]==4,Break[]]],{n,1,Length[w]-1,1}];
```

```
(* Determine node points from Eq. (8.30) *)
  node={}; mss={};
  Do[npt={}; ms1=shapeIns8[etta,nf[[nct]] π];
  ms1=ms1/Max[Abs[ms1]];
    Do[If[ms1[[nv]] ms1[[nv+1]]<0,
    AppendTo[npt,y/.Quiet[FindRoot[
      shapeIns8[y,nf[[nct]] π],{y,etta[[nv]],
      etta[[nv+1]]}]]]]],{nv,1,le-1,1}]; AppendTo[mss,ms1];
    PrependTo[npt,0]; AppendTo[node,npt],{nct,1,4}];
ypot={etas (le-1)+1,0};
(* Plot results on a 2×2 grid *)
GraphicsGrid[{{plt[mss[[1]],1,le,node[[1]],ypot,nf[[1]]],
  plt[mss[[2]],2,le,node[[2]],ypot,nf[[2]]]},
  {plt[mss[[3]],3,le,node[[3]],ypot,nf[[3]]],
  plt[mss[[4]],4,le,node[[4]],ypot,nf[[4]]]}}],
(* Create Sliders *)
Style["In-span Attachment Properties",Bold,11],
{{mi,0.8,"mi"},0.0,3,0.1,
  Appearance->"Labeled",ControlType->Slider},
{{kKi,3.8,"Log10(Ki)"},-4,10,0.1,
  Appearance->"Labeled",ControlType->Slider},
{{etas,0.4,"ηs=ηm"},0.01,0.99,0.01,
  Appearance->"Labeled",ControlType->Slider},
TrackedSymbols:>{mi,kKi,etas},
Initialization:>(Ωe=Range[0.001,25.,0.1];
  etta=Range[0,1,0.01]; le=Length[etta];
(* Functions representing Eq. (8.27) *)
  Q[Ω_,y_]:=(Cos[Ω y]+Cosh[Ω y])/2.;
  R[Ω_,y_]:=(Sin[Ω y]+Sinh[Ω y])/2.;
  S[Ω_,y_]:=(-Cos[Ω y]+Cosh[Ω y])/2.;
  T[Ω_,y_]:=(-Sin[Ω y]+Sinh[Ω y])/2.;
(* Functions representing Eq. (8.26) *)
  d3[Ω_,y_]:=R[Ω,y] T[Ω,y]-Q[Ω,y]^2;
  h3[Ω_,x_,y_]:=T[Ω,x] (T[Ω,1] R[Ω,1-y]-Q[Ω,1] Q[Ω,1-y])+
    S[Ω,x] (R[Ω,1] Q[Ω,1-y]-Q[Ω,1] R[Ω,1-y]);
(* Function representing Eq. (8.25) *)
  ce8[Ω_]:=If[Ki<500.,d3[Ω,1]-(Ki-mi Ω^4) h3[Ω,etas,etas]/
    Ω^3, d3[Ω,1]/Ki-(1.-mi Ω^4/Ki) h3[Ω,etas,etas]/Ω^3];
(* Function representing Eq. (8.29) *)
  shapeIns8[η_,Ω_]:=-T[Ω,η-etas] d3[Ω,1] UnitStep[η-etas]+
    h3[Ω,η,etas];
```

(* Plotting function *)

```
plt[f_,n1_,le_,nod_,ypot_,freq_]:=
 ListLinePlot[f, Axes->False,PlotLabel->Row[{"Ω"n1,"/π=",
  NumberForm[freq,5]}],PlotRange->{{1,le},{-1.5,1.1}},
  Epilog->{{Black,Dashed,Line[{{1,0},{le,0}}]},
  {Inset[Style[Row[{"nodes: ",NumberForm[nod,3]}],10],
  {45,-1.3}]},{PointSize[Medium],Red,Point[ypot]}}];)]
```

8.4.2 Effects of Electrostatic Force on the Natural Frequency and Stability of a Beam

We shall examine a beam clamped at both ends and a cantilever beam when each beam is subjected to an electrostatic force. The electrostatic force field can include a fringe correction factor. In addition, for the beam clamped at both ends two additional effects will be taken into account. One is due to the application of an externally applied in-plane tensile force p_o. The second is to include in-plane stretching due to the transverse displacement.

Consider a beam of constant cross section of length L (m), cross-sectional area A (m^2), moment of inertia I (m^4), Young's modulus E (N·m^{-2}), and density ρ (kg·m^{-3}). The cross section of the beam is rectangular with a depth of h and a width b. The bottom surface of the beam is parallel to a fixed, flat surface that is a distance d_o from it. A voltage of magnitude V_o is applied across the gap d_o, which forms an electrostatic field. If the transverse displacement of the beam is w and no transverse force other than the electrostatic force is applied, then an estimate for the lowest natural frequency assuming small oscillations at frequency ω about a static equilibrium position denoted φ_s is given by [2, Section 4.3.3]

$$\Omega^2 = \sqrt{\frac{1}{m_o}\left[k + 3k_1\varphi_s^2 - E_1^2 V_o^2 \hat{G}_r'(\varphi_s)\right]} \tag{8.31}$$

where

$$\hat{G}_r'(\varphi_s) = \int_0^1 \left(\frac{2Y^2}{(1-\varphi_s Y)^3} + \frac{1.24c_3 Y^2}{(1-\varphi_s Y)^{2.24}}\right) d\eta \tag{8.32}$$

and $\Omega = \omega t_o$ where t_o is defined in Eq. (8.28). In Eq. (8.32), we have for a beam clamped at both ends

$$\begin{aligned} Y(\eta) &= \eta^2(\eta-1)^2, \qquad m_o = 1/630 \\ k &= 4/5 + (2/105)S_o, \qquad k_1 = d_r(2/105)^2 \end{aligned} \tag{8.33}$$

and for a cantilever beam

$$\begin{aligned} Y(\eta) &= \eta^2(\eta^2 - 4\eta + 6), \quad m_o = 104/45 \\ k &= 144/5, \quad k_1 = 0 \end{aligned} \tag{8.34}$$

Furthermore, the following definitions have been introduced

$$d_r = 6\left(\frac{d_o}{h}\right)^2, \quad E_1^2 = \frac{\varepsilon_o b L^4}{2EId_o^3}, \quad S_o = \frac{p_o L^2}{EI}$$
$$c_3 = \left(0.204 + 0.6\left(\frac{h}{b}\right)^{0.24}\right)\left(\frac{d_o}{b}\right)^{0.76} \tag{8.35}$$

In Eq. (8.35), ε_o is the permittivity of free space, which for air is 8.854×10^{-12} F·m^{-1}, and c_3 is the fringe correction factor. When the fringe correction is ignored, $c_3 = 0$. The effects of the displacement-induced in-plane stretching become negligible as $d_r \rightarrow 0$.

In Eq. (8.31), there are two interdependent quantities that have to be determined: φ_s and $\hat{G}_r'(\varphi_s)$. First, we need to determine the value of $E_1 V_o$, denoted $(E_1 V_o)_{\mathrm{PI}}$, at which the system becomes unstable. Corresponding to this value is the maximum static equilibrium value denoted $\varphi_{s,\mathrm{PI}}$ and the maximum displacement denoted $\hat{y}_{\mathrm{PI}}$. For a beam clamped at both ends, $\hat{y}_{\mathrm{PI}} = \varphi_{s,\mathrm{PI}}/16$ and for a cantilever beam $\hat{y}_{\mathrm{PI}} = 3\varphi_{s,\mathrm{PI}}$. The procedure to determine these values is as follows. We first determine $\varphi_{s,\mathrm{PI}}$ from the solution to

$$\hat{G}_r(\varphi_{s,\mathrm{PI}})\left(k + 3k_1\varphi_{s,\mathrm{PI}}^2\right) - \hat{G}_r'(\varphi_{s,\mathrm{PI}})\left(k\varphi_{s,\mathrm{PI}} + k_1\varphi_{s,\mathrm{PI}}^3\right) = 0 \tag{8.36}$$

where

$$\hat{G}_r(\varphi_{s,\mathrm{PI}}) = \int_0^1 \left(\frac{Y}{(1-\varphi_{s,\mathrm{PI}}Y)^2} + \frac{c_3 Y}{(1-\varphi_{s,\mathrm{PI}}Y)^{1.24}}\right) d\eta \tag{8.37}$$

and Y is given by Eq. (8.33) or Eq. (8.34) as the case may be. Then, $(E_1 V_o)_{\mathrm{PI}}$ is determined from

$$(E_1 V_o)_{\mathrm{PI}} = \sqrt{\frac{k\varphi_{s,\mathrm{PI}} + k_1\varphi_{s,\mathrm{PI}}^3}{\hat{G}_r(\varphi_{s,\mathrm{PI}})}} \tag{8.38}$$

Next, we evaluate Eq. (8.32) by first assuming a value for $E_1 V_o$ in the range $0 < E_1 V_o < (E_1 V_o)_{\mathrm{PI}}$. Then, for this value of $E_1 V_o$ the value of φ_s that satisfies

$$k\varphi_s + k_1\varphi_s^3 - E_1^2 V_o^2 \hat{G}_r(\varphi_s) = 0 \tag{8.39}$$

is determined. Substituting φ_s into Eq. (8.32) and the result in turn into Eq. (8.31), we obtain the value of Ω^2.

An interactive environment shown in Figure 8.8 is created to explore the effects that the boundary conditions, the in-plane loading, the fringe correction factor and beam geometry, and displacement-induced in-plane stretching have on the first natural frequency coefficient. This program is computationally intensive and, therefore, has a very long delay from

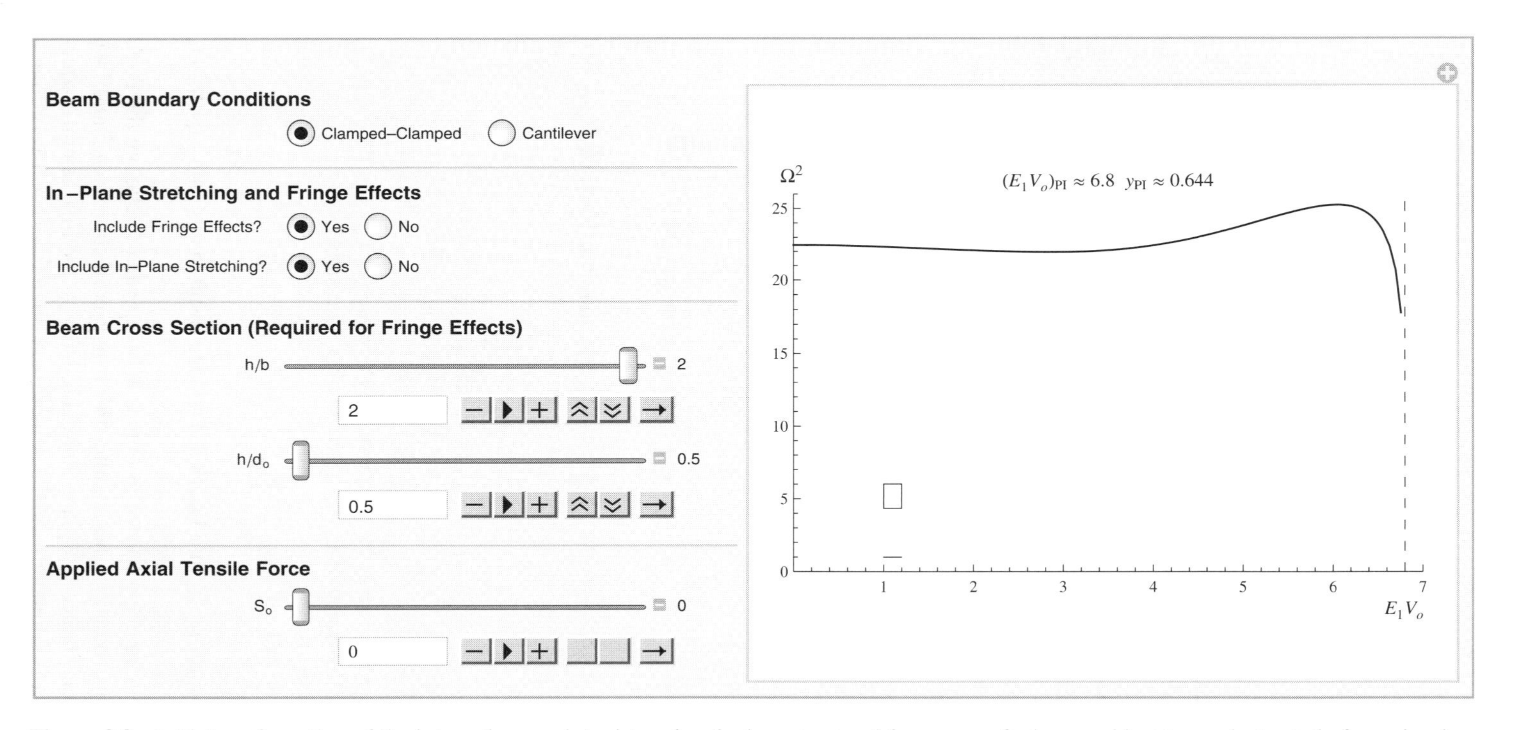

Figure 8.8 Initial configuration of the interactive graph to determine the lowest natural frequency of a beam subject to an electrostatic force, in-plane load, and displacement-induced in-plane force

the time a parameter is changed until the graph is drawn. To overcome this, we have set **ContinuousAction** to **False** and used the general form of the slider so that it is also possible to enter the values directly. Additionally, **FindRoot** internally does some symbolic operations, which cause difficulty when interacting with **NIntegrate** via **gr** and **dgr**. Therefore, as discussed in Sections 5.2 and 5.5, the inputs to these functions are restricted to numerical values by checking the input with **?NumericQ** and by selecting the method used by **NIntegrate** to forego any symbolic operations, which is done by setting **SymbolicProcessing** to **False**.

The program that creates Figure 8.8 is as follows.

```
Manipulate[
            (* Obtain parameters in Eqs. (8.33) to (8.35) *)
  If[fe==1,c3=(0.204+0.6 hb^0.24) (hb/hdo)^0.76,c3=0];
  If[bc==1,mo=1./630.;k=0.8+So 2./105.;
  If[inpl==1,k1=6./hdo^2 (2/105)^2,k1=0],
   mo=104./45.;k=144./5;k1=0];
  If[bc==1,If[So==0,yMax=26,yMax=40],yMax=4];
  If[bc==1,loc=1,loc=0.1];
            (* Create figure for insertion into main figure *)
  If[fe==1,plin=ListLinePlot[{{{0,0},{h/hb,0}},{{0,h/hdo},
   {h/hb,h/hdo},{h/hb,h/hdo+h},{0,h/hdo+h},{0,h/hdo}}},
   Axes->False,PlotRange->{{0,5.1},{-0.1,4.4}},
   ImageSize->Tiny,PlotStyle->{{Black},{Black}}],
   plin=""];
                   (* Solve Eqs. (8.36) and (8.38) *)
  If[bc==1,guess=8.5;pe=11.5,guess=0.08;pe=0.3];
  phiPI=phi/.FindRoot[gr[phi] (k+3. k1 phi^2)
   -dgr[phi] (k phi+k1 phi^3),{phi,guess,0.01,pe}];
  eVPI=Sqrt[(k phiPI+k1 phiPI^3)/gr[phiPI]];
  If[bc==1,yPI=phiPI/16.,yPI=3. phiPI];
                          (* Plot results *)
  Plot[freq2[x],{x,0.0,0.995 eVPI},
   PlotRange->{{0,Ceiling[eVPI]},{0,yMax}},
   MaxRecursion->2,PlotPoints->25,AxesLabel->{"E1Vo","Ω2"},
   PlotLabel->Row[{lab1,NumberForm[eVPI,4],lab2,
   NumberForm[yPI,3]}],Epilog->{{Dashing[Medium],Red,
    Line[{{eVPI,0},{eVPI,yMax}}]},
   Inset[plin,{loc,loc},{0,0}]}],
                 (* Create sliders and radio buttons *)
  Style["Beam Boundary Conditions",Bold],
  {{bc,1," "},{1->"Clamped-Clamped",2->"Cantilever"},
   ControlType->RadioButtonBar},
  Delimiter,
  Style["In-Plane Stretching and Fringe Effects",Bold],
```

```
{{fe,1,"Include Fringe Effects?"},
{1->"Yes",2->"No"},ControlType->RadioButtonBar},
{{inpl,1,"Include In-Plane Stretching?"},
 {1->"Yes",2->"No"},ControlType->RadioButtonBar,
 Enabled->If[bc==1,True,False]},
Delimiter,
Style["Beam Cross Section (Required for Fringe
 Effects)",Bold],
{{hb,2,"h/b"},0.2,2,0.1,Appearance->{"Labeled","Open"},
 Enabled->If[fe==1,True,False],ControlType->Slider,
 ContinuousAction->False},
{{hdo,0.5,"h/do"},0.5,5,0.1,ContinuousAction->False,
 Appearance->{"Labeled","Open"},ControlType->Slider,
 Enabled->If[fe==1,True,False]},
Delimiter,
Style["Applied Axial Tensile Force",Bold],
{{So,0,"So"},0,80,5,ContinuousAction->False,
 Appearance->{"Labeled","Open"},ControlType->Slider,
 Enabled->If[bc==1,True,False]},
ControlPlacement->Left,
Initialization:>{lab1="(E1Vo)PI ≈ "; lab2=" yPI ≈ "; h=1;
```

(* Functions representing Eqs. (8.37) and (8.32) *)

```
gr[phi_?NumericQ]:=Module[{aa},
 If[bc==1,yY=aa^2 (aa-1)^2,yY=(aa^2 (aa^2-4 aa+6))];
 Quiet[NIntegrate[yY/(1-yY phi)^2+c3 yY/
   (1-yY phi)^1.24,{aa,0,1},MaxRecursion->3,
   Method->{Automatic,
   "SymbolicProcessing"->False}]]];
dgr[phi_?NumericQ]:=Module[{aa},
 If[bc==1,yY=aa^2 (aa-1)^2,yY=(aa^2 (aa^2-4 aa+6))];
 Quiet[NIntegrate[2.yY^2/(1-yY phi)^3+1.24 c3 yY^2/
   (1-yY phi)^2.24,{aa,0,1},MaxRecursion->3,
   Method->{Automatic,
    "SymbolicProcessing"->False}]]];
```

(* Function that solves Eq. (8.39) *)

```
phiStatic[eV_?NumericQ]:=(If[bc==1,gues=5.,gues=0.1];
 phix/.Quiet[FindRoot[k phix+k1 phix^3-eV^2 gr[phix],
  {phix,gues}]]);
```

(* Function representing Eq. (8.31) *)

```
freq2[eV_]:=(phiStat=phiStatic[eV];
 Sqrt[(k+3 k1 phiStat^2-eV^2 dgr[phiStat])/mo]);},
TrackedSymbols:>{bc,inpl,fe,hb,hdo,So}]
```

8.4.3 *Response of a Cantilever Beam with an In-Span Attachment to an Impulse Force*

We shall use the results of Section 8.4.1 and the program presented therein to examine the response a cantilever beam of length L to an impulse force at $x = L_1$. At $x = L_s$, $0 \le L_s \le L$, a spring with spring constant k_i is attached. The response of this system is given by [2, Section 3.10.3]

$$y(\eta,\tau) = \sum_{n=1}^{\infty} \frac{Y_n(\eta)Y_n(\eta_1)}{\Omega_n^2 N_n} \sin(\Omega_n^2 \tau) \tag{8.40}$$

where $Y_n(\eta)$ is given by Eq. (8.29), $\tau = t/t_o$, where t_o is given by Eq. (8.28), $\eta_1 = L_1/L$, $0 \le \eta_1 \le 1$, is the location of the impulse, Ω_n are solutions to Eq. (8.25), and

$$N_n = \int_0^1 [Y_n(\eta)]^2 d\eta \tag{8.41}$$

An interactive environment shown in Figure 8.9 is created to explore the effects that various attachments and point of application of an impulse force have on the spatial and temporal response of a cantilever beam. Eight mode shapes are used in Eq. (8.40).

The program that creates Figure 8.9 is as follows.

```
Manipulate[Ki=10^kKi;
                    (* Obtain lowest eight natural frequencies *)
  w=ce8[Ωe]; nf={};
  Do[If[w[[n]] w[[n+1]]<0,
   AppendTo[nf,(x/.Quiet[FindRoot[ce8[x],
    {x,Ωe[[n]],Ωe[[n+1]]}]]))];
   If[Length[nf]==nroot,Break[]]],{n,1,Length[w]-1,1}];
                    (* Evaluate Eq. (8.41) *)
  nsubn=Table[NIntegrate[shapeIns8[x,nf[[n]]]^2,{x,0,1}]+
    mR shapeIns8[1,nf[[n]]]^2,{n,1,nroot}];
                    (* Evaluate Eq. (8.40) *)
  ta=Range[0,tend,tend/40.];
  yet=Table[Table[{τ,η,Total[-shapeIns8[η,nf]*
   shapeIns8[eta1,nf] Sin[nf^2 τ]/(nf^2 nsubn)]},
   {τ,ta}],{η,et}];
  ListPlot3D[Flatten[yet,1],Mesh->{40,20},
   ViewPoint->{-1.3,-2,2},AxesLabel->{"τ","η"," y(η,τ)"}],
                    (* Create sliders and radio buttons *)
  Style["Boundary Attachments at Right End",Bold],
  {{kKR,1.5,"Log10(KR)"},-5,12,0.1,Appearance->"Labeled",
   ControlType->Slider,ContinuousAction->False},
```

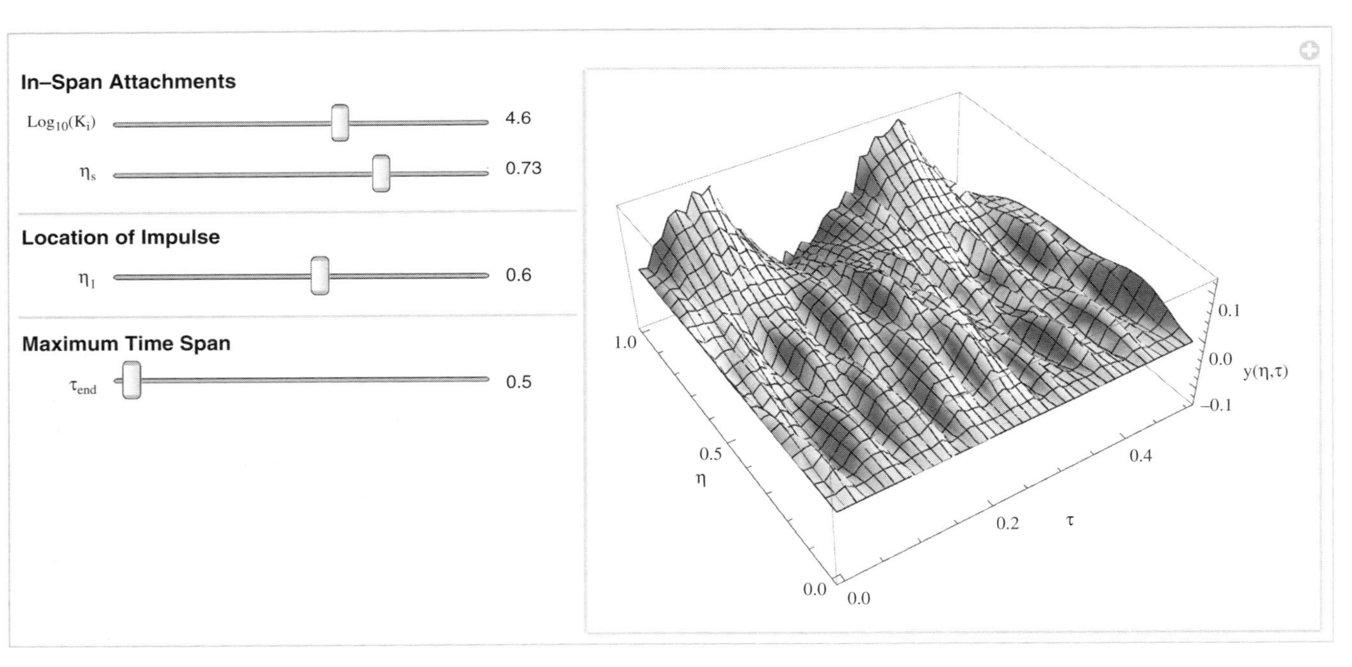

Figure 8.9 Initial configuration of the interactive graph to determine the response of a cantilever beam with an in-span attachment to an impulse force at η_1

```
{{mR,0.3,"m_R"},0,1,0.1,Appearance->"Labeled",
 ControlType->Slider,ContinuousAction->False},
Delimiter,
Style["In-Span Attachments",Bold],
{{inspan,2," "},{1->"No",2->"Yes"},
 ControlType->RadioButtonBar},
Delimiter,
Style["In-span Attachment Properties",Bold],
{{kKi,4.6,"Log_10(K_i)"},-4,10,0.1,Appearance->"Labeled",
 ControlType->Slider,ContinuousAction->False},
{{etas,0.73,"η_s"},0.01,0.99,0.01,Appearance->"Labeled",
 ControlType->Slider,ContinuousAction->False},
Delimiter,
Style["Location of Impulse",Bold],
{{eta1,0.6,"η_1]"},0.1,1.0,0.01,ControlType->Slider,
 Appearance->"Labeled",ContinuousAction->False},
Delimiter,
Style["Maximum Time Span",Bold],
{{tend,0.5,"τ_end"},0.5,3,0.5,ControlType->Slider,
 Appearance->"Labeled",ContinuousAction->False},
```

(* Generate some numerical values *)

```
TrackedSymbols:>{ kKi,etas,tend,eta1 },
ControlPlacement->Left,
Initialization:>(Ω_e=Range[0.01,45.,0.1];
 etta=Range[0,1,0.01]; le=Length[etta]; nroot=8;
 et=Range[0,1,0.05];
```

(* Functions representing Eq. (8.27) *)

```
Q[Ω_,y_]:=(Cos[Ω y]+Cosh[Ω y])/2.;
R[Ω_,y_]:=(Sin[Ω y]+Sinh[Ω y])/2.;
S[Ω_,y_]:=(-Cos[Ω y]+Cosh[Ω y])/2.;
T[Ω_,y_]:=(-Sin[Ω y]+Sinh[Ω y])/2.;
```

(* Functions representing Eq. (8.26) *)

```
d3[Ω_,y_]:=R[Ω,y] T[Ω,y]-Q[Ω,y]^2;
h3[Ω_,x_,y_]:=T[Ω,x] (T[Ω,1] R[Ω,1-y]-Q[Ω,1] Q[Ω,1-y])+
 S[Ω,x] (R[Ω,1] Q[Ω,1-y]-Q[Ω,1] R[Ω,1-y]);
```

(* Functions representing Eq. (8.25) with $m_i = 0$ *)

```
ce8[Ω_]:=If[Ki<500.,d3[Ω,1]-Ki h3[Ω,etas,etas]/Ω^3,
 d3[Ω,1]/Ki-h3[Ω,etas,etas]/Ω^3]]);
```

(* Function representing Eq. (8.29) *)

```
shapeIns8[η_,Ω_]:=-T[Ω,η-etas] d3[Ω,1] UnitStep[η-etas]+
 h3[Ω,η,etas]])]
```

References

[1] B. Balachandran and E. B. Magrab, *Vibrations*, 2nd edn, Cengage Learning, Ontario Canada, 2009.

[2] E. B. Magrab, *Vibrations of Elastic Systems*, Solid Mechanics and Its Applications 184, Springer Science+ Business Media B.V., New York, 2012.

9

Statistics

9.1 Descriptive Statistics

9.1.1 Introduction

Descriptive statistics are used to describe the properties of data; that is, how their values are distributed. One class of descriptors is concerned with the location of where the data tends to aggregate as typically described by the mean, median, or root mean square (rms). A second class of descriptors is concerned with the amount of dispersion of the data as typically described by the variance (or standard deviation) or by its quartiles. A third class of descriptors is the shape of the data as typically indicated by its histogram and its closeness to a known probability distribution. The commands used to determine these statistical descriptors are introduced in this section.

9.1.2 Location Statistics: `Mean[]`, `StandardDeviation[]`, *and* `Quartile[]`

The location statistics that we shall illustrate are the mean, median, and rms values, respectively, which are determined from

```
Mean[dat]
Median[dat]
RootMeanSquare[dat]
```

where **`dat`** is a list of numerical values representing measurements from a random process.
To illustrate these and other statistical functions, we shall consider the following data

```
datex1={111.,103.,251.,169.,213.,140.,224.,205.,166.,202.,
 227.,160.,234.,137.,186.,184.,163.,157.,181.,207.,189.,
 159.,180.,160.,196.,82.,107.,148.,155.,206.,192.,180.,
 205.,121.,199.,173.,177.,169.,93.,182.,127.,126.,187.,
 166.,200.,190.,171.,151.,166.,156.,187.,174.,164.,214.,
```

An Engineer's Guide to Mathematica®, First Edition. Edward B. Magrab.

Companion Website: www.wiley.com/go/magrab

```
139.,141.,178.,177.,243.,176.,186.,173.,182.,164.,162.,
235.,164.,154.,156.,124.,149.,147.,116.,139.,129.,152.,
175.,164.,141.,155.};
```

which will be used here and, subsequently, in several tables and examples. When used in the following examples, we shall employ the shorthand notation

```
datex1={ ...};
```

Thus, to obtain, respectively, the mean, median, and rms values of **datex1**, we have

```
datex1={ ...};  (* Data given above *)
Print["mean = ",Mean[datex1]]
Print["median = ",Median[datex1]]
Print["rms = ",RootMeanSquare[datex1]]
```

which yield

```
mean = 168.663
median = 167.5
rms = 171.969
```

The dispersion measures that we shall illustrate are, respectively, the standard deviation and the quantiles, which are determined from

```
StandardDeviation[dat]
Quantile[dat,q]
Quartiles[dat]
```

where **q** is the qth quantile in **dat**. **Quartiles** gives the $1/4$, $1/2$, and $3/4$ quantiles of **dat**. However, **Quantile** always gives a result that is equal to an element of the **dat**. In this regard, it is different from **Quartiles**. Thus, to obtain the standard deviation and the quartiles using the two different functions, we have

```
datex1={ ...};  (* Data given above *)
Print["std dev = ",StandardDeviation[datex1]]
Print["qtl = ",Quantile[datex1,{1/4,3/4}]]
Print["quart = ",Quartiles[datex1]]
```

which yield

```
std dev = 33.7732
qtl = {149.,187.}
quart = {150.,167.5,187.}
```

Before introducing the shape descriptors, we shall first introduce several continuous distribution functions.

9.1.3 *Continuous Distribution Functions:* **PDF[]** *and* **CDF[]**

If X is a random variable that belongs to probability distribution **dist**, then the probability density function (pdf), cumulative distribution function (cdf), mean value of the distribution (μ), and the variance of the distribution (σ^2), respectively, are obtained from

```
pdf=PDF[dist,x]
cdf=CDF[dist,x]
mu=Mean[dist]
var=Variance[dist]
```

Several distributions commonly used in engineering and their corresponding Mathematica functions are given in Table 9.1. In addition, the formulas for the mean and variance for each distribution are also tabulated. If the arguments of **dist** are symbolic, then the results from these four functions are symbolic expressions. For example, if we use the Weibull distribution, then the following commands, respectively,

```
Print["pdf = ",PDF[WeibullDistribution[α,β],x]]
Print["cdf = ",CDF[WeibullDistribution[α,β],x]]
Print["μ = ",Mean[WeibullDistribution[α,β]]]
Print["σ² = ",Variance[WeibullDistribution[α,β]]]
```

yield, for $x > 0$,

$$\text{pdf} = \frac{1}{\beta}\left(e^{-\left(\frac{x}{\beta}\right)^{\alpha}}\,\alpha\left(\frac{x}{\beta}\right)^{-1+\alpha}\right)$$

$$\text{cdf} = 1 - e^{-\left(\frac{x}{\beta}\right)^{\alpha}}$$

$$\mu = \beta\,\text{Gamma}\left[1+\frac{1}{\alpha}\right]$$

$$\sigma^2 = \beta^2\left(-\text{Gamma}\left[1+\frac{1}{\alpha}\right]^2 + \text{Gamma}\left[1+\frac{2}{\alpha}\right]\right)$$

If α and β are numerical values, then for, say, $\alpha = 2$, $\beta = 3$, the probability that $X = 1.5$ is determined from

```
w=PDF[WeibullDistribution[2,3],1.5]
```

which gives `w = 0.2596`. To obtain a graph of the probability density function of the Weibull distribution for $\alpha = 2$ and $\beta = 3$ and $0 \leq X \leq 6$, we use

```
Plot[PDF[WeibullDistribution[2.,3.],x],{x,0,6},
  AxesLabel->{"x","Probability"}]
```

and obtain Figure 9.1.

Table 9.1 Determination of the symbolic means and variances of selected probability distributions for the random variable X

Distribution	`dist`	`Mean[dist]`*	`Variance[dist]`*
Normal	`NormalDistribution[μ, σ]` μ = mean σ = standard deviation	μ	σ^2
Weibull ($X > 0$)	`WeibullDistribution[α, β, μ]` α = shape parameter β = scale parameter μ = location parameter	$\mu + \beta\,\text{Gamma}\left[1 + \frac{1}{\alpha}\right]$	$\beta^2 \left(-\text{Gamma}\left[1 + \frac{1}{\alpha}\right]^2 + \text{Gamma}\left[1 + \frac{2}{\alpha}\right]\right)$
Rayleigh ($X > 0$)	`RayleighDistribution[λ]` λ = scale parameter	$\lambda \sqrt{\frac{\pi}{2}}$	$\lambda^2 \left(2 - \frac{\pi}{2}\right)$
Exponential ($X > 0$)	`ExponentialDistribution[λ]` λ = parameter	$\frac{1}{\lambda}$	$\frac{1}{\lambda^2}$
Lognormal ($X > 0$)	`LogNormalDistribution[μ, σ]` μ = mean σ = standard deviation	$e^{\mu+\sigma^2/2}$	$e^{2\mu+\sigma^2}\left(-1 + e^{\sigma^2}\right)$
Chi square (χ^2, $X > 0$)	`ChiSquareDistribution[ν]` ν = degrees of freedom	ν	2ν
Student t	`StudentTDistribution[ν]` ν = degrees of freedom	$0 \quad \nu > 1$	$\frac{\nu}{\nu - 2} \quad \nu > 2$
f ratio ($X > 0$)	`FRatioDistribution[n,m]` `n` = degrees of freedom of numerator `m` = degrees of freedom of denominator	$\frac{m}{m - 2} \quad m > 2$	$\frac{2m^2 (m + n - 2)}{(m - 4)(m - 2)^2 n} \quad m > 4$

* The formulas for the mean and variance were obtained by substituting the symbolic form of the statement in the second column into **`Mean`** and **`Variance`** as shown in the text.

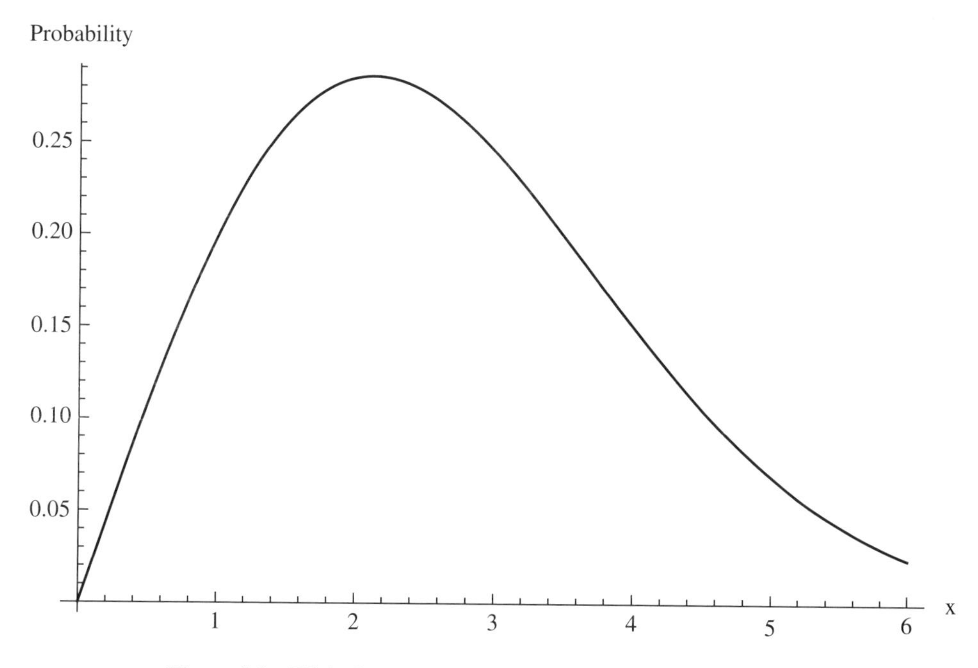

Figure 9.1 Wiebull probability density function for $\alpha = 2$ and $\beta = 3$

9.1.4 *Histograms and Probability Plots*: `Histogram[]` *and* `ProbabilityScalePlot[]`

The shape visualization functions that we shall consider are

```
Histogram[dat,binspec,hspec]
ProbabilityScalePlot[dat,dist]
```

where **`dat`** is a list of numerical values representing measurements from a process and **`dist`** is the name of a distribution in quotation marks. The other quantities in **`Histogram`** are defined in Table 9.2. **`Histogram`** plots a histogram with its bins determined automatically and with any of several types of amplitudes for the y-axis. **`ProbabilityScalePlot`** plots **`dat`** on a probability scale that is specified by the probability distribution **`dist`**. If almost all the data values of **`dat`** lie on or are very close to the reference line that indicates that distribution's variation with respect to X, then the data can be represented by that distribution. Some of the choices for the arguments of **`Histogram`** and **`ProbabilityScalePlot`** are summarized in Table 9.2.

Examples of the effects that these arguments have on the vertical axes of the figures produced by **`Histogram`** are shown in Table 9.3. It is mentioned that when one passes the cursor over the bars in the histogram plot a tooltip appears that provides the height of that bar: either the number of samples or the fraction of the total number of samples depending on the plot options selected.

Table 9.2 Histograms and probability graphs

Plot type	Mathematica function

Histogram*	`Histogram[dat,binspec,hspec,LabelingFunction->pos]`
	`dat` = list of values
	`binspec` = `Automatic` – automatically selects number of bins (default)
	= `n` – specify number of bins
	`hspec` = `"Count"` – plots number of values in each bin (default)
	= `"CumulativeCount"` – plots the number of values in bin and preceding bins
	= `"Probability"` – plots fraction of number of values in each bin with respect to total number of values
	= `"PDF"` – plots probability density function : same as `"Probability"`
	= `"CDF"` – plots cumulative distribution function
	`pos` = position of where the labels are to appear: `Top`, `Bottom`, `Left`, `Right`, and `Above`

Probability	`ProbabilityScalePlot[dat,dist]` `dat` = list of values `dist` = `"Normal"` (default, if omitted) – plots `dat` assuming a normal distribution `dist` = `"Weibull"` – plots `dat` assuming a Weibull distribution `dist` = `"Exponential"` – plots `dat` assuming an exponential distribution `dist` = `"LogNormal"` – plots `dat` assuming a lognormal distribution `dist` = `"Rayleigh"` – plots `dat` assuming a Rayleigh distribution

* The differences in using `"Count"` and `"PDF"` and using `"CumulativeCount"` and `"CDF"` are the units of the *y*-axis; for an equal number of bins, the shape of the histograms in each of these pairs is the same. See Table 9.3.

9.1.5 *Whisker Plot*: `BoxWhiskerChart[]`

Another way to visualize data in the aggregate is with a whisker plot. This type of display is useful when comparing two or more data sets. The command for this function is

```
BoxWhiskerChart[{dat1,dat2, ...},specw,graph]
```

where **datN** are the lists of the data sets, **specw** is a specification for the specific form of the whisker plot, and **graph** are graphics-enhancement instructions. See the *Documentation Center* under the search entry **BoxWhiskerChart** for the details. We shall illustrate the use of **BoxWhiskerChart** for two data sets obtained by splitting **datex1** in half. In addition, we shall use a notched whisker plot with its outliers appearing as blue dots, label each of the data sets, and give different colors to the boxes. Then,

```
datex1={...}; (* See Section 9.1.2 *)
BoxWhiskerChart[{datex1[[1;;40]],datex1[[41;;80]]},
  {"Notched",{"Outliers",Blue}},ChartStyle->{Red,Magenta},
  ChartLabels->{"Set 1","Set 2"}]
```

produces Figure 9.2. It is mentioned that when one passes the cursor over the figure a tooltip appears that provides the details of each whisker plot: maximum value, 75th quartile value,

Table 9.3 Effects of several options of **`Histogram`***

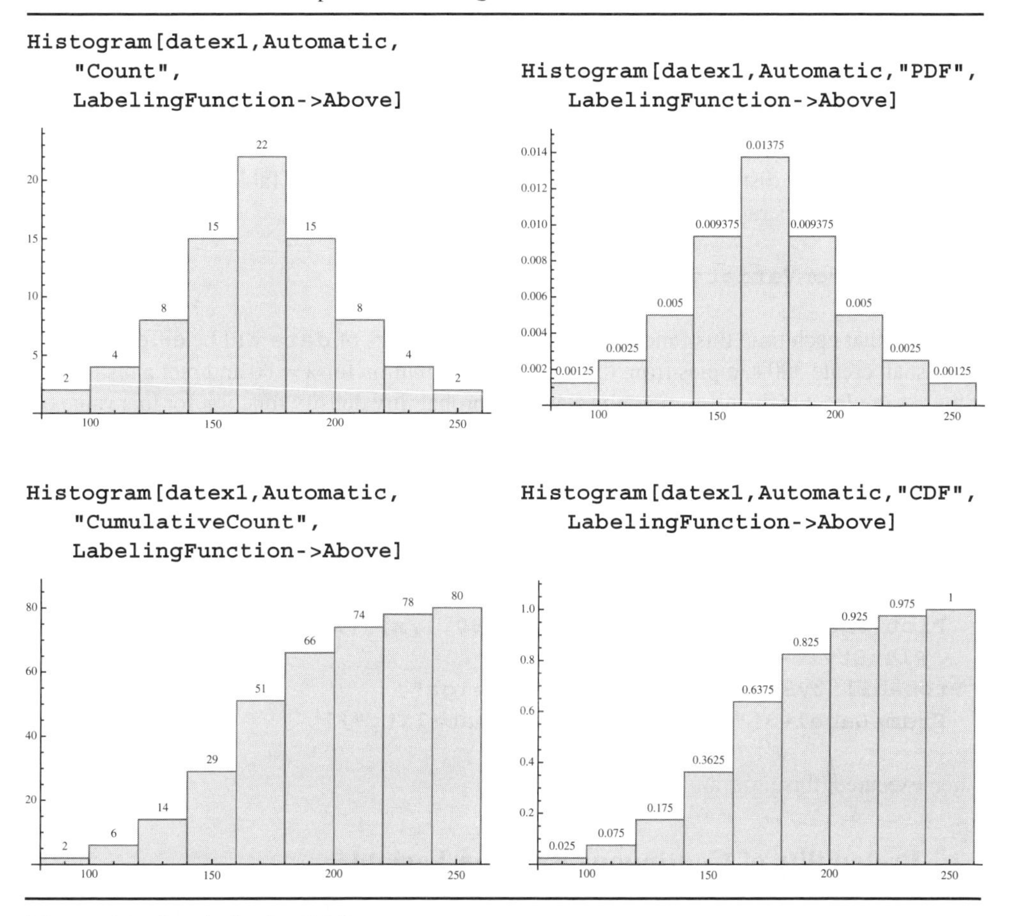

* **`datex1`** is given in Section 9.1.2.

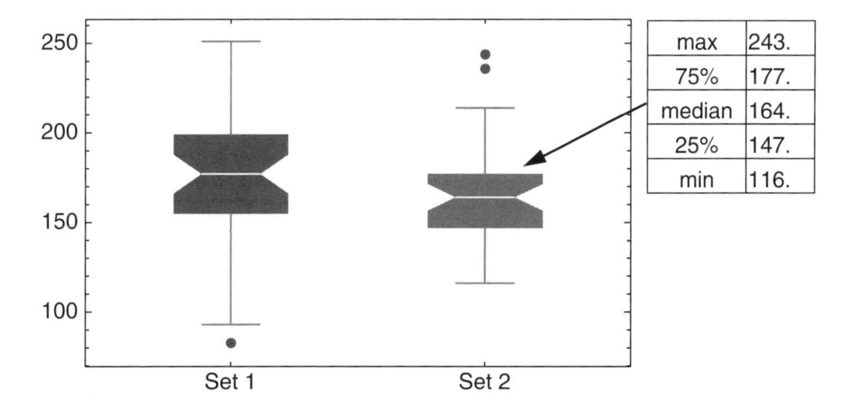

Figure 9.2 Enhanced notched whisker plot with outliers displayed. When the cursor is placed over each whisker plot, a Tooltip is displayed as shown in the figure for Set 2

median, 25th quartile value, and minimum value. The maximum and minimum values include any outliers.

9.1.6 *Creating Data with Specified Distributions*: **RandomVariate[]**

One can create a list of n data values (samples) that has a distribution given by **dist**, where **dist** is given by the distributions appearing in the second column[1] of Table 9.1. The command that performs this operation is

```
data=RandomVariate[dist,n]
```

It is noted that each time this function is executed, the values of **data** will be different.

We shall create 300 samples from the Rayleigh distribution for $\lambda = 80$ and plot a histogram of these samples. On the histogram, the Rayleigh probability density function for this value of λ will be superimposed. In addition, an independent probability plot of the generated data set will be created. Then,

```
data=RandomVariate[RayleighDistribution[80.],300];
Show[Histogram[data,Automatic,"PDF",
 AxesLabel-> {"x","Probability"}],
 Plot[PDF[RayleighDistribution[80.],x],{x,0,300},
  PlotStyle->Black]]
ProbabilityScalePlot[data,"Rayleigh",
 FrameLabel->{"x","Rayleigh Probability"}]
```

When executed, this program yields Figure 9.3.

9.2 Probability of Continuous Random Variables

9.2.1 *Probability for Different Distributions*: **NProbability[]**

The probability $P(X)$ that a continuous random variable X lies in the range $x_1 \leq X \leq x_2$, where x_1 and x_2 are from the set of all possible values of X, is defined as

$$P(x_1 \leq X \leq x_2) = \int_{x_1}^{x_2} f(u)du$$

where $f(x) \geq 0$ for all x, and

$$\int_{-\infty}^{\infty} f(u)du = 1$$

[1] There are many other distribution functions that are available: see the *Documentation Center* under the search entry *guide/ParametricStatisticalDistributions*.

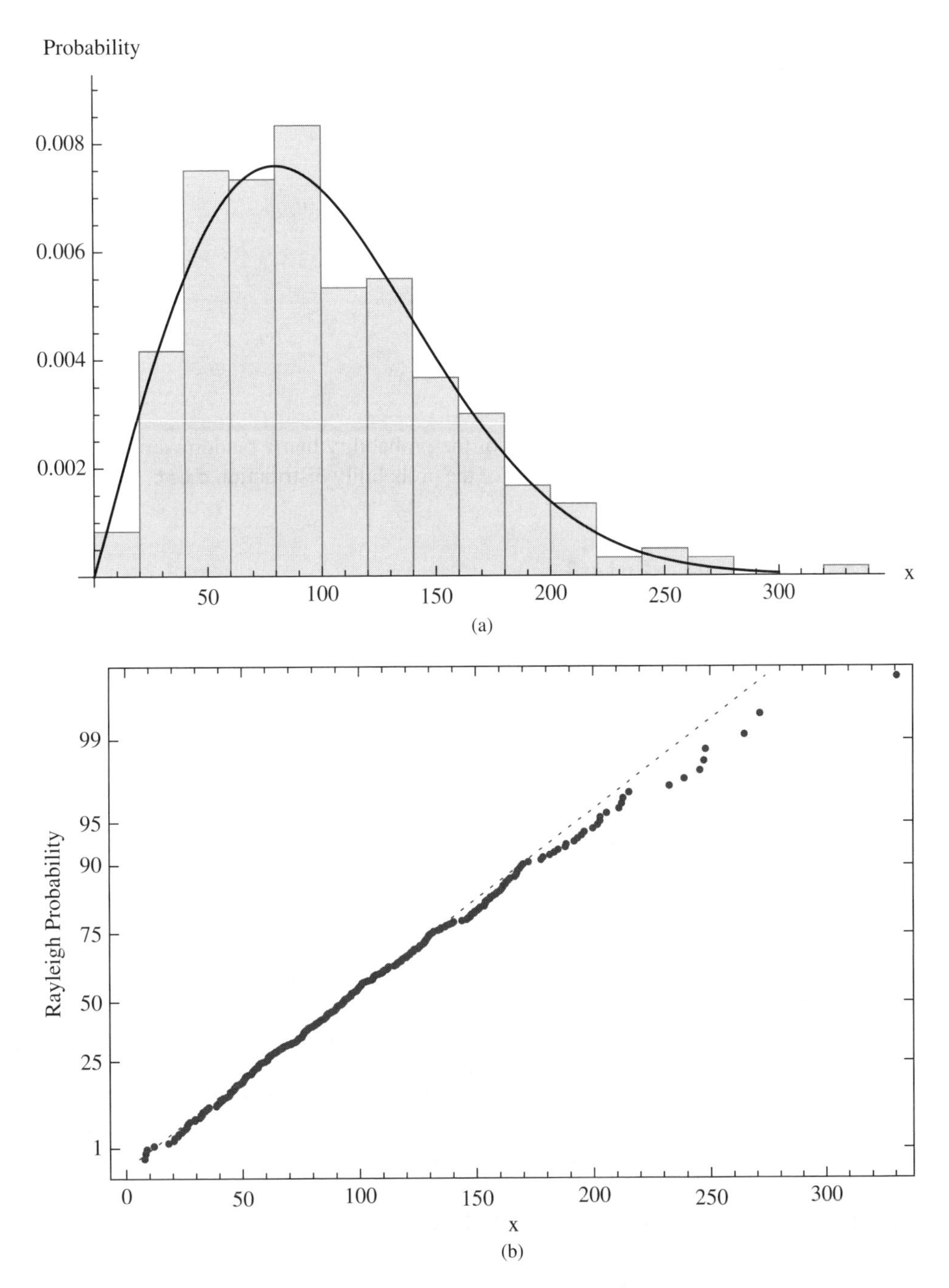

Figure 9.3 (a) Histogram of 300 data samples generated from the Rayleigh distribution compared to the Rayleigh probability density function (solid line); (b) Rayleigh probability plot of the 300 data samples

The quantity $f(x)$ is called the probability density function (pdf) for a continuous random variable.

The cumulative distribution function (cdf) $F(x)$ is

$$F(x) = P(X \le x) = \int_{-\infty}^{x} f(u)du = 1 - \int_{x}^{\infty} f(u)du$$

and, therefore,

$$P(X \ge x) = 1 - F(x)$$

The determination of the numerical value of the probability that a random variable X lies within a specified range when X is a member of the probability distribution **dist**, is obtained from

```
NProbability[rng,x≈dist]
```

where **rng** is the range for which the probability is determined, **dist** is the distribution, and ≈ = | *esc* |dist| *esc* |; that is, ≈ is a representation for the sequential typing of the *esc* key followed by the four letters *dist* that is followed again by the typing of the *esc* key. The mathematical symbol ≈ will not work. Several of the distributions that can be used in this command are listed in the second column of Table 9.1.

NProbability can be used as a table of probabilities for various distributions. For example, if X is a random variable from the Weibull distribution for $\alpha = 1.5$ and $\beta = 3.7$, then the probability that $X \ge 2$ is obtained from

```
p=NProbability[x>=2.,x≈WeibullDistribution[1.5,3.7]]
```

which yields `0.672056`. Thus, there is a 67% likelihood that the next sample will have a value greater than or equal to 2.

The probability from a set of data can be determined when the distribution is known (or can be assumed). For example, if it is assumed that **datex1** of Section 9.1.2 can be approximated by a normal distribution, then we can use **NProbability** to estimate the probability that the next sample to be added to **datex1** lies, say, between $\mu - 1.5\sigma \le X \le \mu + 1.5\sigma$, where μ is the mean and σ is the standard deviation. This probability estimate is obtained with

```
datex1={...}; (* See Section 9.1.2 *)
μ=Mean[datex1];
σ=StandardDeviation[datex1];
p=Probability[σ-1.5 σ<x<μ+1.5 σ, x≈NormalDistribution[μ,σ]]
```

The execution of this program gives `0.866386`; that is, there is an 87% chance that the next sample will lie in the region indicated.

9.2.2 *Inverse Cumulative Distribution Function:* InverseCDF[]

Let us rewrite **NProbability** as

```
p=NProbability[ro(x,a),x≈dist]
```

where **dist** is the distribution and **ro(x,a)** is a relational operator that defines a region of **x**, the random variable, and the limit of that region is determined by **a**. One is often interested in the inverse of this function; that is, for a given distribution **dist** what is the value of **a** for a given value of **p**. The function that answers this question is

```
a=InverseCDF[dist,p]
```

The value of **a** is sometimes called the critical value for these stated conditions. The inverse function essentially replaces tables that typically appear in many statistics books.

To illustrate the use of **InverseCDF**, we shall consider the Student t distribution and determine the value of **a** for $\nu = 15$ and a probability of **p** $= 0.95$. Then

```
a=InverseCDF[StudentTDistribution[15],0.95]
```

which gives `a = 1.75305`. To confirm that this is correct, we use

```
p=NProbability[x<=1.75305,x≈StudentTDistribution[15]]
```

and find that `p = 0.95`.

9.2.3 *Distribution Parameter Estimation:* EstimatedDistribution[] *and* FindDistributionParameters[]

For a given set of data and an assumption about the distribution of the data, it is possible to determine the parameters that govern that distribution. The parameters are determined in either of two ways. The first way, which is suitable for plotting, is

```
fcn=EstimatedDistribution[dat,dist[p1,p2, ...]]
```

whose output is the function

```
dist[n1,n2, ...]
```

In these expressions, **dat** is a list of data, **pN** are the parameters of the assumed distribution **dist**, and **nN** are the numerical values found by **EstimatedDistribution** for distribution **dist**. The values of **nN** are not accessible.

The second way to find the parameters is to use

```
para=FindDistributionParameters[dat,dist[p1,p2, ...]]
```

where the output is of the form

```
{p1->n1,p2->n2, ...}
```

In this case, the individual parameters are accessed by using **p1/.para**, etc.

To illustrate the use of these functions, we shall create 400 samples of data assuming a Weibull distribution and compare the parameters used to obtain these data with those obtained from **FindDistributionParameters**. Then

```
data=RandomVariate[WeibullDistribution[2.1,4.9,60],400];
fcn=EstimatedDistribution[data,WeibullDistribution[α, β, μ]]
para=FindDistributionParameters[data,
  WeibullDistribution[α, β, μ]]]
```

yield, respectively,

```
WeibullDistribution[1.74857,4.26186,60.4793]
```

and

```
{α->1.74857,β->4.26186,μ->60.4793}
```

In the first output, the values of α, β, and μ are inferred from their location in **WeibullDistribution**. In the second output, the values of α, β, and μ are accessible as follows: $\alpha = \alpha$**/.para** = **1.74857**, $\beta = \beta$**/.para** = **4.26186**, and $\mu = \mu$**/.para** = **60.4793**.

Example 9.1

Histograms

We shall use **FindDistributionParameters, EstimatedDistribution**, and **RandomVariate** to create a figure that plots a histogram of some computer-generated data for the Weibull distribution, the probability density function using the estimated parameters determined from these data, and the probability density function that created the data. In addition, we shall place in the figure's label the values of the estimated parameters, the mean and standard deviation of the data, and the mean and standard deviation as computed from the third and fourth columns of Table 9.1 corresponding to the Weibull distribution. Then, the program is as follows.

```
(* Compute various statistical parameters *)
data=RandomVariate[WeibullDistribution[2.1,4.9,60],400];
meen=Mean[data];
stddev=StandardDeviation[data];
para=FindDistributionParameters[data,
  WeibullDistribution[α,β,μ]]
(* Create plot labels *)
α=α/.para;  β=β/.para;  μ=μ/.para;
fcn=EstimatedDistribution[data,WeibullDistribution[α,β,μ]]
```

```
al=Style[Row[{"α = ",NumberForm[α,4]}],12];
bet=Style[Row[{"β = ",NumberForm[β,4]}],12];
mu=Style[Row[{"μ = ",NumberForm[μ, 4]}],12];
avgd=Style[Row[{"Mean (data) = ",NumberForm[meen,4]}],12];
stdd=Style[Row[{"Std dev (data) = ",
 NumberForm[stddev,4]}],12];
avgc=Style[Row[{"Mean (est dist) = ",
 NumberForm[μ+β Gamma[1+1/α],4]}],12];
stdc=Style[Row[{"Std dev (est dist) = ",
 NumberForm[β Sqrt[-Gamma[1+1/α]^2+Gamma[1+2/α]],4]}],12];
                          (* Plot results *)
Show[Histogram[data,Automatic,"PDF",
  PlotLabel->Column[{Row[{al,"  ",bet,"  ",mu}],
   Row[{avgd,"  ",stdd}],Row[{avgc,"  ",stdc}]}],
   AxesLabel->{"x","Probability"}],
  Plot[PDF[WeibullDistribution[2.1,4.9,60],x],{x,60,75}],
  Plot[PDF[fcn,x],{x,60,75},
   PlotStyle->{Black,Dashing[Large]}]]
```

The execution of this program produces Figure 9.4.

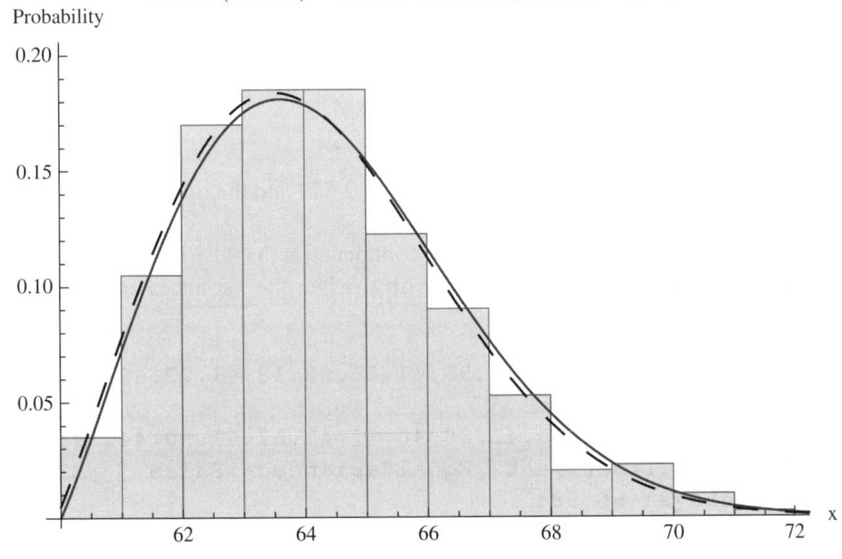

Figure 9.4 Computed parameters for a Weibull distribution and the comparison of a histogram of the data with the probability density functions for the estimated parameters (dashed line) and the original data (solid line)

9.2.4 *Confidence Intervals:* ···**CI[]**

Let θ be a numerical value of a statistic (e.g., the mean, variance, difference in means) of a collection of n samples. We are interested in determining the values of l and u such that the following is true

$$P(l \leq \theta \leq u) = 1 - \alpha$$

where $0 < \alpha < 1$. This means that we will have a probability of $1-\alpha$ of selecting a collection of n samples that will produce an interval that contains the true value of θ. The interval

$$l \leq \theta \leq u$$

is called the $100(1-\alpha)\%$ two-sided confidence interval for θ.

The confidence limits depend on the distribution of the samples and on whether or not the standard deviation of the population is known. Several commonly used relationships to determine these confidence limits are summarized in Table 9.4. The quantities $\bar{x}$ and s^2, respectively, are the mean and variance for the sample values that are determined from **Mean** and **Variance**.

To illustrate several of these commands, we shall first determine the confidence interval at the 90% level of the mean of **datex1** given in Section 9.1.2 when the variance is unknown. Then, from Case 1 of Table 9.4,

```
Needs["HypothesisTesting`"]
datex1={...}; (* See Section 9.1.2 *)
StudentTCI[Mean[datex1],
  StandardDeviation[datex1]/Sqrt[Length[datex1]],
  Length[datex1],ConfidenceLevel->0.90]
```

yields

```
{162.379,174.946}
```

Thus, the lower confidence limit at the 90% level is 162.379 and the upper confidence limit is 174.946.

For a second example, we shall determine the confidence at the 98% level for the difference in the means of two sets of data, **set1** and **set2**, when the variances are unknown and unequal. Then, from Case 3 of Table 9.4,

```
set1={41.60,41.48,42.34,41.95,41.86,42.18,41.72,42.26,
  41.81,42.04};
set2={39.72,42.59,41.88,42.00,40.22,41.07,41.90,44.29};
MeanDifferenceCI[set1,set2,EqualVariances->False,
  ConfidenceLevel->0.98]
```

yields

```
{-1.29133,1.72183}
```

Table 9.4 Determination of the confidence intervals of several statistical measures

Case	Assumptions	Statistic	Command*
1	Mean with σ^2 unknown (Student *t* test)	$\bar{x}$	`StudentTCI[mu,sig/Sqrt[n],n,ConfidenceLevel->cl]` `mu` = mean of sample `sig` = standard deviation of samples `n` = number of samples `cl` = confidence level; if omitted, `cl` = 0.95
2	Difference in means with $\sigma_1^2 = \sigma_2^2$ unknown (Student *t* test)	$\bar{x}_1 - \bar{x}_2$	`MeanDifferenceCI[set1,set2,EqualVariances->True, ConfidenceLevel->cl]` `set1` = list of values for data set 1 `set2` = list of values for data set 2 `cl` = confidence level; if omitted, `cl` = 0.95
3	Difference in means with $\sigma_1^2 \neq \sigma_2^2$ unknown (Student *t* test)	$\bar{x}_1 - \bar{x}_2$	`MeanDifferenceCI[set1,set2,EqualVariances->False, ConfidenceLevel->cl]` `set1` = list of values for data set 1 `set2` = list of values for data set 2 `cl` = confidence level; if omitted, `cl` = 0.95
4	Variance (χ^2 test)	s^2	`VarianceCI[dat,ConfidenceLevel->cl]` `dat` = list of values `cl` = confidence level; if omitted, `cl` = 0.95
5	Ratio of variances (*f* ratio test)	$\frac{s_1^2}{s_2^2}$	`VarianceRatioCI[set1,set2,ConfidenceLevel->cl]` `set1` = list of values for data set 1 `set2` = list of values for data set 2 `cl` = confidence level; if omitted, `cl` = 0.95

* To run these commands, the hypothesis testing package using `Needs["HypothesisTesting`"]` is required.

Table 9.5 Hypothesis testing of means and variances

Case	H_0	H_1	Mathematica functions*
1	$\mu = \mu_0$	$\mu \neq \mu_0$	`LocationTest[dat,mu0]` `dat` = list of values `mu0` = comparison mean to which the mean of `dat` is compared
2	$\mu_1 = \mu_2$	$\mu_1 \neq \mu_2$	`LocationTest[{set1,set2}]` `set1` = list of values for data set 1 `set2` = list of values for data set 2
3	$\sigma^2 = \sigma_0^2$	$\sigma^2 \neq \sigma_0^2$	`VarianceTest[dat,var0]` `dat` = list of values `var0` = comparison variance to which the variance of `dat` is compared
4	$\sigma_1^2 = \sigma_2^2$	$\sigma_1^2 \neq \sigma_2^2$	`VarianceTest[{set1,set2}]` `set1` = list of values for data set 1 `set2` = list of values for data set 2

* To run these functions, the hypothesis testing package using `Needs["HypothesisTesting`"]` must be executed.

9.2.5 *Hypothesis Testing:* `LocationTest[]` *and* `VarianceTest[]`

Let θ be a numerical value of a statistic (e.g., the mean, variance, difference in means) of a collection of n samples. Suppose that we are interested in determining whether this parameter is equal to θ_o. In the hypothesis-testing procedure, we postulate a hypothesis, called the null hypothesis and denoted H_0, and then based on the parameter θ form an appropriate test statistic q_0. For testing mean values, q_0 would be a t statistic, for variances it would be χ^2, and for the ratio of variances it would be the f ratio statistic. We then compare the test statistic to a value that corresponds to the magnitude of the test statistic that one can expect to occur naturally, q. Based on the respective magnitudes of q_0 and q, the null hypothesis is either accepted or rejected. If the null hypothesis is rejected, the alternative hypothesis denoted H_1 is accepted. This acceptance or rejection of H_0 is based on a quantity called the p-value, which is the smallest level of significance that would lead to the rejection of the null hypothesis. The percentage confidence level is 100(1 − p-value)%. Therefore, the smaller the p-value, the less plausible is the null hypothesis and the greater confidence we have in H_1.

In Table 9.5, we have summarized several hypothesis-testing commands that are useful in engineering. The output of these commands is the p-value.

We shall illustrate these commands with the following examples. Let us determine if there is a statistically significant difference at the 95% confidence level between the mean of `datex1` given in Section 9.1.2 and a mean value of 174; that is, $\mu_0 = 174$. This is determined from Case 1 of Table 9.5. Thus,

```
datex1={...}; (* See Section 9.1.2 *)
p=LocationTest[datex1,174]
```

which yields that `p = 0.161423`. Since we can only be 100(1 − 0.161) = 83.6% confident, we do not reject H_0. Recall from Section 9.1.2 that the mean value of these data is 168.7.

However, if $\mu_0 = 181$, then we would find that **p = 0.00161** and we are now $100(1 - 0.00161) = 99.84\%$ confident that the mean of **datex1** is different from a mean of 181.

For a second example, we determine whether the variances of data sets **set1** and **set2** of Section 9.2.4 are different. In this case, we use Case 4 of Table 9.5. Then,

```
set1={...};  set2={...} (* See Section 9.2.4 *)
p=VarianceTest[{set1,set2}]
```

gives that **p = 0.0000653785** and we reject H_0 with 99.993% confidence level that the variances are equal.

9.3 Regression Analysis: LinearModelFit[]

9.3.1 Simple Linear Regression[2]

Regression analysis is a statistical technique for modeling and investigating the relationship between two or more variables. A simple linear regression model has only one independent variable. If the input to a process is x and its response y, then a linear model is

$$y = y(x) = \beta_0 + \beta_1 x + \beta_2 x^2 + \cdots \tag{9.1}$$

where any regression model that is linear in the parameters β_j is a linear regression model, regardless of the shape of the curve y that it generates. If there are n values of the independent variable x_i and n corresponding measured responses y_i, $i = 1, 2, \ldots, n$, then estimates of y are obtained from

$$\hat{y} = \hat{y}(x) = \hat{\beta}_0 + \hat{\beta}_1 x + \hat{\beta}_2 x^2 + \cdots \quad x_{min} \le x \le x_{max}$$

where x_{min} is the minimum value of x_i, x_{max} is the maximum value of x_i, and $\hat{\beta}_k$ are estimates of β_k.

The function to determine $\hat{\beta}_k$ is

```
modl=LinearModelFit[coord,{x,x^2,...},x,
  ConfidenceLevel->cl]
```

where **coord = {{x1,y1},{x2,y2},...}**, **x** is the independent variable, **{x,x^2,...}** matches the form of the independent variables in Eq. (9.1), and **cl** is the confidence level, which, if omitted is equal to 0.95. The output of **LinearModelFit** is the symbolic object of the form

```
FittedModel[b0+b1 x+b2 x^2+...]
```

---

[2] For a complete listing and explanation of the very large number of options associated with the commands introduced in this section, enter *guide/StatisticalModelAnalysis* in the *Documentation Center* search area.

where **bK** will be numerical values for the estimates $\hat{\beta}_k$. The access to numerous quantities that resulted from the statistical procedure used by **LinearModelFit** is obtained from

```
modl["Option"]
```

where the choices for **Option** can be found in the *Documentation Center* under the search entry *LinearModelFit.*

Several of the more commonly used choices for **Option** are summarized in Table 9.6 for the case where

$$y = \beta_0 + \beta_1 x$$

and the data set

```
xx={2.38,2.44,2.70,2.98,3.32,3.12,2.14,2.86,3.50,3.20,2.78,
    2.70,2.36,2.42,2.62,2.80,2.92,3.04,3.26,2.30};
yy={51.11,50.63,51.82,52.97,54.47,53.33,49.90,51.99,55.81,
    52.93,52.87,52.36,51.38,50.87,51.02,51.29,52.73,
    52.81,53.59,49.77};
pts=Table[{xx[[n]],yy[[n]]},{n,1,Length[xx]}];
```

We shall create a figure that plots **pts**, the fitted line, and the 90% confidence bands. Assuming that the above three definitions have been executed, the rest of the program is

```
modl=LinearModelFit[pts,{x},x,ConfidenceLevel->0.90];
bands=modl["MeanPredictionBands"];
xm=Min[xx];  xmx=Max[xx];
Show[Plot[modl[x],{x,xm,xmx},PlotStyle->Thick,
   AxesLabel->{"x","y(x)"}],
 ListPlot[pts,PlotMarkers->Automatic],
 Plot[{bands[[1]],bands[[2]]}/.x->f,{f,xm,xmx},
 PlotStyle->Dashing[Medium]]]
```

The results are shown in Figure 9.5.

To determine if the residuals between the fitted curve and the original data are normally distributed, we plot them on a probability plot. If the residuals are close to the line representing a normal distribution, we can say that the fit is good. Thus,

```
ProbabilityScalePlot[modl["FitResiduals"],"Normal",
 PlotMarkers->Automatic,
 FrameLabel->{"Residual","Normal Probability"}]
```

produces Figure 9.6, which indicates that the fit is good.

Table 9.6 Several output quantities from `LinearModelFit` for simple linear regression

Execution of*

`modl=LinearModelFit[pts,{x},x,ConfidenceLevel->0.90]`

yields `FittedModel[41.7498+3.73663 x]`

Various characteristics of model are accessed with `modl["Option"]` as indicated in the columns below

Option description	"Option"	opt=modl["Option"]
R^2 – Coefficient of determination	"RSquared"	0.877424
Pure function	"Function"	`41.7498+3.73663 #1 &` Example: `opt[2.8]` yields `52.2124`
Confidence interval at 0.90 (in this case) on fitted model as a function of x	"MeanPredictionBands"	$\{41.75 + 3.74\,x - 1.73\sqrt{0.86 - 0.605\,x + 0.108\,x^2},$ $41.75 + 3.74\,x + 1.73\sqrt{0.86 - 0.605\,x + 0.108\,x^2}\}$ Example: `opt/.x->2.8` yields `{52.0017,52.4231}`
Confidence intervals at 0.90 (in this case) on fit parameters presented in tabular form	"ParameterConfidenceIntervalTable"	(see table below)
Residuals	"FitResiduals"	List of $\hat{y} - y$ evaluated at each value of x in `pts`
Fit parameters	"BestFitParameters"	`{41.7498,3.73663}` Example: β_0 = `opt[[1]] = 41.7498` and β_1 = `opt[[2]] = 3.73663`

Output of "ParameterConfidenceIntervalTable":

	Estimate	Standard Error	Confidence Interval
1	41.7498	0.9271	{40.1422, 43.3574}
x	3.73662	0.3292	{3.1658, 4.3075}

* Form of `pts` is given in the text.

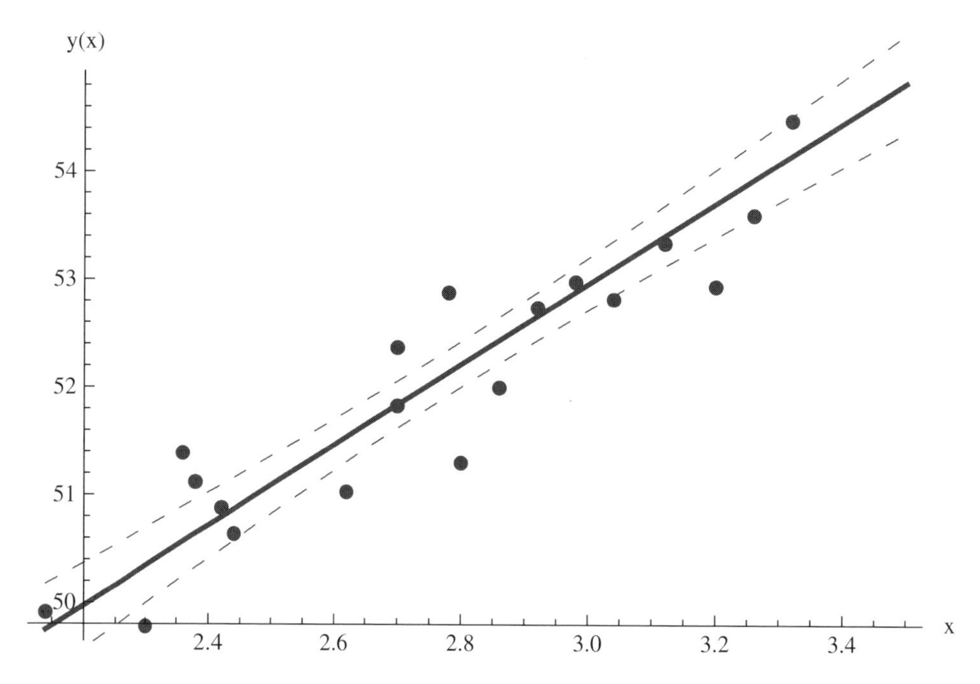

Figure 9.5 Straight line fit to data values shown and the 90% confidence bands of the fitted line

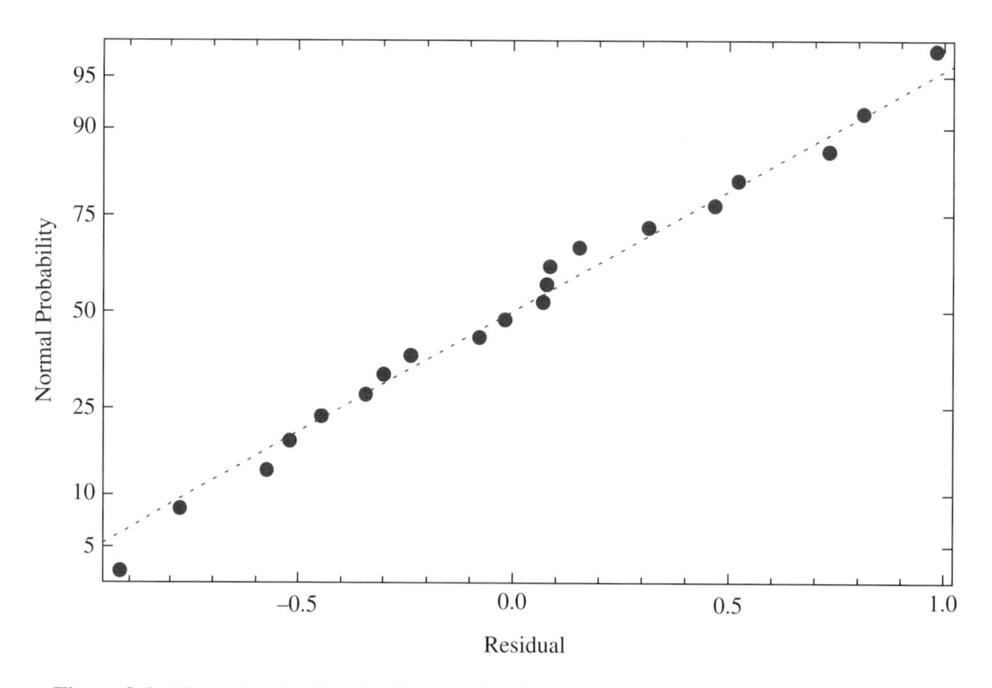

Figure 9.6 Normal probability distribution plot of the residuals for the fit shown in Figure 9.5

Table 9.7 Data for the multiple linear regression example

k	y_k	$x_{1,k}$	$x_{2,k}$	$x_{3,k}$	k	y_k	$x_{1,k}$	$x_{2,k}$	$x_{3,k}$
1	0.22200	7.3	0.0	0.0	14	0.10100	7.3	2.5	6.8
2	0.39500	8.7	0.0	0.3	15	0.23200	8.5	2.0	6.6
3	0.42200	8.8	0.7	1.0	16	0.30600	9.5	2.5	5.0
4	0.43700	8.1	4.0	0.2	17	0.09230	7.4	2.8	7.8
5	0.42800	9.0	0.5	1.0	18	0.11600	7.8	2.8	7.7
6	0.46700	8.7	1.5	2.8	19	0.07640	7.7	3.0	8.0
7	0.44400	9.3	2.1	1.0	20	0.43900	10.3	1.7	4.2
8	0.37800	7.6	5.1	3.4	21	0.09440	7.8	3.3	8.5
9	0.49400	10.0	0.0	0.3	22	0.11700	7.1	3.9	6.6
10	0.45600	8.4	3.7	4.1	23	0.07260	7.7	4.3	9.5
11	0.45200	9.3	3.6	2.0	24	0.04120	7.4	6.0	10.9
12	0.11200	7.7	2.8	7.1	25	0.25100	7.3	2.0	5.2
13	0.43200	9.8	4.2	2.0	26	0.00002	7.6	7.8	20.7

9.3.2 Multiple Linear Regression

Multiple linear regression is similar to that for simple linear regression except that the form for the input to **LinearModelFit** is slightly different. Consider the output of a process $y = y(x_1, x_2, \ldots, x_n)$, where x_k are the independent input variables to that process. Then, for a specific set of values $x_{k,m}$, $m = 1, 2, \ldots$ the output is y_m. A general linear regression model for this system is of the form

$$y = \beta_0 + \beta_1 f_1 + \beta_2 f_2 + \cdots + \beta_l f_l$$

where $f_l = x_k^p x_j^s$, $p, s = 0, 1, 2, \ldots$, are independently chosen integers, but do not include the case $p = s = 0$.

Then, for multiple regression analysis, an estimate of y and β_k, denoted, respectively, $\hat{y}$ and $\hat{\beta}_k$, are obtained from

```
modlm=LinearModelFit[coord,{f1,f2, ...},{x1,x2, ...},
  ConfidenceLevel->cl]
```

where `coord={{x11,x21, ... ,y1},{x12,x22, ... ,y2}, ...}`.

To illustrate multiple regression analysis, consider the data shown in Table 9.7. We shall fit these data with the following model

$$y = \beta_0 + \beta_1 x_1 + \beta_2 x_2 + \beta_3 x_3 + \beta_4 x_1 x_2 + \beta_5 x_1 x_3 + \beta_6 x_2 x_3 + \beta_7 x_1^2 + \beta_8 x_2^2 + \beta_9 x_3^2$$

These data are converted to the form shown for **coord** above as follows

```
y={0.22200,0.39500,0.42200,0.43700,0.42800,0.46700,0.44400,
  0.37800,0.49400,0.45600,0.45200,0.11200,0.43200,0.10100,
  0.23200,0.30600,0.09230,0.11600,0.07640,0.43900,0.09440,
  0.11700,0.07260,0.04120,0.25100,0.00002};
x11={7.3,8.7,8.8,8.1,9.0,8.7,9.3,7.6,10.0,8.4,9.3,7.7,9.8,
  7.3,8.5,9.5,7.4,7.8,7.7,10.3,7.8,7.1,7.7,7.4,7.3,7.6};
x21={0.0,0.0,0.7,4.0,0.5,1.5,2.1,5.1,0.0,3.7,3.6,2.8,4.2,
  2.5,2.0,2.5,2.8,2.8,3.0,1.7,3.3,3.9,4.3,6.0,2.0,7.8};
x31={0.0,0.3,1.0,0.2,1.0,2.8,1.0,3.4,0.3,4.1,2.0,7.1,2.0,
  6.8,6.6,5.0,7.8,7.7,8.0,4.2,8.5,6.6,9.5,10.9,5.2,20.7};
coord=Table[{x11[[n]],x21[[n]],x31[[n]],y[[n]]},
  {n,1,Length[y]}];
```

Executing these statements, the multiple regression model is determined from

```
modlm=LinearModelFit[coord,{x1,x2,x3,x1 x2,x1 x3,x2 x3,
  x1^2,x2^2,x3^2},{x1,x2,x3},ConfidenceLevel->0.95];
```

which, upon using **Normal[modlm]**, gives the following expression for an estimate for y

$$-1.76936 + 0.420798\, x_1 - 0.0193246\, x_1^2 + 0.222453\, x_2 - 0.0198764\, x_1 x_2 -0.00744853 x_2^2 - 0.127995\, x_3 + 0.00915146\, x_1 x_3 - 0.00257618\, x_2 x_3 +0.000823969 x_3^2$$

This expression can be converted to a pure function with

```
fcn=modlm["Function"]
```

which gives

```
-1.76936+0.420798 #1-0.0193246 #1^2+0.222453 #2-0.0198764 #1 #2
 -0.00744853 #2^2-0.127995 #3+0.00915146 #1 #3+
 0.00257618 #2 #3+0.000823969 #3^2 &
```

The coefficient of determination is obtained from

```
R2=modlm["RSquared"]
```

which yields `R2 = 0.916949`. A table of the confidence intervals on the estimates $\hat{\beta}_k$ is obtained from

```
modlm["ParameterConfidenceIntervalTable"]
```

which produces

	Estimate	Standard Error	Confidence Interval
1	-1.76936	1.28698	{-4.49763, 0.958902}
x_1	0.420798	0.294173	{-0.20282, 1.04442}
x_2	0.222453	0.130742	{-0.0547082, 0.499614}
x_3	-0.127995	0.0702452	{-0.276909, 0.0209179}
$x_1 x_2$	-0.0198764	0.0120374	{-0.0453946, 0.00564188}
$x_1 x_3$	0.00915146	0.00762128	{-0.00700493, 0.0253078}
$x_2 x_3$	0.00257618	0.00703927	{-0.0123464, 0.0174988}
x_1^2	-0.0193246	0.0167968	{-0.0549322, 0.0162831}
x_2^2	-0.00744853	0.0120477	{-0.0329886, 0.0180915}
x_3^2	0.000823969	0.0014411	{-0.00223102, 0.00387896}

In this table, the column labeled *Estimate* gives the values of $\hat{\beta}_k$ corresponding to the value f_k, which are given in the left column; the right column contains the corresponding confidence intervals on their respective estimates.

We shall now create a figure that shows the surface for $x_3 = 6$ and its confidence interval surfaces. The surface is only valid within the respective maximum and minimum values of x_1, x_2, and x_3. These values determine the plot limits on these quantities. The surface for $\hat{y}$ is obtained with **Normal[modlm]** and the confidence interval surfaces are obtained from

```
bands=modlm["MeanPredictionBands"];
```

Then,

```
fgn=Normal[modlm];
bands=modlm["MeanPredictionBands"];
Show[Plot3D[fgn/.{x1->s,x2->p,x3->6},{s,7.1,10.3},
 {p,0,7.8},PlotRange->{-0.6,0.8},PlotStyle->Opacity[0.5],
  Mesh->None,ViewPoint->{1.8,-1.5,0.9},
  AxesLabel->{"x1","x2"," y(x1,x2,6)"}],
  Plot3D[bands/.{x1->s,x2->p,x3->6},{s,7.1,10.3},{p,0,7.8},
  PlotStyle->Opacity[0.5],Mesh->None]]
```

produces Figure 9.7.

To determine if the residuals are normally distributed, we obtain the probability plot shown in Figure 9.8, by using

```
ProbabilityScalePlot[modlm["FitResiduals"],"Normal",
 PlotMarkers->Automatic,
 FrameLabel->{"Residual","Normal Probability"}]
```

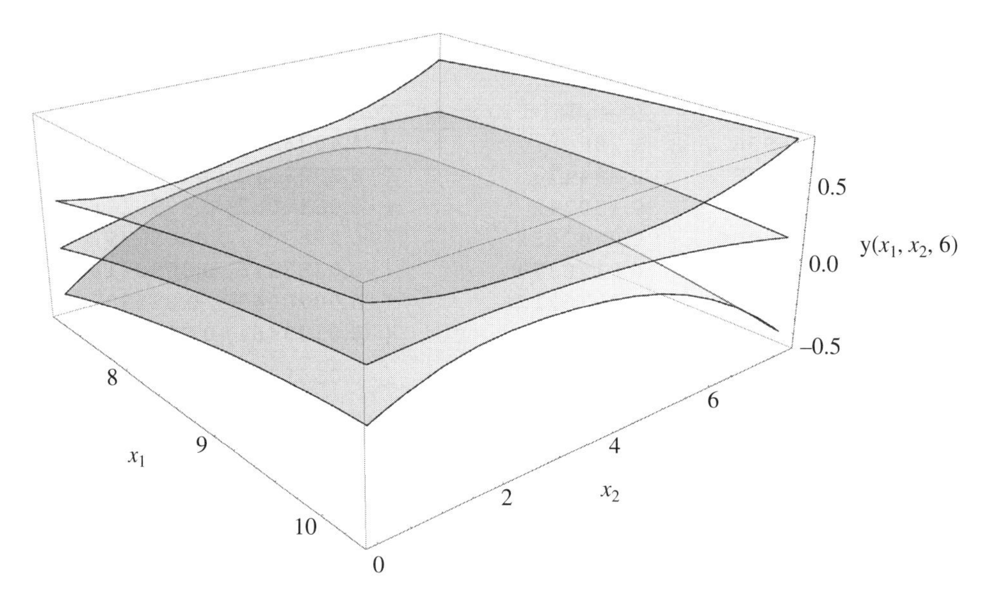

Figure 9.7 Multiple regression fitted surface for x_3 = 6 and its confidence interval surfaces

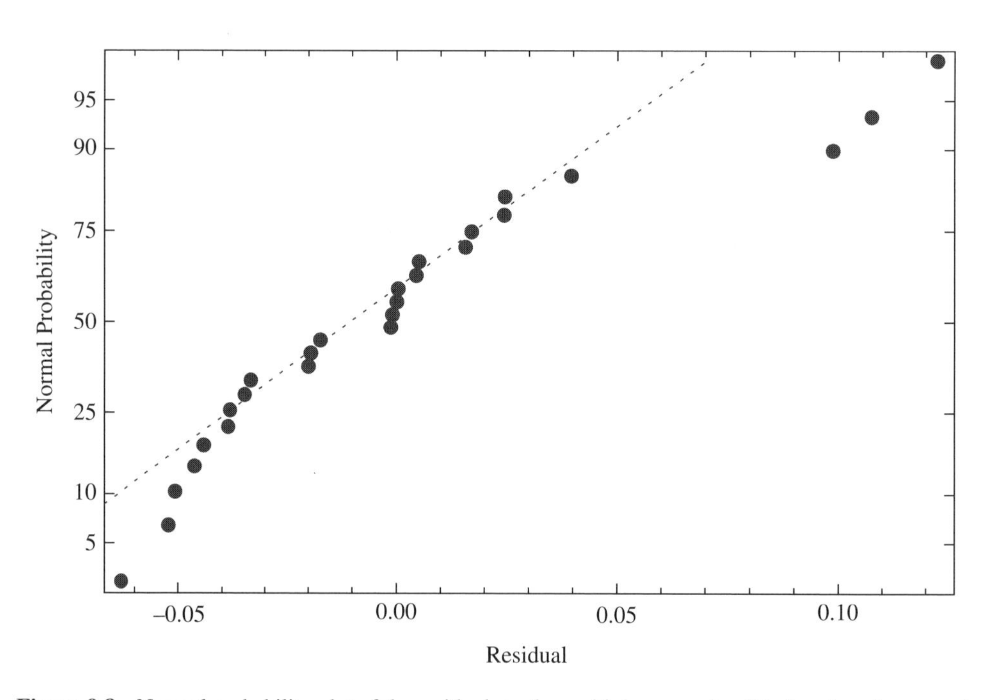

Figure 9.8 Normal probability plot of the residuals to the multiple regression fitted surface for x_3 = 6

9.4 Nonlinear Regression Analysis: NonLinearModelFit[]

Nonlinear regression analysis is performed with **NonLinearModelFit**, which is the statistical version of **FindFit**. **FindFit** only provides the model's coefficient, whereas **NonLinearModelFit** additionally provides several statistical estimates, the same statistical estimates that **LinearModelFit** provides. The form for **NonLinearModelFit** is

```
modl=NonLinearModelFit[coord,{expr,con},par,var,
  ConfidenceLevel->cl]
```

where **coord** = **{{x1,y1, ... ,f1},{x2,y2, ... ,f2}, ...}**, **expr** is an expression that is a function of the parameters **par** = **{{p1,g1},{p2,g2}, ...}**, where **gN** are the (optional) guesses for each parameter, **var** = **{x,y, ...}** are the independent variables, and **con** are the constraints on the parameters. If there are no constraints, then this quantity is omitted and if no initial guess is provided for a parameter that **gM** is omitted. The quantity **cl** is the confidence level; if omitted, a value of 0.95 is used.

We shall illustrate the use of this function by using the data in Table 5.2 of Exercise 5.33 and assuming that the data can be modeled with the following function

$$y = \tan^{-1}\left[\frac{a\cot x\sin^2 x - b\cot x}{c + d\cos(2x)}\right]$$

where the constants a, b, c, and d are to be determined. For the initial guess, we shall use $a = c = d = 2.0$ and $b = 0.1$ and we shall assume a confidence level of 0.9. The following program will plot the fitted curve, the data values, and the confidence bands as shown in Figure 9.9 and plot the residuals as shown in Figure 9.10. Several of the different entities that can be accessed from the implementation of **NonLinearModelFit** for this example are summarized in Table 9.8.

```
dat={{0.01,0},{0.1141,0.09821},{0.2181,0.1843},
  {0.3222,0.2671},{0.4262,0.3384},{0.5303,0.426},
  {0.6343,0.5316},{0.7384,0.5845},{0.8424,0.6527},
  {0.9465,0.6865},{1.051,0.8015},{1.155,0.8265},
  {1.259,0.7696},{1.363,0.7057},{1.467,0.4338},{1.571,0}};
nlModel=NonlinearModelFit[dat,
  ArcTan[(c+d Cos[2 x]),(a Cot[x] Sin[x]^2-b Cot[x])],
  {{a,2},{b,0.1},{c,2},{d,2}},x,ConfidenceLevel->0.90];
bands=nlModel["MeanPredictionBands"];
Show[Plot[Normal[nlModel],{x,0.01,π/2},
   AxesLabel->{"x","y(x)"}],
  ListPlot[dat,PlotMarkers->Automatic],
  Plot[{bands[[1]],bands[[2]]}/.x->g,{g,0.01,π/2.},
   PlotStyle->{Red,Dashing[Medium]}]]
ListPlot[nlModel["FitResiduals"],Filling->Axis,
  AxesLabel->{"x","Residual"}]
```

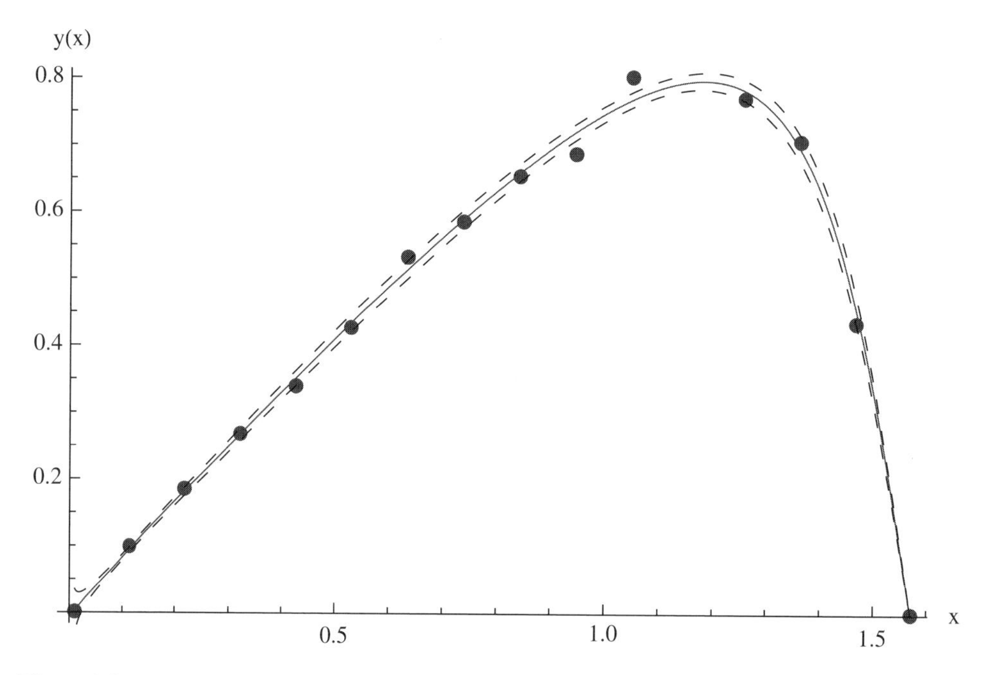

Figure 9.9 Nonlinear curve fit to data values shown and the 90% confidence bands of the fitted curve

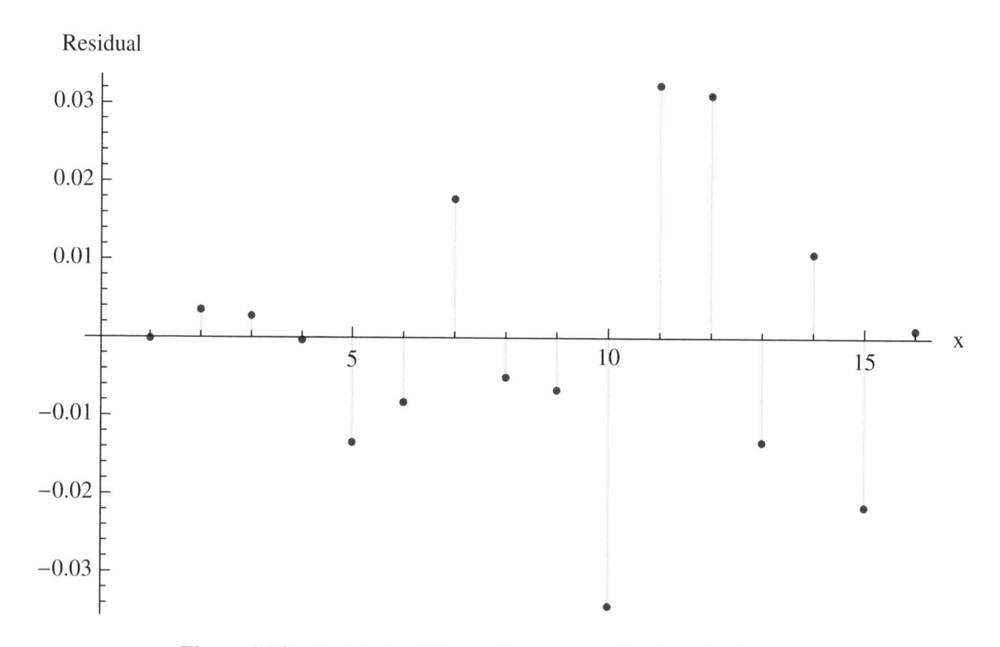

Figure 9.10 Residuals of the nonlinear curve fit given in Figure 9.9

Table 9.8 Several output quantities from `NonLinearModelFit`

Execution of*

```
nlModel=NonlinearModelFit[dat,ArcTan[(c+d Cos[2 x]),(a Cot[x] Sin[x]^2-b Cot[x])],
  {{a,2},{b,0.1},{c,2},{d,2}},x,ConfidenceLevel->0.90]
```

yields

```
FittedModel[ArcTan[8.89463+6.34702 Cos[2 x],-0.00125361 Cot[x]+12.755 Cos[x] Sin[x]]
```

Characteristics of model accessed with `nlModel["Option"]` as indicated in the columns below.

Option description	"Option"	opt=nlModel["Option"]
R^2 (Coefficient of determination)	`"RSquared"`	`0.998986`
Pure function	`"Function"`	`ArcTan[8.89463+6.34702 Cos[2 #1], -0.00125361 Cot[#1]+ 12.755 Cos[#1] Sin[#1]]&` Example: `opt[π/4.]` yields `0.621954`
Confidence interval at 0.90 (in this case) on fitted model as a function of x	`"MeanPredictionBands"`	Formulas are too large to include; however, the results can be evaluated at, say, $x = \pi/4$ as follows: `opt/.x->π/4.` yields `{0.608569,0.635339}`
Confidence intervals at 0.90 (in this case) on fit parameters in tabular form	`"ParameterConfidenceIntervalTable"`	(see table below)
Residuals	`"FitResiduals"`	List of $\hat{y} - y$ evaluated at each value of x in `dat`

Fit parameters	`"BestFitParameters"`	`{a→12.755,b→0.00125361,c→8.89463,d→6.34702}`
		Then, `{a,b,c,d}/.opt` gives
		`{12.755,0.00125361,8.89463,6.34702}`

	Estimate	Standard Error	Confidence Interval
a	12.755	0.104995	{12.568, 12.942}
b	0.00125	0.002971	{-0.0040, 0.00655}
c	8.89463	0.071459	{8.7673, 9.022}
d	6.34702	0.127471	{6.1198, 6.5742}

*Form of `dat` is given in the text.

9.5 Analysis of Variance (ANOVA) and Factorial Designs: **ANOVA[]**

Analysis of variance (ANOVA) is a statistical analysis procedure that can determine whether a change in the output of a process is caused by a change in one or more inputs to the process or is due to a natural (random) variation in the process. In addition, the procedure can also determine if the various inputs to the process interact with each other or if they are independent of each other. The results of ANOVA are typically used to maximize or minimize the output of the process.

Let there be v_k, $k = 1, 2, \ldots, K$ inputs to the process and each v_k assumes L different values called levels. Each v_k, is called a main effect. Each level is identified by an integer that ranges from 1 to L. Corresponding to each level input to the process is the output f. In the physical experiment, the actual values are used to obtain f; however, in the ANOVA analysis the actual values are not used, only the integer designation of its level. We shall designate the value of the input v_k at its lth level as l_k, where $1 \le l_j \le L$. Therefore, the output is $f = f(l_1, l_2, \ldots, l_K)$. When all possible combinations l_j are considered, the experiment is considered a full factorial experiment; that is, an analysis where all the main effects and all the interactions of the main effects are considered. In addition, this full combination of levels can be replicated M times so that in general $f = f(l_{1m}, l_{2m}, \ldots, l_{Km})$, $1 \le m \le M$.

To employ ANOVA, the *Analysis of Variance* package must first be loaded with

```
Needs["ANOVA`"]
```

Then the analysis is performed with

```
ANOVA[dat,modl,{v1,v2, ... ,vK}]
```

where **vk, k** = 1,2,... ,K are the variable names of the main effects, **modl** is the model to be used, and **dat** are the data. Using the notation introduced above, the data have the form **{{l11,l21, ... ,lK1,f1},{l12,l22, ... lK2,f2}, ... }**, where **fn** is the value of the output corresponding to the respective combination of levels. For a full factorial experiment, **modl** is given by **{v1,v2, ... ,vK,All}** where **All** indicates that all combinations of the main effects are to be considered. The effect of **All** will become clear in the examples that follow.

We shall illustrate the use of **ANOVA** and the interpretation of the results with the following two examples.

Example 9.2

Two-Factor ANOVA

Consider the data in Table 9.9, which is for a process with two factors A and B, each taken at three levels. In addition, each combination of levels of these factors is replicated four times. The data given in Table 9.9 are placed in the appropriate form resulting in the array

```
dat={{1,1,130},{1,1,155},{1,1,74},{1,1,180},
  {1,2,34},{1,2,40},{1,2,80},{1,2,75},
  {1,3,20},{1,3,70},{1,3,82},{1,3,58},
  {2,1,150},{2,1,188},{2,1,159},{2,1,126},
  {2,2,136},{2,2,122},{2,2,106},{2,2,115},
```

Table 9.9 Data for Example 9.2

		Factor *B*		
		1	2	3
Factor *A*	1	130, 155, 74, 180	34, 40, 80, 75	20, 70, 82, 58
	2	150, 188, 159, 126	136, 122, 106, 115	25, 70, 58, 45
	3	138, 110, 168, 160	174, 120, 150, 139	96, 104, 82, 60

```
{2,3,25},{2,3,70},{2,3,58},{2,3,45},
{3,1,138},{3,1,110},{3,1,168},{3,1,160},
{3,2,174},{3,2,120},{3,2,150},{3,2,139},
{3,3,96},{3,3,104},{3,3,82},{3,3,60}};
```

where the first value of each triplet is the level of A, the second value the level of B, and the third value the output of the process. Then the analysis of variance is performed with

```
Needs["ANOVA`"]
ANOVA[dat,{A,B,All},{A,B}]
```

which outputs the following ANOVA table

		DF	SumOfSq	MeanSq	FRatio	PValue
	A	2	10683.7	5341.86	7.91137	0.00197608
	B	2	39118.7	19559.4	28.9677	1.9086×10^{-7}
ANOVA->	AB	4	9613.78	2403.44	3.55954	0.0186112
	Error	27	18230.8	675.213		
	Total	35	77647.			

and the following mean values of the main factors and their interactions at each level

	All	105.528
	A[1]	83.1667
	A[2]	108.333
	A[3]	125.083
	B[1]	144.833
	B[2]	107.583
	B[3]	64.1667
CellMeans→	A[1] B[1]	134.75
	A[1] B[2]	57.25
	A[1] B[3]	57.5
	A[2] B[1]	155.75
	A[2] B[2]	119.75
	A[2] B[3]	49.5
	A[3] B[1]	144.
	A[3] B[2]	145.75
	A[3] B[3]	85.5

From the ANOVA table, it is seen that the main factors and their interactions are statistically significant at better than the 98% level. Consequently, from an examination of the cell means, if the objective is to find the combination of parameters that produces the maximum output, then one should run the process with factor A at level 2, denoted `A[2]`, and factor B at level 1, denoted `B[1]`. The output of the process at these levels will be, on average, 155.8. If the interaction term was not statistically significant, then all these interactions would have been ignored; they would be considered a random occurrence. The first value in `CellMeans` designated `All` is the overall mean of the data; that is, `N[Mean[dat[[All,3]]]] = 105.528`.

Example 9.3

Four-Factor Factorial Analysis

Consider the data in Table 9.10, which is for a process with four factors U, V, W, and Y, each taken at two levels. In addition, each combination of levels of these factors is replicated two times. The data given in Table 9.10 are placed in the appropriate form resulting in the array

```
datfac={
 {1,1,1,1,159},{1,1,1,1,163},{2,1,1,1,168},{2,1,1,1,175},
 {1,2,1,1,158},{1,2,1,1,163},{2,2,1,1,166},{2,2,1,1,168},
 {1,1,2,1,175},{1,1,2,1,178},{2,1,2,1,179},{2,1,2,1,183},
 {1,2,2,1,173},{1,2,2,1,168},{2,2,2,1,179},{2,2,2,1,182},
 {1,1,1,2,164},{1,1,1,2,159},{2,1,1,2,187},{2,1,1,2,189},
 {1,2,1,2,163},{1,2,1,2,159},{2,2,1,2,185},{2,2,1,2,191},
 {1,1,2,2,168},{1,1,2,2,174},{2,1,2,2,197},{2,1,2,2,199},
 {1,2,2,2,170},{1,2,2,2,174},{2,2,2,2,194},{2,2,2,2,198}};
```

Then the analysis of variance is performed with

```
Needs[ANOVA`"] (* Not needed if already executed *)
ANOVA[datfac,{U,V,W,Y,All},{U,V,W,Y}]
```

which outputs the following ANOVA table

		DF	SumOfSq	MeanSq	FRatio	PValue
	U	1	2312.	2312.	241.778	4.45067×10^{-11}
	V	1	21.125	21.125	2.20915	0.156633
	W	1	946.125	946.125	98.9412	2.95785×10^{-8}
	Y	1	561.125	561.125	58.6797	9.69219×10^{-7}
	UV	1	0.125	0.125	0.0130719	0.910397
	UW	1	3.125	3.125	0.326797	0.575495
	UY	1	666.125	666.125	69.6601	3.18663×10^{-7}
ANOVA →	VW	1	0.5	0.5	0.0522876	0.822026
	VY	1	12.5	12.5	1.30719	0.269723
	WY	1	12.5	12.5	1.30719	0.269723
	UVW	1	4.5	4.5	0.470588	0.502537

Table 9.10 Data for Example 9.3

Factors and their levels				Response ($y_{m,j}$)	
U	V	W	Y	$j=1$	$j=2$
1	1	1	1	159	163
2	1	1	1	168	175
1	2	1	1	158	163
2	2	1	1	166	168
1	1	2	1	175	178
2	1	2	1	179	183
1	2	2	1	173	168
2	2	2	1	179	182
1	1	1	2	164	159
2	1	1	2	187	189
1	2	1	2	163	159
2	2	1	2	185	191
1	1	2	2	168	174
2	1	2	2	197	199
1	2	2	2	170	174
2	2	2	2	194	198

```
UVY     1   2.       2.       0.20915     0.653583
UWY     1   0.       0.       0.          1.
VWY     1   0.125    0.125    0.0130719   0.910397
UVWY    1   21.125   21.125   2.20915     0.156633
Error   16  153.     9.5625
Total   31  4716.
```

From this table, it is seen that main factors U, W, and Y and the interaction UY have a statistically meaningful effect on the output. Since the list of mean values of the main factors and their interactions is quite long, we shall only list those associated with the statistically meaningful effects. These mean values at each level are

```
U[1]        166.75
U[2]        183.75
W[1]        169.813
W[2]        180.688
Y[1]        171.063
Y[2]        179.438
U[1]Y[1]    167.125
U[1]Y[2]    166.375
U[2]Y[1]    175.
U[2]Y[2]    192.5
```

From these mean values, it is seen that the maximum response will be obtained when factor U is at its high level (**U[2]**) and factor Y is at its high level (**Y[2]**).

9.6 Functions Introduced in Chapter 9

A list of functions introduced in Chapter 9 is given in Table 9.11.

Table 9.11 Commands introduced in Chapter 9

Command	Usage
`ANOVA`	Performs an analysis of variance
`BoxWhiskerChart`	Creates a box whisker chart
`CDF`	Gives the cumulative distribution function for a specified distribution
`EstimatedDistribution`	Estimates the parameters of a specified distribution for a set of data
`FindDistributionParameters`	Estimates the parameters of a specified distribution for a set of data
`Histogram`	Plots a histogram
`InverseCDF`	Determines the inverse of a specified `CDF`
`LinearModelFit`	Performs a simple or a multiple linear regression analysis
`LocationTest`	Performs hypothesis tests on means
`Mean`	Obtains the mean of a list of values
`MeanDifferenceCI`	Determines the confidence interval between the means of two lists of values
`Median`	Obtains the median of a list of values
`NonlinearModelFit`	Determines the parameters of a nonlinear model assumed described a list of values
`NProbability`	Determines the probability of an event for a specified probability distribution
`PDF`	Gives the symbolic expression or the numerical value of the probability density function for a specified distribution
`Probability`	Gives the probability of an event for a specified distribution
`ProbabilityScalePlot`	Creates a probability plot of a list of values for a specified distribution
`Quartile`	Give the specified quartile for a list of values
`RandomVariate`	Generates a list of values that have a specified distribution
`RootMeanSquare`	Determines the root mean square of a list of values
`StudentTCI`	Gives the confidence interval of the mean of a list of values
`StandardDeviation`	Obtains the standard deviation of a list of values
`Variance`	Obtains the variance of a list of values
`VarianceTest`	Used to test the hypothesis that the variances of two lists of values are equal

10

Control Systems and Signal Processing

10.1 Introduction

A control system is often employed to provide a physical system with the ability to meet specified performance goals. In order to design such a system, one usually creates a model of the physical system and a model of the control system so that the combined system can be analyzed and the appropriate control system characteristics chosen. Mathematica provides a collection of commands that allows one to model the system, analyze the system, and plot the characteristics of the system in different ways. In this chapter, we shall demonstrate the usage of several commands that can be employed to design control systems. In addition, we shall illustrate several commands that can be used in various aspects of signal processing and spectral analysis: filters and windows.

10.2 Model Generation: State-Space and Transfer Function Representation

10.2.1 Introduction

Before illustrating the various Mathematica commands that can be used to represent control systems, we shall introduce a permanent magnet motor as a physical system to be modeled and controlled. This system will be used as the specific linear system when many of the commands are introduced. The governing equations for one such system are [1]

$$
\begin{aligned}
&L\frac{di}{dt} + k_m\frac{d\theta}{di} + Ri = v \\
&J\frac{d^2\theta}{di^2} + \zeta\frac{d\theta}{di} - k_g i = 0
\end{aligned}
\tag{10.1}
$$

An Engineer's Guide to Mathematica®, First Edition. Edward B. Magrab.

Companion Website: www.wiley.com/go/magrab

where $v = v(t)$ is the input voltage to the motor windings, $i = i(t)$ is the current in the motor coil, $\theta = \theta(t)$ is the angular position of the rotor, R is the motor resistance, L is the inductance of the winding, k_m is the conversion coefficient from current to torque, k_g is the back electromotive force generator constant, ζ represents the motor friction, and J is the mass moment of inertia of the system and its load.

10.2.2 *State-Space Models:* `StateSpaceModel[]`

Equation (10.1) can be converted to a system of first-order differential equations with the definitions

$$\begin{aligned} x_1(t) &= \theta(t) \\ x_2(t) &= \frac{d\theta}{dt} \\ x_3(t) &= i(t) \end{aligned}$$

Then Eq. (10.1) becomes the following system of first-order equations

$$\begin{aligned} \dot{x}_1 &= \frac{dx_1}{dt} = x_2 \\ \dot{x}_2 &= \frac{dx_2}{dt} = -\frac{\zeta}{J}x_2 + \frac{k_g}{J}x_3 \\ \dot{x}_3 &= \frac{dx_3}{dt} = -\frac{k_m}{L}x_2 - \frac{R}{L}x_3 + \frac{v}{L} \end{aligned}$$

which can be written in matrix form as

$$\{\dot{x}\} = [A]\{x\} + \{B\}\{u\}$$

where $\{x\}$ is the state vector, $\{u\}$ is the input vector, and

$$\{x\} = \begin{Bmatrix} x_1 \\ x_2 \\ x_3 \end{Bmatrix}, \quad \{\dot{x}\} = \begin{Bmatrix} \dot{x}_1 \\ \dot{x}_2 \\ \dot{x}_3 \end{Bmatrix}, \quad \{u\} = \begin{Bmatrix} 0 \\ 0 \\ v \end{Bmatrix}$$

$$[A] = \begin{bmatrix} 0 & 1 & 0 \\ 0 & -\zeta/J & k_g/J \\ 0 & k_m/L & -R/L \end{bmatrix}, \quad [B] = \begin{Bmatrix} 0 \\ 0 \\ 1/L \end{Bmatrix}$$

The matrix $[A]$ is called the state matrix and the matrix $[B]$ the input matrix.

In addition, we define an output vector $\{y\}$ as follows

$$\{y\} = [C]\{x\} + \{D\}\{u\}$$

where, for the system under consideration,

$$[C] = \{1 \quad 0 \quad 0\}, \qquad \{D\} = \{0\}$$

It is recalled that $x_1 = \theta$. The equations

$$\{\dot{x}\} = [A]\,\{x\} + \{B\}\,\{u\}$$
$$\{y\} = [C]\,\{x\} + \{D\}\,\{u\}$$

are the state-space equations for a linear time-invariant system.

The state-space representation for this linear system for the formulation given above is obtained with

```
StateSpaceModel[{a,b,c,d}]
```

where **a, b, c**, and **d** are the matrices and vectors as defined above. Then, for the system represented by Eq. (10.1),

```
a={{0,1,0},{0,-ζ/J,kg/J},{0,-km/L,-R/L}};
b={{0},{0},{1/L}};
c={{1,0,0}};
d={{0}};
ssM=StateSpaceModel[{a,b,c,d}]
```

which displays

$$\left(\begin{array}{ccc|c} 0 & 1 & 0 & 0 \\ 0 & -\zeta/J & k_g/J & 0 \\ 0 & -k_m/L & -;R/L & 1/L \\ \hline 1 & 0 & 0 & 0 \end{array}\right)^{s} \tag{10.2}$$

The state-space model can also be obtained directly from Eq. (10.1) by using

```
StateSpaceModel[eqs,x,u,t]
```

where **eqs** is a list of the differential equations, **x** is a list of the dependent variables and their derivatives up to the $n-1$ derivative in each dependent variable, **u** is a list of the input functions, and **t** is the independent variable. Then, for Eq. (10.1), the state-space equations can be obtained from

```
StateSpaceModel[{L i'[t]+km θ'[t]+R i[t]==v[t],
  J θ''[t]+ζ θ '[t]-kg i[t]==0},{θ[t],θ'[t],i[t]},{v[t]},
  {θ[t]},t]
```

which produces Eq. (10.2).

10.2.3 *Transfer Function Models:* `TransferFunctionModel[]`

Equation (10.1) can be converted to a transfer function model by first taking the Laplace transform of these equations assuming zero initial conditions and then solving for the ratio of the Laplace transform of the output variable and the Laplace transform of the input function. Thus, the Laplace transform of Eq. (10.1) with zero initial conditions yields

$$\begin{aligned} &k_m s\Theta(s) + (Ls + R)\, I(s) = V(s) \\ &\left(Js^2 + \zeta s\right)\Theta(s) - k_g I(s) = 0 \end{aligned} \tag{10.3}$$

The solution to Eq. (10.3) is

$$\begin{aligned} &\frac{\Theta(s)}{V(s)} = \frac{k_g}{sD} \\ &\frac{I(s)}{V(s)} = \frac{Js + \zeta}{D} \\ &D = (Ls + R)\,(Js + \zeta) + k_m k_g \end{aligned} \tag{10.4}$$

The transfer function model is created with

```
TransferFunctionModel[tf,s]
```

where `tf` is the transfer function in terms of the variable `s`. Thus, for the transfer function $\Theta(s)/V(s)$ given above,

```
tfM=TransferFunctionModel[kg/(s ((R+L s) (J s+ζ)+kg km)),s]
```

which displays

$$\left(\frac{k_g}{s\left((L\,s+R)\,\left(J\,s+\zeta\right)+k_m\,k_g\right)}\right)^{\mathcal{T}} \tag{10.5}$$

This transfer function representation can also be obtained from the state-space representation obtained previously. In this case, the command argument is

```
TransferFunctionModel[ssMod]
```

where `ssMod` is the state-space model obtained from `StateSpaceModel`. Thus, for the system under consideration,

```
Simplify[TransferFunctionModel[ssM]]
```

creates the same result as that shown in Eq. (10.5).

It is seen that a direct way to arrive at the transfer function is to use the sequence

```
tfMod=TransferFunctionModel[StateSpaceModel[eqs,x,u,t]]
```

Thus, for the system given by Eq. (10.1), we have

```
tfM=TransferFunctionModel[StateSpaceModel[
  {L i'[t]+k_m θ'[t]+R i[t]==v[t],
   J θ''[t]+ζ θ'[t]-k_g i[t]==0},{θ[t],θ'[t],i[t]},
   {v[t]},{θ[t]},t]]//Simplify
```

which yields Eq. (10.5).

The transfer function model can be converted to a state-space model with

```
StateSpaceModel[tfMod]
```

where **tfMod** is a transfer function model. Thus, using our previous results,

```
ssM1=StateSpaceModel[tfM]
```

displays

$$\left(\begin{array}{ccc|c} 0 & 1 & 0 & 0 \\ 0 & 0 & 1 & 0 \\ 0 & -\frac{R\zeta + k_g k_m}{J L} & -\frac{R}{L} - \frac{\zeta}{J} & 1 \\ \hline \frac{k_g}{J L} & 0 & 0 & 0 \end{array}\right)^{S}$$

which is in a different form than that given by Eq. (10.2) and is a result of the state-space representation not being unique. However, **TransferFunctionModel[ssM1]** recovers Eq. (10.5).

10.3 Model Connections – Closed-Loop Systems and System Response: SystemsModelFeedbackConnect[] and SystemsModelSeriesConnect[]

There are several commands that allow one to connect transfer function objects to form a closed-loop transfer function model. Two of the most commonly used are as follows. Consider two systems **sy1** and **sy2** that are transfer function objects. If **sy2** is in a feedback loop with **sy1**, then this system is represented by

```
SystemsModelFeedbackConnect[sy1,sy2,fbk]
```

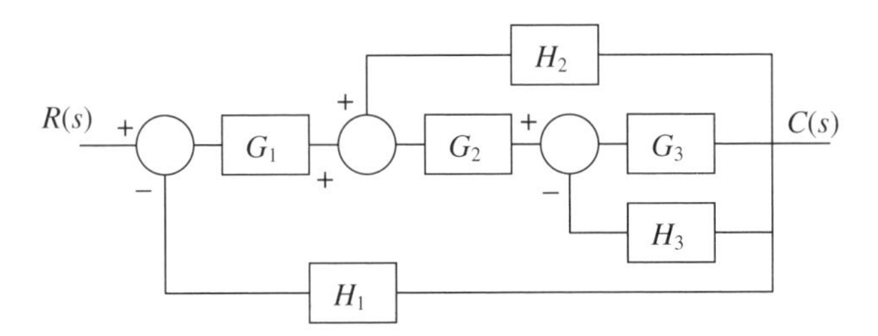

Figure 10.1 Block diagram of interconnected transfer function elements representing a control system

where **fbk** = −1 is used to indicate negative feedback and **fbk** = 1 is used for positive feedback. When omitted, the default value is −1. If, in addition, **sy2** is omitted, then it is assumed that unity negative feedback is being used.

When these two systems are cascaded, that is, they are in series, then the systems are combined using

```
SystemsModelSeriesConnect[sy1,sy2]
```

To illustrate the use of these two commands, we consider the system block diagram shown in Figure 10.1. The transfer function representing the ratio $C(s)/R(s)$ is obtained from

```
TF[h_]:=TransferFunctionModel[h]
sy1=SystemsModelFeedbackConnect[TF[G3],TF[H3]];
sy2=SystemsModelSeriesConnect[TF[G2],sy1];
sy3=SystemsModelFeedbackConnect[sy2,TF[H2],1];
sy4=SystemsModelSeriesConnect[sy3,TF[G1]];
CR=SystemsModelFeedbackConnect[sy4,TF[H1]]//Simplify
```

where we have created the function **TF** to improve the readability of the program. The execution of the program gives

$$\frac{G_1\, G_2\, G_3}{1 + G_1\, G_2\, G_3\, H_1 - G_2\, G_3\, H_2 + G_3\, H_3}$$

To further illustrate the use of **SystemsModelFeedbackConnect**, we examine the transfer function given by Eq. (10.5) and make it into a closed-loop system with unity feedback. However, before doing so, we shall introduce a function that can be used to obtain the output response of a system when the input **v** is specified. The command is

```
OutputResponse[systfss,v,{t,tmin,tmax}]
```

when an interpolating function is desired and

```
OutputResponse[systfss,v,t]
```

when a symbolic solution function is desired. This symbolic function can be used to determine some of the characteristics of the system's response such as rise time and percentage overshoot: see Example 10.1.

The system **systfss** can be either a transfer function model or a state-space model. The state-space model is used when additionally initial conditions are to be specified or initial conditions are specified and **v** = 0. The quantities **tmin** and **tmax**, respectively, indicate the minimum and maximum values of the time interval of interest. The input **v** is typically **DiracDelta** to determine the impulse response, **UnitStep** to determine the response to a step input, and **τ-(τ-1) UnitStep[τ-1]** to determine the response to a ramp, where $\tau = t/t_o$ and t_o is the duration of the ramp.

We shall now determine the closed-loop response of the transfer function model given by Eq. (10.5) to a unit step function and display the result. It is assumed that the system has the following parameters: L = 0.01 H, R = 6.0 Ohms, ζ = 0.005 N·m·s·rad^{-1}, k_m = 0.09 V·s·rad^{-1}, J = 0.02 kg·m^2, and k_g = 18.0 N·m·A^{-1}. Then,

```
L=0.01; R=6.; ζ=0.005; k_m=0.09; J=0.02; k_g=18; tend=1.5;
sysnf=TransferFunctionModel[k_g/(s ((R+L s) (J s+ζ)+k_g k_m)),s];
tfbk=SystemsModelFeedbackConnect[sysnf[R,L,k_g,k_m,J,ζ]];
sys1out=OutputResponse[tfbk,UnitStep[t],{t,0,tend}];
Plot[sys1out,{t,0,tend},PlotRange->All,
  AxesLabel->{"τ","θ(τ)"}]
```

which displays the results shown in Figure 10.2.

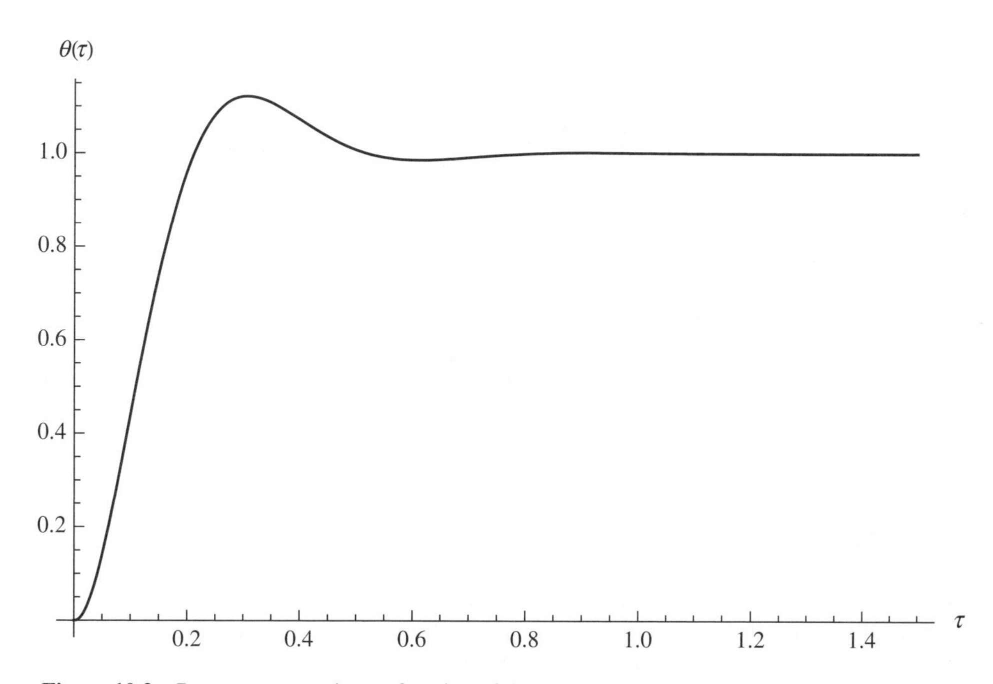

Figure 10.2 Response to a unit step function of the system given by Eq. (10.5) with unit feedback

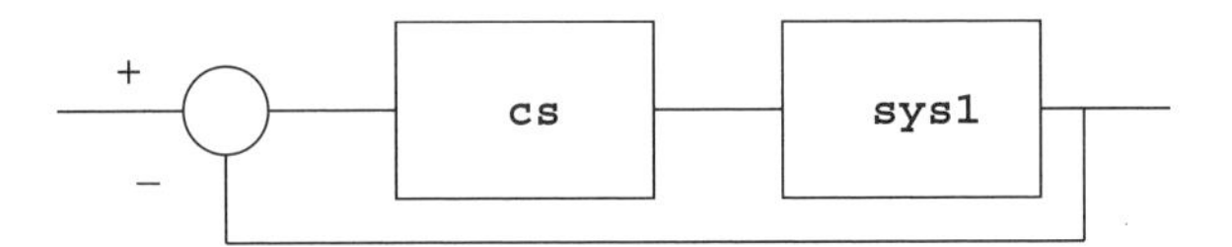

Figure 10.3 Closed-loop system with unity feedback

Another useful command for creating stable closed-loop systems is **PIDTune**. This command allows one to place a general PID controller and any of its special cases in series with a system (denoted **sys1**) as shown in Figure 10.3. The command automatically selects the parameters of the PID system to reject disturbances and to follow as closely as possible changes in the input signal. The model of this system is given by

```
PIDTune[sys1,{"controller","tuningrule"},"output"]
```

where **sys1** is a transfer function model. The option **"controller"** is one of several types: **"P", "PI", "PID"**, and a few others. The default controller is **"PI"**. The option **"tuningrule"** specifies the method that will be used to provide a good estimate of the parameters of the controller that result in stable operation. However, the response may still not satisfy specific transient response characteristics or steady-state characteristics. Therefore, the results from **PIDTune** may in some cases be a starting point from which one determines the final values of the controller's parameters. There are 14 tuning rules available; however, not all tuning rules are applicable with all selections of **"controller"**. The default tuning rule is **"ZieglerNichols"**.

The option **"output"** is used to select one of several output quantities that are available from the function. If this option is omitted, then the transfer function of the **"controller"** selected is the output. Another output that can be selected is the transfer function of the entire system, which is obtained by using **"ReferenceOutput"**.

Example 10.1

PID Control System

We shall determine the response of a PID controller in series with the transfer function model given by Eq. (10.5); that is, the configuration shown in Figure 10.3 with **sys1** given by Eq. (10.5) and **cs** given by a PID controller or one of its special cases. It is again assumed that the system has the following parameters: $L = 0.01$ H, $R = 6.0$ Ohms, $\zeta = 0.005$ N·m·s·rad^{-1}, $k_m = 0.09$ V·s·rad^{-1}, $J = 0.02$ kg·m^2, and $k_g = 18.0$ N·m·A^{-1}. We shall consider a PID controller with two different tuning rules: **"ZieglerNichols"** and **"KappaTau"** and compare the resulting responses to that without the PID controller, which is shown in Figure 10.2. The program is

```
L=0.01; R=6.; ζ=0.005; km=0.09; J=0.02; kg=18; tend=1.5;
sysnf=TransferFunctionModel[kg/(s ((R+L s) (J s+ζ)+kg km)),s];
tfbk=SystemsModelFeedbackConnect[sysnf];
sysnone=OutputResponse[tfbk,UnitStep[t],{t,0,tend}];
zn=PIDTune[sysnf,"PID","ReferenceOutput"];
syszn=OutputResponse[zn,UnitStep[t],{t,0,tend}];
```

```
kt=PIDTune[sysnf,{"PID","KappaTau"},"ReferenceOutput"];
syskt=OutputResponse[kt,UnitStep[t],{t,0,tend}];
Plot[{sysnone,syszn,syskt},{t,0,tend},PlotRange->All,
  PlotStyle->{Black,{Black,Dashed},{Black,Thick}},
  AxesLabel->{"t","θ(t)"},
  PlotLegends->Placed[{"No PID","ZieglerNichols",
   "KappaTau"},{0.8,0.3}]]
```

which produces the results shown in Figure 10.4.

Some of the properties of the resulting responses can be obtained from the preceding results. To compare, for example, the maximum values of the responses of these three closed-loop systems, we use the preceding results and **NMaximum** as follows. Assuming that the previous program has just been run,

```
nzmax=First[NMaximize[{syszn[[1]],tend>t>0},t]];
ktmax=First[NMaximize[{syskt[[1]],tend>t>0},t]];
nonemax=First[NMaximize[{sysnone[[1]],tend>t>0},t]];
Print["Amax (no controller): ", nonemax]
Print["Amax (Ziegler-Nichols controller): ", nzmax]
Print["Amax (kappa tau controller): ", ktmax]
```

displays

```
Amax (no controller): 1.12146
Amax (Ziegler-Nichols controller): 1.37641
Amax (kappa tau controller): 1.06625
```

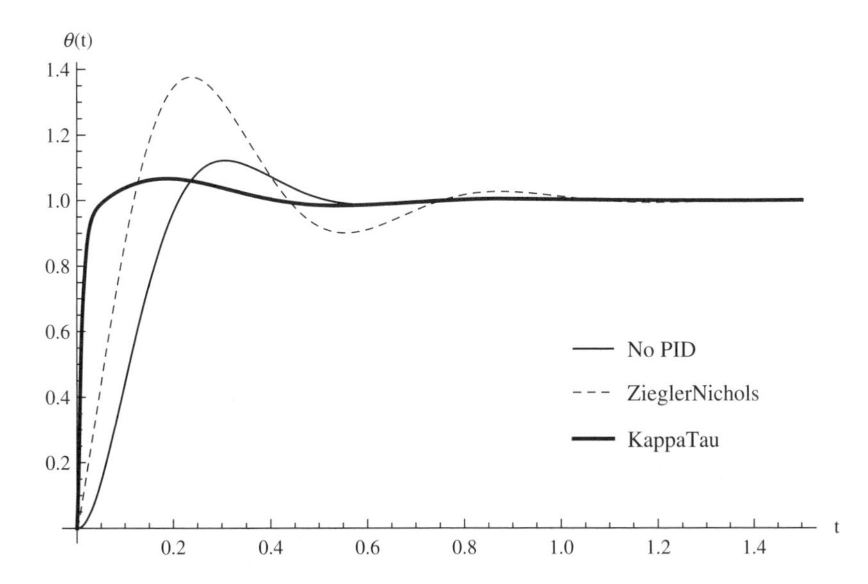

Figure 10.4 Response of the control system shown in Figure 10.3 to a unit step input when **sys1** is given by Eq. (10.5)

The rise time of the system response can also be obtained. Rise time is the time that it takes a system subjected to a unit step input to go from 10% of its final (steady-state) value to 90% of its final value. From the results shown in Figure 10.4, the final value is 1. In order to greatly improve execution time, we shall obtain the rise times using a two-step process. The first step is to create symbolic expressions for the three output responses. These expressions are obtained as follows.

```
L=0.01; R=6.; ζ=0.005; km=0.09; J=0.02; kg=18; tend=1.5;
sysnf=TransferFunctionModel[kg/(s ((R+L s) (J s+ζ)+kg km)),s];
tfbk=SystemsModelFeedbackConnect[sysnf];
symnone=Expand[Chop[Simplify[OutputResponse[tfbk,
  UnitStep[t],t],Assumptions->t>0]]][[1]]
zn=PIDTune[sysnf,"PID","ReferenceOutput"];
symzn=Expand[Chop[Simplify[
  OutputResponse[zn,UnitStep[t],t],Assumptions->t>0]]][[1]]
kt=PIDTune[sysnf,{"PID","KappaTau"},"ReferenceOutput"];
symkt=Expand[Chop[Simplify[
  OutputResponse[kt,UnitStep[t],t],Assumptions->t>0]]][[1]]
```

The results displayed for **symnone, symzn**, and **symkt** are copied into the following program to create three functions that will then be used by **FindRoot** to obtain the rise times. The program is

```
nonee[t_]:=0.000456784-0.000456784 E^(-586.444 t)-
  0.999543 E^(-6.90308 t) Cos[10.2866 t]+
  0.999543 Cos[10.2866 t]^2-
  0.696807 E^(-6.90308 t) Sin[10.2866 t]+
  0.999543 Sin[10.2866 t]^2
znn[t_]:=-0.0143311+0.01469 E^(-577.808 t)-
  0.000358844 E^(-14.0792 t)-
  1.01433 E^(-4.18153 t) Cos[9.87179 t]+
  1.01433 Cos[9.87179 t]^2+
  0.429655 E^(-4.18153 t) Sin[9.87179 t]+
  1.01433 Sin[9.87179 t]^2
tkk[t_]:=0.926068+0.349449 E^(-465.891 t)-
  1.27552 E^(-126.49 t)-
  0.0739324 E^(-3.93465 t) Cos[8.89399 t]+
  0.0739324 Cos[8.89399 t]^2+
  0.13197 E^(-3.93465 t) Sin[8.89399 t]+
  0.0739324 Sin[8.89399 t]^2
t90=t/.FindRoot[0.9-nonee[t],{t,0.1,0.03,0.2}];
t10=t/.FindRoot[0.1-nonee[t],{t,0.05,0.001,0.2}];
Print["Rise time (no controller): ",t90-t10]
t90=t/.FindRoot[0.9-znn[t],{t,0.05,0.03,0.2}];
t10=t/.FindRoot[0.1-znn[t],{t,0.01,0.001,0.2}];
```

```
Print["Rise time (Ziegler-Nichols controller): ",t90-t10]
t90=t/.FindRoot[0.9-tkk[t],{t,0.02,0.001,0.2}];
t10=t/.FindRoot[0.1-tkk[t],{t,0.005,0.001,0.2}];
Print["Rise time (Kappa-Tau controller): ",t90-t10]
```

Execution of this program displays

```
Rise time (no controller): 0.141746
Rise time (Ziegler-Nichols controller): 0.0911663
Rise time (Kappa-Tau controller): 0.0218086
```

10.4 Design Methods

10.4.1 Root Locus: RootLocusPlot[]

The root locus procedure is used to determine the location of the roots of an open-loop or closed-loop system when a parameter, such as gain, is varied. A plot of the root locus is obtained with

```
RootLocusPlot[sys,{p,pmin,pmax},PlotPoints->npts,
  PoleZeroMarkers->{Automatic,"ParameterValues"->var}]
```

where **sys** is the transfer function of the system, **p** is the parameter in **sys** that is to be varied over the range **pmin** < **p** < **pmax**. The option **PlotPoints** specifies that **npts** points are to be used to obtain the root locus plot. The option **PoleZeroMarkers** allows one to place markers at **p** = **var** along the locus curves, where **var** is a single value or a list of values. When one places the cursor over these points, the value of **p** is shown. By default, **Automatic** plots a pole with an "×" and a zero with an "o". The values of these poles and zeros are obtained as indicated subsequently.

To illustrate the use of **RootLocusPlot**, we shall obtain a root locus plot of the closed-loop system shown in Figure 10.3, where **cs** is now a lead controller whose transfer function is given by

$$\frac{k_o(3s+50)}{s+50} \tag{10.6}$$

Then, the program that obtains a root locus plot for this system, which is shown in Figure 10.5, is

```
L=0.01; R=6.; ζ=0.005; km=0.09; J=0.02; kg=18.; tend=1.5;
tf=TransferFunctionModel[kg/(s ((R+L s) (J s+ζ)+kg km)),s];
lead=TransferFunctionModel[ko (3 s+50)/(s+50),s];
sys1=SystemsModelSeriesConnect[lead,tf];
sys=SystemsModelFeedbackConnect[sys1];
RootLocusPlot[sys,{ko,0,15},PlotRange->{{-60,1},{-20,20}},
```

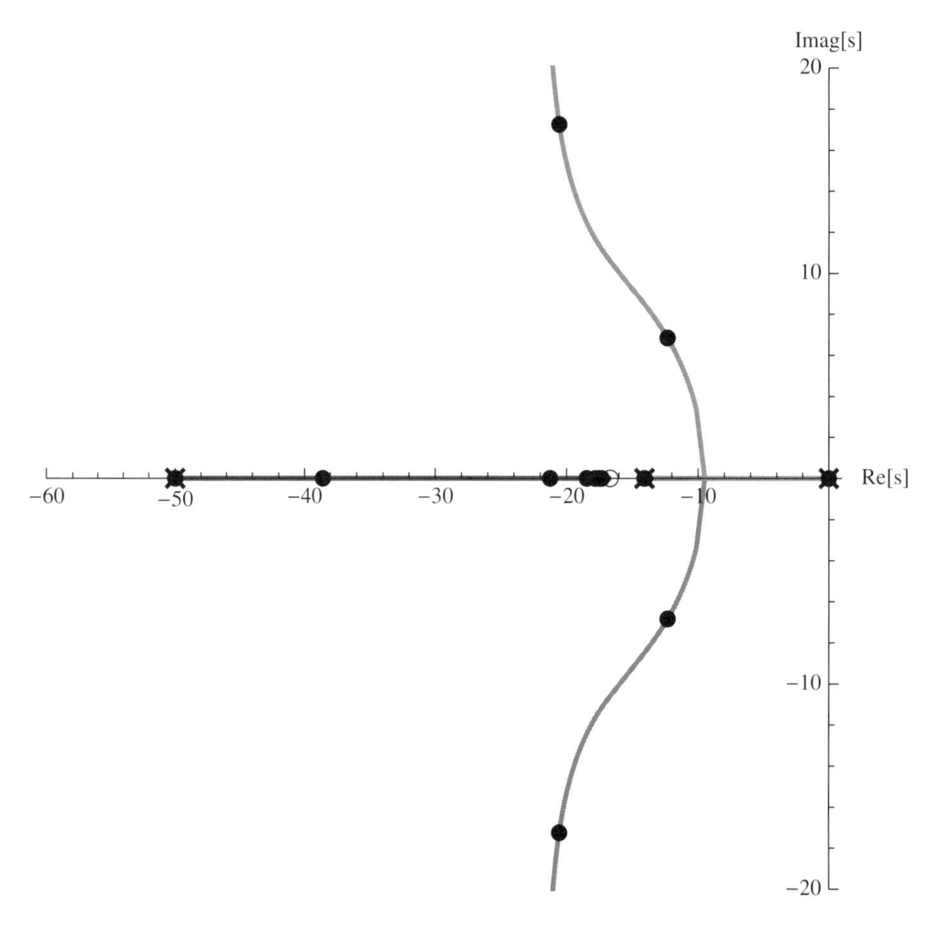

Figure 10.5 Root locus plot of the system shown in Figure 10.3 with **cs** replaced with the transfer function given by Eq. (10.6)

```
PlotPoints->250,AspectRatio->1,
AxesLabel->{"Re[s]","Imag[s]"},PoleZeroMarkers->
{Automatic,"ParameterValues"->Range[0,3,0.5]}]
```

It is mentioned again that when one passes the cursor over the points, the corresponding value of k_o is displayed. Also, we have used **PlotRange** to exclude the most negative pole so that the important part of the root locus plot has ample visual resolution. Setting the aspect ratio to 1 further aids in achieving this visual resolution.

The values of the poles appearing in this figure can be obtained by using

```
Flatten[TransferFunctionPoles[sys1]]
```

which yields

```
{-586.176,-50.,-14.0743,0}
```

The zeros of the open-loop transfer function are obtained with

```
Flatten[TransferFunctionZeros[sys1]]
```

which gives

```
{-16.6667}
```

It is shown in the *Documentation Center* under **RootLocusPlot** how **Manipulate** can be used to create an interactive graphic to explore the various aspects of the root locus curves as a function k_o.

10.4.2 Bode Plot: BodePlot[]

A Bode plot is a plot of the amplitude in dB of the frequency response of a system and on a separate plot its phase response, usually presented in degrees. A Bode plot is obtained with

```
BodePlot[sys,StabilityMargins->ft]
```

where **sys** is either a transfer function model or a state-space model, and the option **StabilityMargins** displays on the graph a vertical line indicating the gain and phase margins at the frequencies at which they are determined.

The values of the phase and gain margins can be obtained from

```
gpm=GainPhaseMargins[sys]
```

or individually from

```
gm=GainMargins[sys]
```

and

```
gp=PhaseMargins[sys]
```

In these expressions,

```
gpm={{{wg1,g1},{wg2,g2}, ...},{{wp1,p1},{wp2,p2}, ...}}
gm={{wg1,g1},{wg2,g2}, ...}
pm={{wp1,p1},{wp2,p2}, ...}
```

where **wgn** is the crossover frequency of the gain margin ratio **gn** and **wpn** is the crossover frequency of the phase margin **pn** in radians.

We shall obtain the Bode plots for the system shown in Figure 10.3 with **sys1** given by Eq. (10.5) and **cs** given by Eq. (10.6) with $k_o = 3.0$ and compare the Bode plots with those in which **cs** is absent. We shall include the display of the stability margins in the Bode plots and also list their values, which will be converted to dB and degrees. Lastly, we shall display the response of these two systems to a unit step input. **Tooltip** will also be employed so that placing the cursor over each curve will display the transfer function associated with that curve. The program is

```
L=0.01; R=6.; ζ=0.005; km=0.09; J=0.02; kg=18.;
ko=3; tend=0.7;
tf=TransferFunctionModel[kg/(s ((R+L s) (J s+ζ)+kg km)),s];
lead=TransferFunctionModel[ko (3s+50)/(s+50),s];
syslead=SystemsModelFeedbackConnect[
  SystemsModelSeriesConnect[lead,tf]];
sysno=SystemsModelFeedbackConnect[tf];
BodePlot[{Tooltip[syslead],Tooltip[sysno]},
  PlotStyle->{{},Dashed}, StabilityMargins->True,
  StabilityMarginsStyle->{{},Dashed}]
orx=OutputResponse[syslead,UnitStep[t],{t,0,tend}];
ory=OutputResponse[sysno,UnitStep[t],{t,0,tend}];
Plot[{orx,ory},{t,0,tend},PlotRange->All,
  AxesLabel->{"t","θ(t)"},
  PlotStyle->{Black,{Black,Dashed}}]
gmlead=GainMargins[syslead];
pmlead=PhaseMargins[syslead];
Map[{#[[1]],#[[2]]/Degree} &,pmlead]
Map[{#[[1]],20.Log10[#[[2]]]} &,gmlead]
```

which produces the Bode plots shown in Figure 10.6 and the output response to a unit step shown in Figure 10.7. In addition, the program displays the phase margins

```
{{30.0758,104.239},{0.,180.}}
```

which are in degrees and the gain margin

```
{{166.353,26.3221}}
```

which is expressed in dB.

10.4.3 Nichols Plot: NicholsPlot[]

A Nichols plot is a plot of the open-loop or closed-loop system's phase on the x-axis expressed in degrees versus the open-loop or closed-loop system's modulus (gain) expressed in dB on the y-axis. It is a very useful tool in frequency-domain analysis. In Mathematica, the creation of a Nichols plot is shown in Table 10.1. As is seen in the table, one must exercise several options to get the plot to look like a classical Nichols chart. It is mentioned that passing the cursor over a grid line displays its value. The values of ω shown as large points on the Nichols curves are as follows: the smallest value of ω in the list **w** appears at the topmost location and

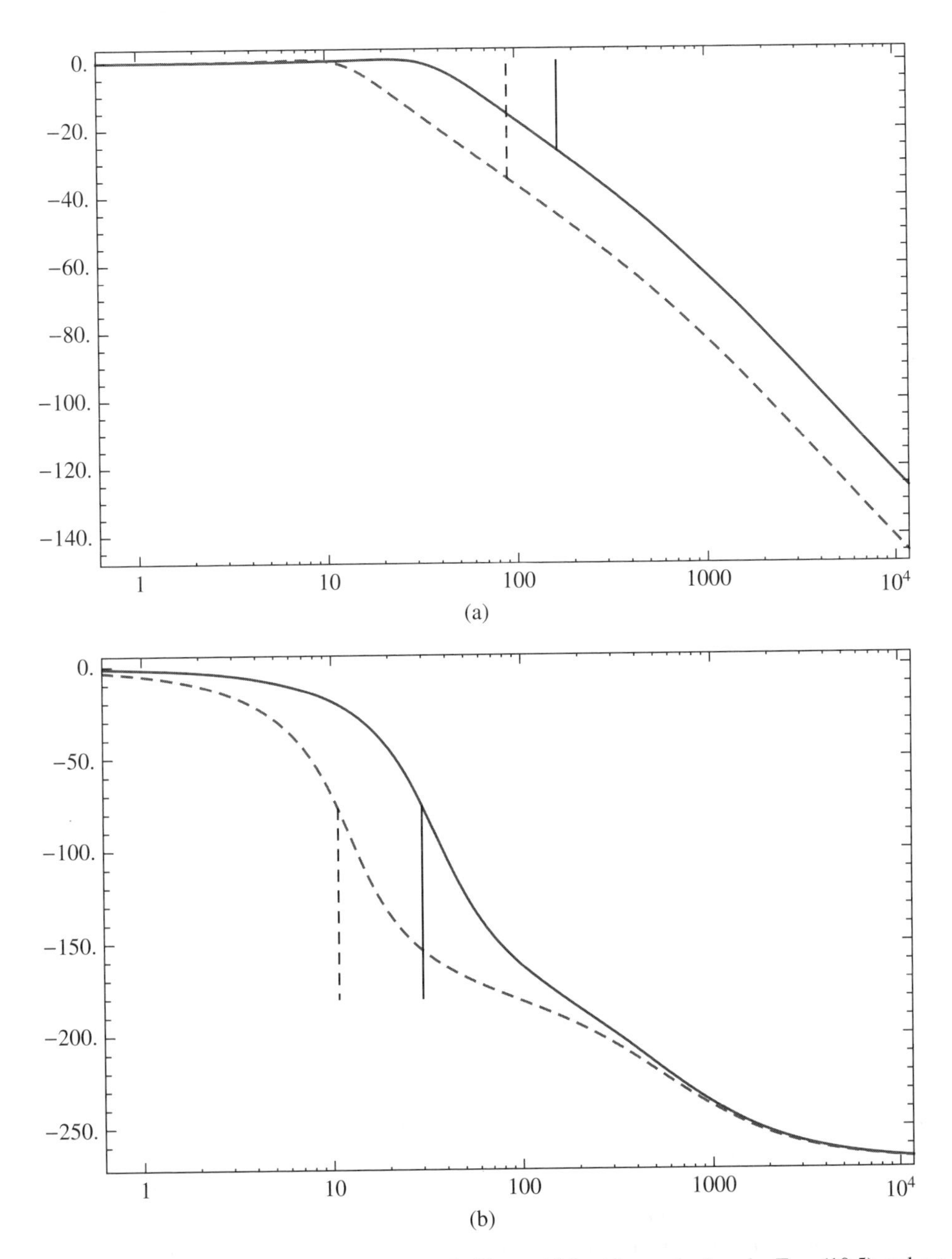

Figure 10.6 Bode plots for the system shown in Figure 10.3 with **sys1** given by Eqs. (10.5) and **cs** given by Eq. (10.6) with $k_o = 3.0$ (solid line) and the Bode plots for the system in which **cs** is absent (dashed lines): (a) amplitude (b) phase

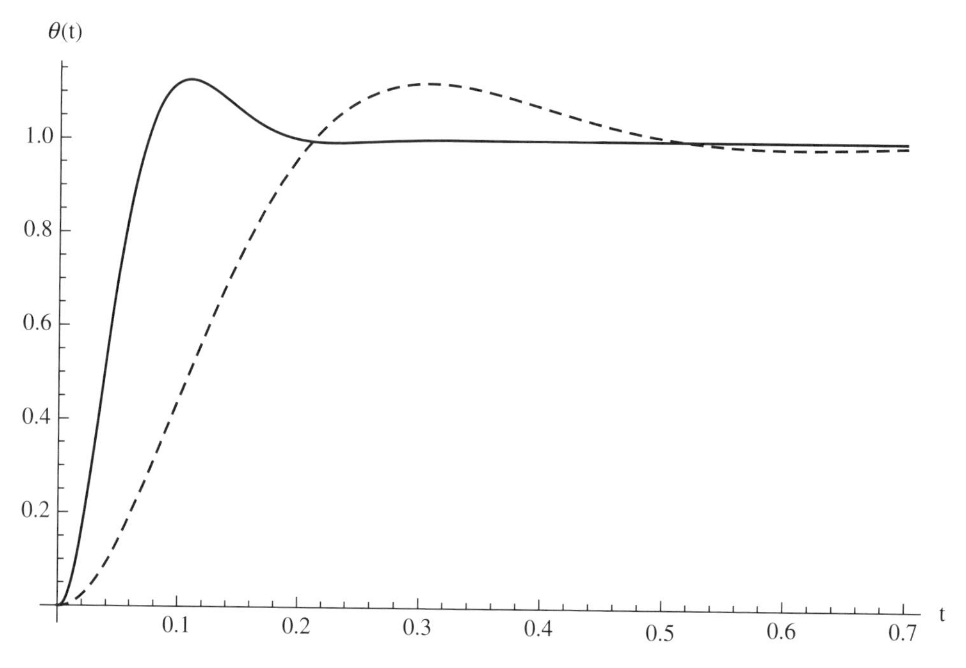

Figure 10.7 Output responses to a unit step input for the two systems shown in Figure 10.6

the maximum value of ω at the lowest position. The thick horizontal line emanating from the origin at (−180,0) is the phase margin and the thick vertical line is the corresponding gain margin. Their values can be confirmed with **`GainMargins`** and **`PhaseMargins`**.

10.5 Signal Processing

10.5.1 Filter Models: `ButterworthFilterModel[]`, `EllipticFilterModel[], ...`

Mathematica 9 has commands that create the transfer functions of analog filters using different filter models. The models that we shall demonstrate are the Butterworth, elliptic, Chebyshev1, and the Chebyshev2 filters. For each of these models, one can select whether the filter is low pass, high pass, band pass, or band stop. These filter models and types will be examined by displaying the effects that each has on a signal composed of three sinusoidal waves of unit amplitude and different frequencies.

The four commands that create these four transfer functions are defined by the parameters shown in Figure 10.8. In these figures, $g_s < 1$ is the stop-band attenuation, $g_p \le 1$ is the pass band attenuation, and $g_s < g_p$. The corresponding pass-band frequencies and stop-band frequencies are in rad·s^{-1}.

The transfer function of a Butterworth filter is obtained with

```
ButterworthFilterModel[arg]
```

Table 10.1 Creation of a Nichols plot

```
(*Creation of open-loop system *)
L=0.01; R=6.; ζ=0.005; km=0.09; J=0.02; kg=18.; ko=3;
tf=TransferFunctionModel[kg/(s ((R+L s) (J s+ζ)+kg km)),s];
lead=TransferFunctionModel[ko (3s+50)/(s+50),s];
syslead=SystemsModelSeriesConnect[lead,tf];
(*Quantities used in NicholsPlot *)
w={1,10,20,30,40,100,300}; rt={{Red,Thick},{Red,Thick}};
pts=Table[{w[[n]],Directive[{PointSize[Large]}]},{n,1,Length[w]}];
lb={Style["Open-loop phase (°)",12],Style["Open-loop gain (dB)",12]};
```

```
NicholsPlot[syslead,Frame->True,
  FrameLabel->lb]
```

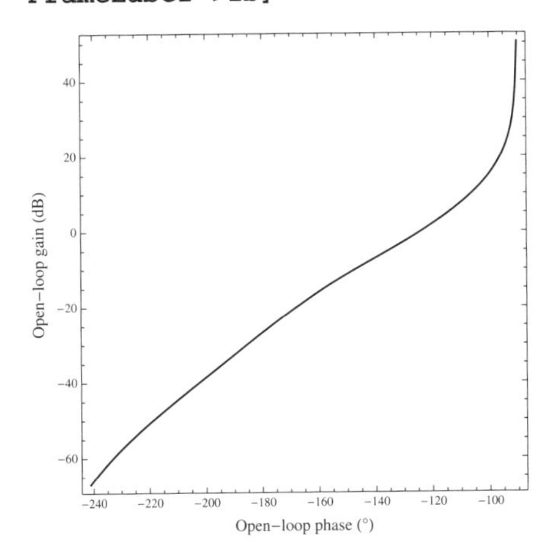


```
NicholsPlot[syslead,Frame->True,
  FrameLabel->lb, Mesh->{pts}]
```

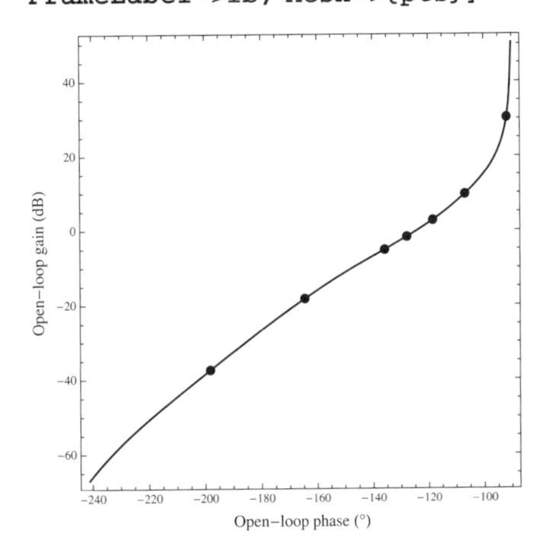


```
NicholsPlot[syslead,{1,300},
  Frame->True,FrameLabel->lb,
  Mesh->{pts}]
```

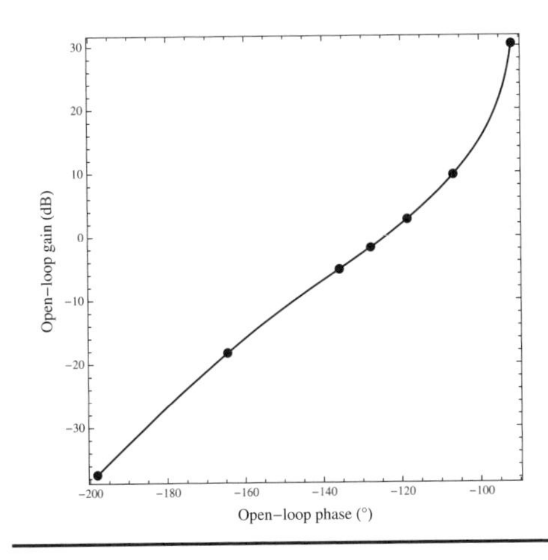


```
NicholsPlot[syslead,{1,300},
  Frame->True,FrameLabel->lb,
  NicholsGridLines->Automatic,
  Mesh->{pts}]
```

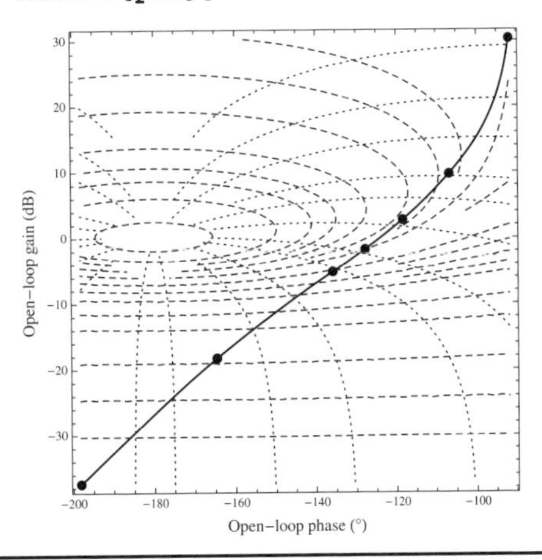

*

(continued)

Table 10.1 (*Continued*)

```
NicholsPlot[syslead,{1,300},Frame->True,FrameLabel->lb,Mesh->{pts},
  NicholsGridLines->Automatic,AxesOrigin->{-200,-38},
  StabilityMargins->True,StabilityMarginsStyle->rt]
```

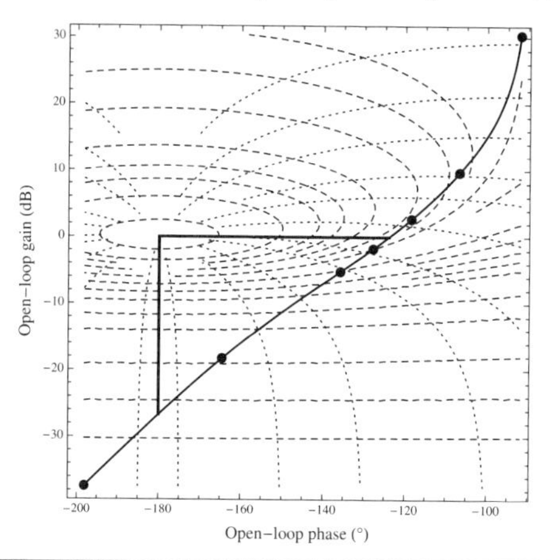

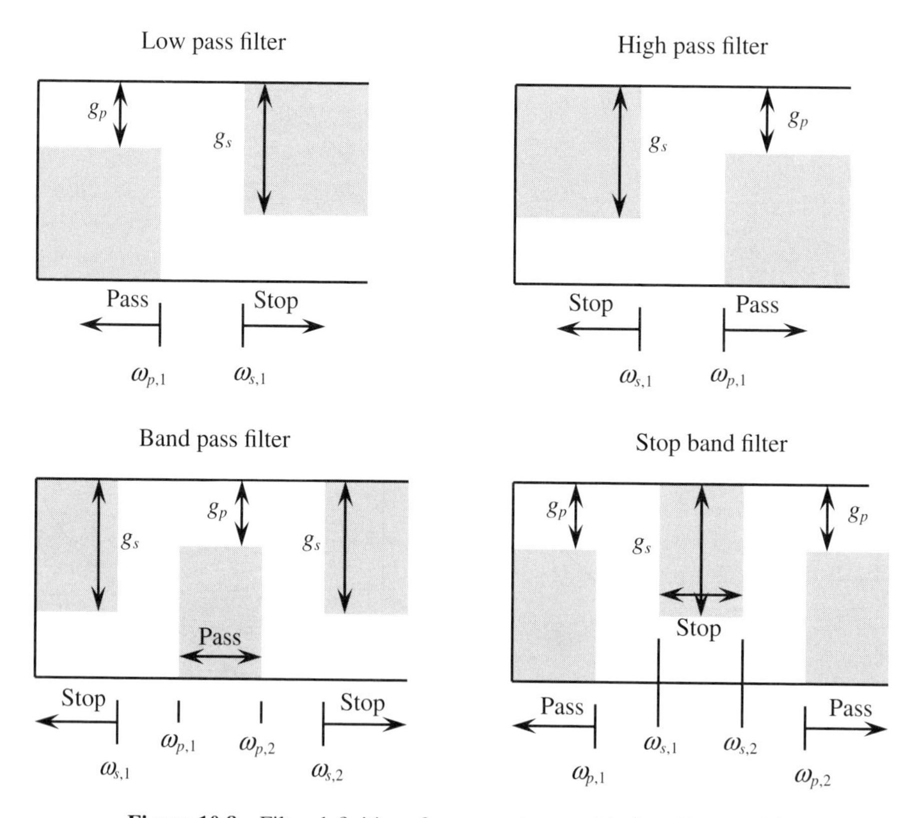

Figure 10.8 Filter definitions for parameters used in four filter models

where `arg` has one of the following four definitions:

Low-pass filter

```
{"Lowpass",{wp1,ws1},{ap,as}}
```

High-pass filter

```
{"Highpass",{ws1,wp1},{as,ap}}
```

Band-pass filter

```
{"Bandpass",{ws1,wp1,wp2,ws2},{as,ap}}
```

Stop-band filter

```
{"Bandstop",{wp1,ws1,ws2,wp2},{ap,as}}
```

In these definitions,

```
as=-20 Log10[gs]
ap=-20 Log10[gp]
```

If only a low-pass filter is to be specified, `arg` is given by `{n,wc}`, where `n` is the filter order and `wc` is the filter's cutoff frequency.

The remaining filter models can be obtained using

```
EllipticFilterModel[arg]
Chebyshev1FilterModel[arg]
Chebyshev2FilterModel[arg]
```

where `arg` is any one of the four definitions given above.

Example 10.2

Effects of Filters on Sinusoidal Signals

To illustrate the use of these filter commands, we shall construct an interactive graphic whereby one can display the frequency response function for any of the four filter models for any of the four filter types. The resulting filter frequency response function will be displayed along with their band-stop and band-pass regions as shown in Figure 10.8. In addition, the effects of the filter's frequency characteristics on an input signal composed of three sinusoids of the form

$$s_{in}(t) = \sin(10t) + \sin(25t) + \sin(60t)$$

will be displayed two ways. The first way will be to include the amplitudes of the frequency components of $s_{in}(t)$ after they have passed through the filter. The relative magnitudes of these

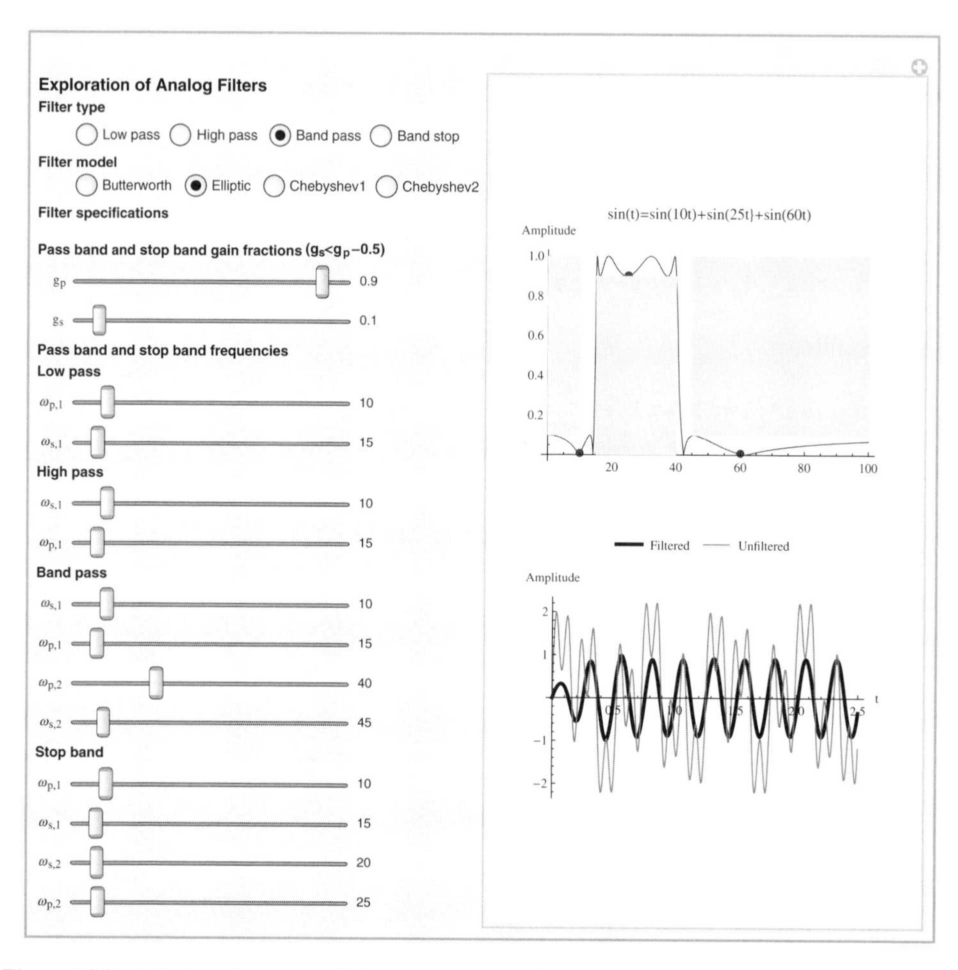

Figure 10.9 Initial configuration of the interactive graphic to explore the effects of analog filters on sinusoidal signals

amplitudes expressed in dB will be made available with **Tooltip**. The second way will be to compare the time waveform $s_{in}(t)$ with that from the output of the filter. The program that creates Figure 10.9 follows. In the program, several safeguards have been employed to ensure that the various pass-band and stop-band frequencies are always in ascending order. These safeguards permit one to use the sliders in any order. In addition, **Sequence** has been used to improve the readability of the program, as has the function **gg**.

```
Manipulate[
  (* Create sequences to improve readability of subsequent code *)
  seq=Sequence[100.,0.25,Appearance->"Labeled",
   ControlType->Slider];
  seq1=Sequence[0.05,0.95,0.05,Appearance->"Labeled",
```

```
 ControlType->Slider];
col=Sequence[Blue,Opacity[0.1]];
If[gstop>=gpass,gstop=gpass-0.05];
```

(* Define `arg` and create pass band and stop regions *)

```
spc=Which[
 ftype==1, If[fs1lp<=fp1lp,fs1lp=fp1lp+2.];
 rec1={col,Rectangle[{0.1,0},{fp1lp,gpass}]};
 rec2={col,Rectangle[{fs1lp,gstop},{100,1}]};
 reg={rec1,rec2};
 {"Lowpass",{fp1lp,fs1lp},{-20. Log10[gpass],
   -20. Log10[gstop]}},
 ftype==2, If[fp1hp<=fs1hp,fp1hp=fs1hp+2.];
 rec1={col,Rectangle[{fs1hp,gstop},{0,1}]};
 rec2={col,Rectangle[{fp1hp,gpass},{100,0}]};
 reg={rec1,rec2};
 {"Highpass",{fs1hp,fp1hp},{-20. Log10[gstop],
   -20. Log10[gpass]}},
 ftype==3, If[fp1bp<=fs1bp,fp1bp=fs1bp+2.];
 If[fp2bp<=fp1bp,fp2bp=fp1bp+2.];
 If[fs2bp<=fp2bp,fs2bp=fp2bp+2.];
 rec1={col,Rectangle[{0,1},{fs1bp,gstop}]};
 rec2={col,Rectangle[{fp1bp,0},{fp2bp,gpass}]};
 rec3={col,Rectangle[{fs2bp,1},{100,gstop}]};
 reg={rec1,rec2,rec3};
 {"Bandpass",{fs1bp,fp1bp,fp2bp,fs2bp},
   {-20. Log10[gstop],-20. Log10[gpass]}},
 ftype==4,If[fs1sb<=fp1sb,fs1sb=fp1sb+2.];
 If[fs2sb<=fs1sb,fs2sb=fs1sb+2.];
 If[fp2sb<=fs2sb,fp2sb=fs2sb+2.];
 rec1={col,Rectangle[{0,0},{fp1sb,gpass}]};
 rec2={col,Rectangle[{fs1sb,1},{fs2sb,gstop}]};
 rec3={col,Rectangle[{fp2sb,0},{100,gpass}]};
 reg={rec1,rec2,rec3};
 {"Bandstop",{fp1sb,fs1sb,fs2sb,fp2sb},
   {-20. Log10[gpass],-20. Log10[gstop]}}];
```

(* Select filter model *)

```
ftf=Which[
 fmodel==1,ButterworthFilterModel[spc],
 fmodel==2,EllipticFilterModel[spc],
 fmodel==3,Chebyshev1FilterModel[spc],
 fmodel==4,Chebyshev2FilterModel[spc]];
out=OutputResponse[ftf,(Sin[wo t]+Sin[2.5 wo t]+
 Sin[6 wo t]) UnitStep[t],{t,ts,tend}];
ampout={Flatten[{wo,Abs[ftf[I wo]]}],
```

```
  Flatten[{2.5 wo,Abs[ftf[I 2.5 wo]]}],
  Flatten[{6. wo,Abs[ftf[I 6 wo]]}]};
```

(* Create tool tip *)

```
ttip=Table[Tooltip[ampout[[n]],
  Column[{Row[{"ω=",ampout[[n,1]]}],
  Row[{"Attn=",NumberForm[
    20. Log10[ampout[[n,2]]],3]," (dB)"}]}]],{n,1,3}];
```

(* Plot results *)

```
GraphicsColumn[{Show[
  Plot[Abs[ftf[I w]],{w,0.1,100},PlotRange->All,
   Epilog->reg,AxesOrigin->{0,0},
   AxesLabel->{"ω","Amplitude"}, PlotLabel->lab],
  ListPlot[ttip,Filling->Axis,PlotMarkers->Automatic,
   PlotRange->{{0,100},{0,1.05}}]],
  Plot[{out,Sin[wo t]+Sin[2.5 wo t]+Sin[6 wo t]},
   {t,ts,tend},PlotRange->All,
   PlotStyle->{{Black,Thick},Gray},
   PlotLegends->Placed[{"Filtered",
     "Unfiltered"},Top],ImageSize->250,
    AxesLabel->{"t","Amplitude"}]}],
```

(* Create sliders and buttons *)

```
Style["Exploration of Analog Filters",12,Bold],
Style["Filter type",10,Bold],
{{ftype,3,""},{1->"Low pass",2->"High pass",
 3->"Band pass",4->"Band stop"},
 ControlType->RadioButtonBar},
Style["Filter model",10,Bold],
{{fmodel,2,""},{1->"Butterworth",2->"Elliptic",
 3->"Chebyshev1",4->"Chebyshev2"},
 ControlType->RadioButtonBar},
Style["Filter specifications",10,Bold],
" ",
Style["Pass band and stop band gain fractions
  (gs<gp-0.5)",10,Bold],
{{gpass,0.9,"gp"},0.05,0.95,0.05,Appearance->"Labeled",
  ControlType->Slider},
{{gstop,0.1,"gs"},0.05,0.95,0.05,Appearance->"Labeled",
 ControlType->Slider},
Style["Pass band and stop band frequencies",10,Bold],
Style["Low pass",10,Bold],
{{fp1lp,10,"ωp,1"},1.,seq,Enabled->gg[ftype,1]},
{{fs1lp,15,"ωs,1"},fp1lp+0.25,seq,Enabled->gg[ftype,1]},
Style["High pass",10,Bold],
```

```
{{fs1hp,10,"ωs,1 "},1.,seq,Enabled->gg[ftype,2]},
{{fp1hp,15,"ωp,1"},fs1hp+0.25,seq,Enabled->gg[ftype,2]},
Style["Band pass",10,Bold],
{{fs1bp,10,"ωs,1"},1.,seq,  Enabled->gg[ftype,3]},
{{fp1bp,15,"ωp,1"},fs1bp+0.25,seq,Enabled->gg[ftype,3]},
{{fp2bp,40,"ωp,2"},fp1bp+0.25,seq,Enabled->gg[ftype,3]},
{{fs2bp,45,"ωs,2"},fp2bp+0.25,seq,Enabled->gg[ftype,3]},
Style["Stop band",10,Bold],
{{fp1sb,10,"ωp,1"},1.,seq,Enabled->gg[ftype,4]},
{{fs1sb,15,"ωs,1"},fp1sb+0.25,seq,Enabled->gg[ftype,4]},
{{fs2sb,20,"ωs,2"},fs1sb+0.25,seq,Enabled->gg[ftype,4]},
{{fp2sb,25,"ωp,2"},fs2sb+0.25,seq,Enabled->gg[ftype,4]},
TrackedSymbols:>{ftype,fmodel,gpass,gstop,fp1lp,fs1lp,
  fp1hp,fs1hp,fs1bp,fp1bp,fp2bp,fs2bp,fs1sb,fp1sb,
  fp2sb,fs2sb},
Initialization:>{wo=10; ampin={{wo,1},{2.5 wo,1},
  {6 wo,1}}; lab="sin(t)=sin(10t)+sin(25t}+sin(60t)";
  ts=0.; tend=2.5;
  gg[typ_,n_]:=If[typ==n,pp=True,pp=False]}]
```

10.5.2 *Windows:* **HammingWindow[], HannWindow[], ...**

In practice, one frequently takes the Fourier transform of signals over a portion of its total duration. This type of transform is called the short-time Fourier transform (STFT). Essentially, one has multiplied the signal by a rectangular window of unit amplitude and duration t_d. This window is called a Dirichlet window or a rectangular window or a boxcar window and can be represented by the difference of two unit step functions; that is, $u(t) - u(t-t_d)$. A result of the STFT using the Dirichelt window is to introduce side lobes that sometimes can mask useful information. These side lobes appear when the window does not coincide with the period of a sine wave or a multiple of the periods of a signal with several sine waves. Many window functions have been developed to mitigate this effect. We shall examine, in addition to the rectangular window, five other windows: the Hamming window, the Hann window, the Blackman window, the Kaiser window, and the Nuttall window. The equations that describe each of these functions can be found in the *Documentation Center* along with about 20 other window functions.[1]

Before illustrating the window functions, we consider a signal $f(t)$ of duration t_d composed of two sine waves, one with frequency αf_o and unit amplitude and the other with frequency βf_o and amplitude A_2, where $0.5 \leq \alpha, \beta \leq 2.0$, and $0 \leq A_2 \leq 1$. Then,

$$f(t) = \sin(2\pi\alpha f_o t) + A_2 \sin(2\pi\beta f_o t) \tag{10.7}$$

[1] Enter *guide/WindowFunctions* in the *Documentation Center* search area.

A window function denoted $w(t)$ is introduced, where $w(t)$ is equal to zero everywhere except for $t_s \le t \le t_s + t_w$, t_s is the time at which one starts to sample the signal, and $t_w < t_d$. Then the signal is given by

$$s_w(t) = f(t)w(t) \tag{10.8}$$

We are interested in the magnitude of the Fourier transform of Eq. (10.8), which is denoted $|S_w(\omega)|$, where ω is the frequency in rad·s^{-1}. The transform will be obtained with **Fourier**, which was introduced in Section 5.8. Hence, numerically the signal $s_w(t)$ is being sampled every $\Delta t = 1/f_s$ seconds, where for convenience it is assumed that $f_s = \gamma f_o$, where $\gamma > 2$. In addition, it is assumed that $t_w = m\gamma\Delta t$, where m is an integer; thus the sampled function contains $n_T = m\gamma$ samples. It we let $t_s = n_s\Delta t$ and $t = (n_s + n)\Delta t$, where n_s is an integer including zero and $1 \le n \le n_T$, then the sampled waveform is

$$\begin{aligned} s_{w,n}(t) &= f\left((n_s + n - 1)\Delta t\right) w\left((n-1)\Delta t\right) \quad 1 \le n \le n_T \\ &= \left\{\sin\left(2\pi\alpha(n_s + n - 1)/\gamma\right) + A_2 \sin\left(2\pi\beta(n_s + n - 1)/\gamma\right)\right\} w\left((n-1)\Delta t\right) \end{aligned} \tag{10.9}$$

Notice that the window function is independent of n_s.

The following commands are used to access the selected window functions

```
wt=DirichletWindow[arg]
wt=HammingWindow[arg]
wt=HannWindow[arg]
wt=BlackmanWindow[arg]
wt=KaiserWindow[arg]
wt=NuttallWindow[arg]
```

where **arg** is the independent variable defined over the region $-1/2 \le$ **arg** $\le 1/2$. Therefore, in order to apply these functions to Eq. (10.9), **arg** is replaced by **(n-1)/(n$_T$-1)-1/2**, where $1 \le n \le n_T$.

Example 10.3

Effects of Windows on Spectral Analysis

We shall construct an interactive interface to show the effects of these window functions on $|S_{w,n}(\omega_n)|$, where ω_n is the nth discrete frequency. It is assumed that $f_o = 10$ Hz and $\gamma = 50$. The program that creates the interactive graphic that appears in Figure 10.10 is as follows.

```
Manipulate[nt=g m; arg=(n-1)/(nt-1)-1/2;
                    (* Select window *)
  win=Which[wf==1,DirichletWindow[arg],
    wf==2,HammingWindow[arg],
    wf==3,HannWindow[arg],
    wf==4,BlackmanWindow[arg],
```

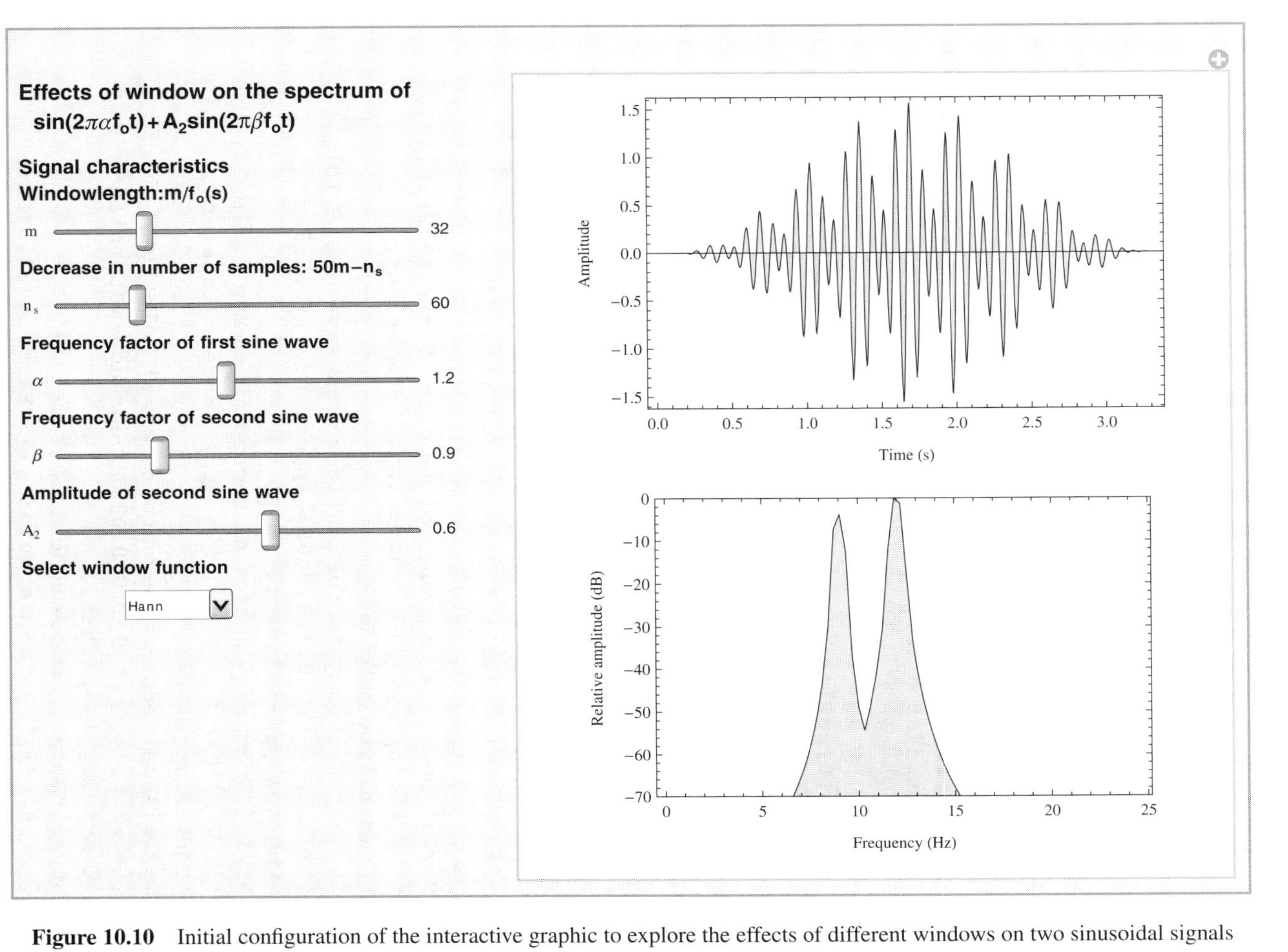

Figure 10.10 Initial configuration of the interactive graphic to explore the effects of different windows on two sinusoidal signals

```
  wf==5,KaiserWindow[arg],
  wf==6,NuttallWindow[arg]];
    (* Create windowed time-domain data and format for plotting *)
 dat=Table[(Sin[2 π α (ns+n-1)/g]+
  a2 Sin[β 2 π (ns+n-1)/g ]) win,{n,1,nt}];
 timns=Table[{(ns+n-1) dt,dat[[n]]},{n,1nt}];

(* Perform Fourier transform and convert to dB and format for plotting *)
 ft=Fourier[dat,FourierParameters->{1,-1}];
 dwmax=2. Max[Abs[ft]]/nt;
 datf=Table[{(k-1)/(nt dt),
  20. Log10[2. Abs[ft[[k]]]/nt/dwmax]},{k,1,nt/2}];
 last=Round[25 m/fo];
                    (* Display results *)
 lp1=ListLinePlot[timns, Frame->True,FrameLabel->lab1];
 lp2=ListLinePlot[datf[[1;;last]],Filling->Axis,
  PlotRange->{-70,0},Frame->True,FrameLabel->lab2];
 GraphicsColumn[{lp1,lp2}],
                    (* Create sliders *)
 Style["Effects of windows on the spectrum of",Bold,12],
 Style["  sin(2παfot)+A2sin(2πβfot)",Bold,12],
 " ",
 Style["Signal characteristics",Bold,11],
 Style["Window length: m/fo (s)",Bold,10],
 {{m,32,"m"},4,128,4,Appearance->"Labeled",
  ControlType->Slider},
 Style["Decrease in number of samples: 50m-ns]",Bold,10],
 {{ns,60,"ns"},0,300,2,Appearance->"Labeled",
  ControlType->Slider},
 Style["Frequency factor of first sine wave",Bold,10],
 {{α,1.2,"α"},0.5,2.0,0.025,Appearance->"Labeled",
  ControlType->Slider},
 Style["Frequency factor of second sine wave",Bold,10],
 {{β,0.9,"β"},0.5,2.0,0.025,Appearance->"Labeled",
  ControlType->Slider},
 Style["Amplitude of second sine wave",Bold,10],
 {{a2,0.6,"A2"},0.0,1.0,0.01,Appearance->"Labeled",
  ControlType->Slider},
 Style["Select window function",Bold,10],
 {{wf,3,""},{1->"Rectangular",2->"Hamming",3->"Hann",
  4->"Blackman",5->"Kaiser",6->"Nuttall"},
  ControlType->PopupMenu},
 TrackedSymbols:>{ns,m,a2,α,wf,β},
```

```
Initialization:>{fo=10.; g=50; dt=1/(g fo);
 lab1={"Time (s)","Amplitude"};
 lab2={"Frequency (Hz)","Relative amplitude (dB)"}}]
```

10.5.3 Spectrum Averaging

We shall build on the results of the preceding section and examine the averaging of the spectral content of sinusoidal signals in the presence of noise. Two types of averaging of periodic signals will be considered; root mean square averaging and vector averaging. Root mean square averaging averages the magnitudes of the spectral content on a frequency-by-frequency basis. This type of averaging tends to smooth the spectral magnitudes of the signals of interest, but doesn't improve the signal-to-noise ratio of the spectrum. For vector averaging, which is also called synchronous averaging, the real parts and imaginary parts of each frequency component are averaged separately. This type of averaging requires a periodically occurring event to trigger the acquisition of the signal. Equation (10.9) is such a signal so that vector averaging can be demonstrated using this signal. This type of averaging can greatly suppress the noise.

The signal that we shall be considering is given by Eq. (10.9), but modified as follows to include additive noise $r(t)$ and account for the kth segment of the signal

$$s_{wr,n}^{(k)}(t) = \{\sin(2\pi\alpha(n_s + n - 1)/\gamma) + A_2 \sin(2\pi\beta(n_s + n - 1)/\gamma) + r((n-1)\Delta t)\} w((n-1)\Delta t) \qquad 1 \le n \le n_T \tag{10.10}$$

and the superscript k indicates the kth segment of the signal of length $n_T\Delta t = n_T/(\gamma f_o)$. We are interested in the manipulation of the components of the Fourier transform of Eq. (10.10) of the kth segment, which are denoted as $S_{wr,n}^{(k)}(\omega_n)$, where ω_n is the nth discrete frequency in rad·s^{-1}. For root mean square averaging, the magnitude is given by

$$\bar{S}_{rms,n}(\omega_n) = \frac{1}{k_{avg}} \sum_{k=1}^{k_{avg}} \left| S_{wr,n}^{(k)}(\omega_n) \right|$$

and for vector averaging by

$$\bar{S}_{vector,n}(\omega_n) = \frac{1}{k_{avg}} \left| \sum_{k=1}^{k_{avg}} S_{wr,n}^{(k)}(\omega_n) \right|$$

where k_{avg} is the number of segments averaged.

Example 10.4

Spectrum Averaging

Using the program given in Example 10.3 as the starting point, the following program illustrates the effects of windows on the root mean square average of signals and the vector average of signals. It is assumed that $\gamma = 50$ and $f_o = 10$ Hz. It creates the interactive graphic shown in Figure 10.11.

```
Manipulate[nt=γ m; arg=(n-1)/(nt-1)-1/2;
                    (* Selected window specified *)
  win=Table[Which[wf==1,DirichletWindow[(n-1)/(nt-1)-1/2],
   wf==2,HammingWindow[(n-1)/(nt-1)-1/2],
   wf==3,HannWindow[(n-1)/(nt-1)-1/2],
   wf==4,BlackmanWindow[(n-1)/(nt-1)-1/2],
   wf==5,KaiserWindow[(n-1)/(nt-1)-1/2],
   wf==6,NuttallWindow[(n-1)/(nt-1)-1/2]],{n,1,nt}];
  ftavg=ConstantArray[0,nt];
  dat1=Table[(Sin[2 πα (ns+n-1)/γ]+
    a2 Sin[β 2 π (ns+n-1)/γ]),{n,1,nt}];
(* Average spectral components according to chosen averaging method *)
  Do[ran=RandomReal[{-rd,rd},nt];
   dat=(dat1+ran) win;
   ft=Fourier[dat,FourierParameters->{1,-1}];
   If[avgtyp==1,ftavg=ftavg+ft,ftavg=ftavg+Abs[ft]],
  {kk,1,kavg}];
  dwmax=If[avgtyp==1,2. Max[Abs[ftavg]/kavg]/nt,
   2. Max[ftavg/kavg]/nt];
  datf=If[avgtyp==1,
   Table[{(k-1)/(nt dt),
    20. Log10[2. Abs[ftavg[[k]]]/kavg/nt/dwmax]},
    {k,1,nt/2}],
   Table[{(k-1)/(nt dt),
    20. Log10[2. ftavg[[k]]/kavg/nt/dwmax]},
    {k,1,nt/2}]];
  last=Round[25 m/fo];
                    (* Display results *)
  ListLinePlot[datf[[1;;last]],Filling->Axis,
   PlotRange->{-70,0},Frame->True,FrameLabel->lab2],
                    (* Create buttons and sliders *)
  Style["Effects of windows on the spectrum of",Bold,12],
  Style["  sin(2παfot)+A2sin(2πβfot)+rdnoise(t)",Bold,12],
  " ",
  Style["Averaging type",Bold,10],
  {{avgtyp,1,""},{1->"Vector",2->"Rms"},
   ControlType->RadioButtonBar},
  Style["Number of averages",Bold,10],
  {{kavg,10,"kavg"},1,50,1,
```

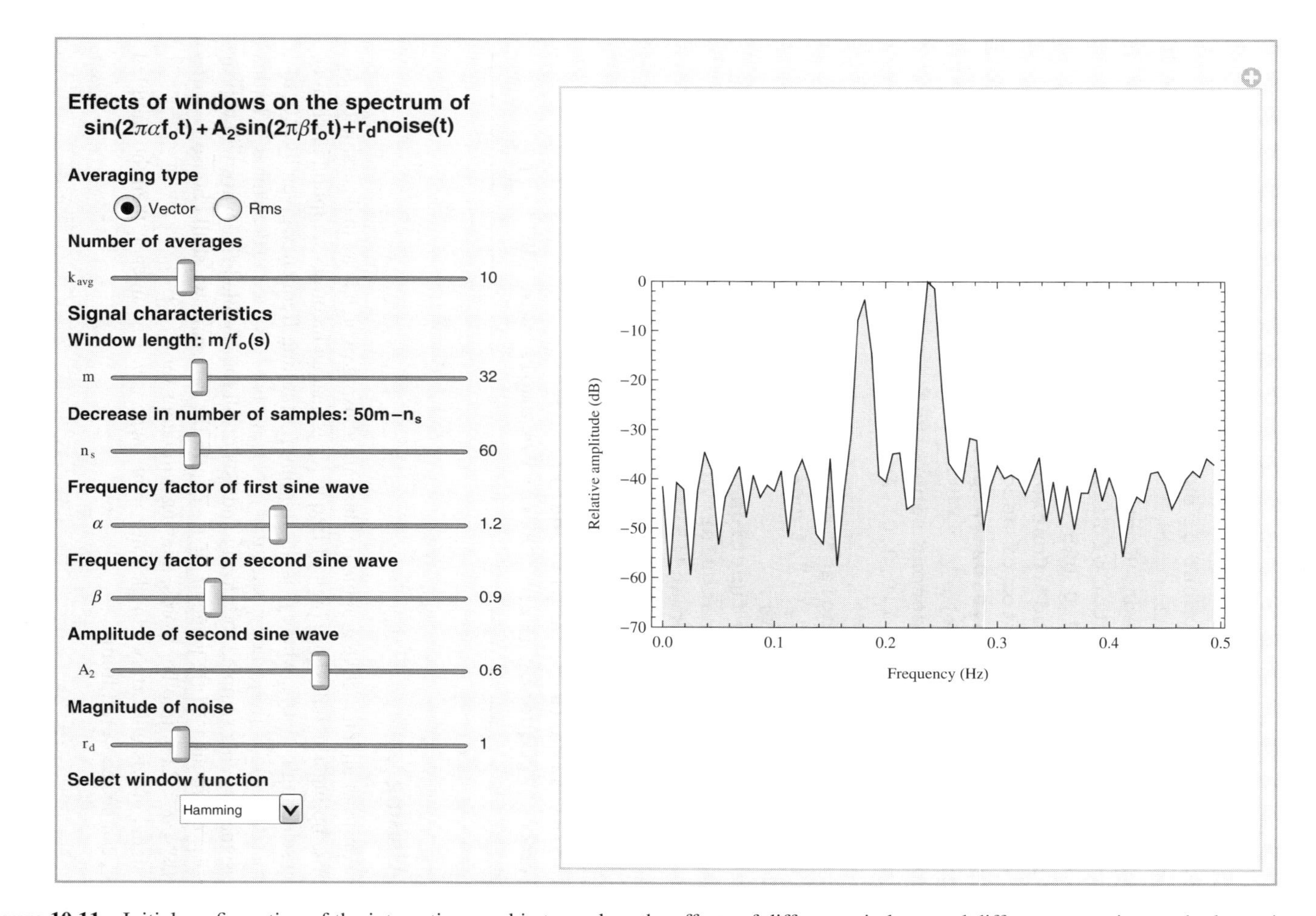

Figure 10.11 Initial configuration of the interactive graphic to explore the effects of different windows and different averaging methods on sinusoidal signals in the presence of noise

```
  Appearance->"Labeled",ControlType->Slider},
 Style["Signal characteristics",Bold,11],
 Style["Window length: m/fo(s)",Bold,10],
 {{m,32,"m"},4,128,4,Appearance->"Labeled",
  ControlType->Slider},
 Style["Decrease in number of samples: 50m-ns",Bold,10],
 {{ns,60,"ns"},0,300,2,
  Appearance->"Labeled", ControlType->Slider},
 Style["Frequency factor of first sine wave",Bold,10],
 {{α,1.2,"α"},0.5,2.0,0.025,
  Appearance->"Labeled",ControlType->Slider},
 Style["Frequency factor of second sine wave",Bold,10],
 {{β,0.9,"β"},0.5,2.0,0.025,Appearance->"Labeled",
  ControlType->Slider},
 Style["Amplitude of second sine wave",Bold,10],
 {{a2,0.6,"A2]"},0.0,1.0,0.01,Appearance->"Labeled",
  ControlType->Slider},
 Style["Magnitude of noise",Bold,10],
 {{rd,1,"rd"},0.0,6.0,0.5,Appearance->"Labeled",
  ControlType->Slider},
 Style["Select window function",Bold,10],
 {{wf,2,""},{1->"Rectangular",2->"Hamming",
  3->"Hann",4->"Blackman",5->"Kaiser",6->"Nuttall"},
  ControlType->PopupMenu},
 TrackedSymbols:>{ns,m,a2,α,wf,β,rd,kavg,avgtyp},
 Initialization:>{fo=10.; γ=50; dt=1/(γ fo);
  lab1={"Time (s)","Amplitude"};
  lab2={"Frequency (Hz)","Relative amplitude (dB)"}}]
```

10.6 Aliasing

We shall construct an interactive graphic to illustrate aliasing, which is an undesirable consequence of sampling a signal whose highest frequency is f_c at an incorrect sampling interval t_s; that is, by selecting $t_s \geq 1/f_s$, where f_s is the minimum sampling frequency given by $f_s > \alpha f_c$. Consider a sine wave of frequency f_o such that $f_o > f_c$, where f_c is used to determine f_s. In this case, the signal is being sampled too slowly and as a result the sampled sine wave will appear as a sine wave with aliased frequency $f_a = f_s - f_o$. In the program, we shall choose $f_s = \alpha f_o$, where $1.09 < \alpha < 2.06$. In other words, aliasing will occur for $\alpha \leq 2$.

If the signal $g(t)$ is uniformly sampled at $t = nt_s$, where n is an integer, then the sampled waveform can be recovered from

$$g(t) = \sum_{n=-k}^{k} \frac{g(nt_s)\sin(2\pi f_o(t - nt_s))}{2\pi f_o(t - nt_s)} \tag{10.11}$$

For our purposes, it has been empirically found that

$$k = \left\lceil \frac{5}{f_o t_s} \right\rceil$$

gives good results. This quantity is determined by using **`Ceiling`**.

Example 10.5

Aliasing

We shall create an interactive graphic shown in Figure 10.12, which demonstrates aliasing by using the preceding results. The program that produces this figure is as follows.

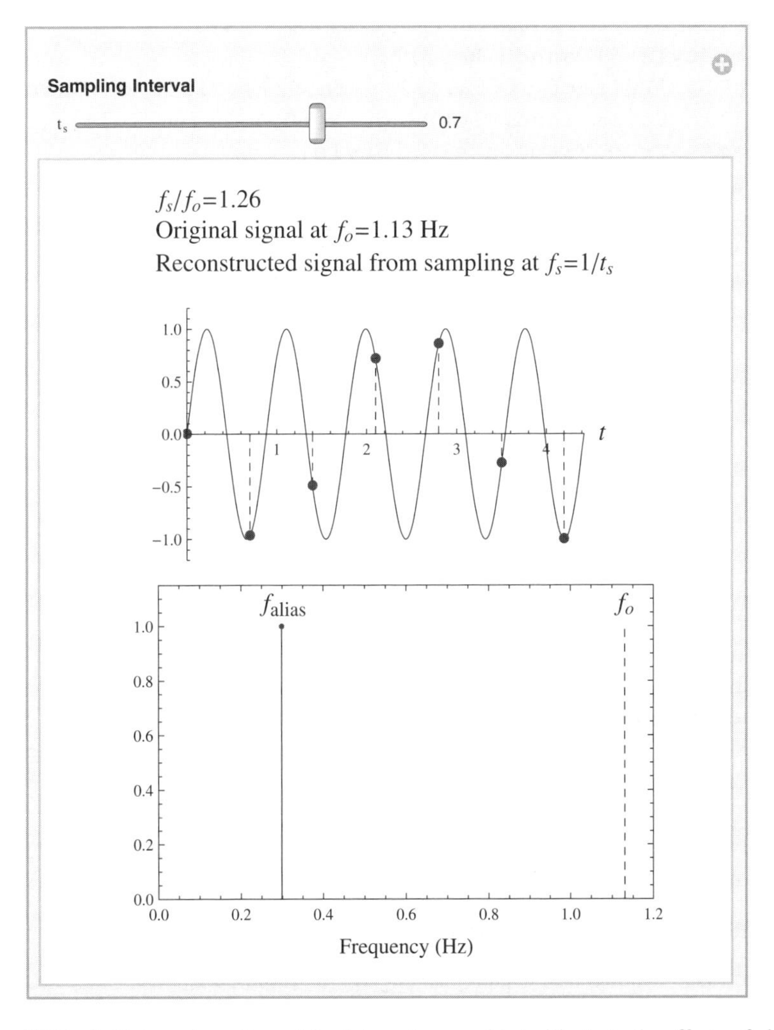

Figure 10.12 Initial configuration of the interactive graphic to illustrate the effects of aliasing

```
Manipulate[a1=ts f0;   nmax=Ceiling[5/f0/ts]; fa=(1-a1)/ts;
  sam=Table[{n ts,Sin[2 n π f0 ts]},{n,0,nmax}];
                    (* Plot label *)
  tab=Column[{Style[Row[{"fs/fo=",NumberForm[1/a1,3]}],14],
   seq}];
                    (* Display results *)
  GraphicsColumn[{Show[Plot[Sin[2 π f0 t],{t,0,5/f0},
    PlotRange->{{0,5/f0},{-1.2,1.2}},PlotStyle->Blue,
    PlotLabel->tab,AxesLabel->{Style["t",14,Italic],""}],
   ListPlot[sam,Filling->Axis,PlotMarkers->Automatic,
    FillingStyle->{Red,Dashed,Dashed}],
   Plot[rest[t,ts,2 nmax,f0],{t,0,nmax ts},
    PlotStyle->{Dashed,Red}]],
   ListPlot[If[fa<f0,{{fa,1}},{{fa,0}}],Filling->Axis,
    FillingStyle->Red,PlotRange->{{0,1.2},{0,1.15}},
    Frame->True,FrameLabel->{Style[
     "Frequency (Hz)",12],""},
    Epilog->{Inset[Style["fo",Blue,14],{f0,1.08}],
    If[fa<fo,Inset[Style["falias",Red,14],{fa,1.08}]],
    {Blue,Dashed,Line[{{f0,0},{f0,1.0}}]}}]},
    ImageSize->Medium],
                    (* Create slider *)
  Style["Sampling Interval",Bold,10],
  {{ts,0.7,"ts"},0.43,0.81,0.01,Appearance->"Labeled",
   ControlType->Slider},
  TrackedSymbols:>{ts},
      (* Plotting label and function representing Eq. (10.11) *)
  Initialization:>{fo =1.13;
   rest[t_,ts_,nmax_,f0_]:=Total[Table[
    Sin[2 n π f0 ts] Sinc[π (t-n ts)/ts],
     {n,-nmax,nmax,1}]];
   seq=Sequence[Style["Original signal at fo=1.13 Hz",
    Blue,14],Style["Reconstructed signal from sampling
    at fs=1/ts",Red,14]]}]
```

10.7 Functions Introduced in Chapter 10

A list of functions introduced in Chapter 10 is given in Table 10.2.

Table 10.2 Commands introduced in Chapter 10

Command	Usage
`BlackmanWindow`	Represents a Blackman window function
`BodePlot`	Creates a Bode plot from a state-space model or a transfer function model for a time-invariant linear system
`ButterworthFilterModel`	Creates a Butterworth low-pass, high-pass, band-pass, and band-stop filter model
`Chebyshev1FilterModel`	Creates a Chebyshev1 low-pass, high-pass, band-pass, and band-stop filter model
`Chebyshev2FilterModel`	Creates a Chebyshev2 low-pass, high-pass, band-pass, and band-stop filter model
`DirichletWindow`	Represents a Dirichlet (rectangular) window function
`EllipticFilterModel`	Creates an elliptic low-pass, high-pass, band-pass, and band-stop filter model
`HammingWindow`	Represents a Hamming window function
`HannWindow`	Represents a Hann window function
`KaiserWindow`	Represents a Kaiser window function
`NicholsPlot`	Creates a Nichols plot from a state-space model or a transfer function model
`NuttallWindow`	Represents a Nuttall window function
`OutputResponse`	Obtains the numerical output response from a state-space model or a transfer function model for a specified input
`PIDTune`	Creates a PID controller for direct use with a linear time-invariant system
`RootLocusPlot`	Creates a root locus plot from a state-space model or a transfer function model for a time-invariant linear system
`StateSpaceModel`	Creates a standard state-space model or converts a transfer function model to a state-space model
`SystemsModelFeedbackConnect`	Connects the output of system to its input using negative or positive feedback
`SystemsModelSeriesConnect`	Connects two state-space models or two transfer function models in series
`TransferFunctionModel`	Creates a transfer function model or converts a state-space model to a transfer function model

Reference

[1] D. K. Anand and R. B. Zmood, *Introduction to Control Systems*, 3rd edn, Butterworth-Heinemann Ltd. Oxford England, 1995.

11

Heat Transfer and Fluid Mechanics

11.1 Introduction

Several topics in heat transfer and fluid mechanics are examined numerically and interactive environments are developed to explore the characteristics of the different systems. In particular, the following applications are considered:

Conduction heat transfer –

- transient heat transfer in slabs, cylinders, and spheres subject to various initial conditions, boundary conditions, and a constant source;
- ablation of a tumor, which is modeled as two concentric spheres;
- efficiency of longitudinal and radial fins.

Convection heat transfer –

- natural convection from a heated vertical plate.

Radiation heat transfer –

- view factor between two parallel rectangular surfaces.

Internal viscous flow –

- laminar flow in horizontal pipes;
- flow in three reservoirs.

External flow –

- pressure distribution around a Joukowski airfoil;
- surface profile of nonuniform flow in open channels.

In addition, an interactive graphic for heat conduction in a slab has been demonstrated in Example 5.13 and for fluid mechanics, the air entrapment of liquid jets has been given in Example 5.14, the flow around a cylinder with circulation appears in Example 6.6, and the streamlines of flow around an ellipse have been demonstrated in Example 7.6.

An Engineer's Guide to Mathematica®, First Edition. Edward B. Magrab.

Companion Website: www.wiley.com/go/magrab

11.2 Conduction Heat Transfer

11.2.1 One-Dimensional Transient Heat Diffusion in Solids

The flow of heat in three geometric shapes will be examined: a slab of uniform thickness $2L_o$; a very long solid circular cylinder of radius $x = L_o$; and a solid sphere of radius $x = L_o$. For the slab, it is assumed that the surface temperature is the same on both sides of the center plane. For each of these solids, it will be assumed that their initial temperature is $f(x)$ and that the heat is dissipated from their respective surfaces at $x = L_o$ by convection into an environment whose temperature varies with time. It will also be assumed that heat is generated at a rate of $g(x,t)$ W·m^{-3}. If the temperature in these solids as a function of location and time is denoted $T = T(x,t)$, then the governing equation for these three solids in nondimensional form is

$$\frac{\partial \theta}{\partial \tau} = \frac{1}{\eta^n}\frac{\partial}{\partial \eta}\left(\eta^n \frac{\partial \theta}{\partial \eta}\right) + G(\eta, \tau) \tag{11.1}$$

where for a slab $n = 0$, for a cylinder $n = 1$, and for a sphere $n = 2$. In addition,

$$\eta = \frac{x}{L_o}, \quad \tau = \frac{\alpha t}{L_o^2}, \quad \theta(\eta, \tau) = \frac{T(\eta, \tau) - T_\infty}{T_i - T_\infty}, \quad G(\eta, \tau) = \frac{g(\eta, \tau)L_o^2}{k(T_i - T_\infty)}$$

where α is the thermal diffusivity, k is the thermal conductivity, T_∞ is the ambient temperature of the medium surrounding the solid, and T_i is the initial temperature of the solid.

The initial condition can be expressed as

$$\theta(\eta, 0) = \frac{f(\eta) - T_\infty}{T_i - T_\infty} = F(\eta)$$

At the boundary $\eta = 0$, it is assumed that

$$\frac{\partial \theta(0, \tau)}{\partial \eta} = 0$$

At $\eta = 1$, three boundary conditions will be considered. The first boundary condition is one in which a time-dependent heat flux is applied to the boundary. In this case,

$$\frac{\partial \theta(1, \tau)}{\partial \eta} = Q_o q(\tau)$$

where

$$Q_o = \frac{q_x'' L_o}{k(T_i - T_\infty)}$$

and q_x'' is the heat flux per unit area.

The second boundary condition is one in which there is heat transfer by convection. In this case, if we express the heat convection as

$$h\left[T_\infty + (T_i - T_\infty)c(\tau)\right]$$

then the boundary condition can be written as

$$\theta(1,\tau) + \frac{1}{Bi}\frac{\partial\theta(1,\tau)}{\partial\eta} = c(\tau)$$

where $Bi = hL_o/k$ is the Biot number.

The third boundary condition is one in which the temperature on the surface varies with time. In this case, if we express the temperature as

$$T_\infty + (T_i - T_\infty)b(\tau)$$

then the boundary condition can be written as

$$\theta(1,\tau) = b(\tau)$$

We shall create an interactive graphic that determines the temperature distribution $\theta(\eta,\tau)$ for $G(\eta,\tau) = G_o$, a constant, and for the following different combinations of initial conditions and boundary conditions at $\eta = 1$. It will be noted that in certain cases a rapidly decaying exponential function has been introduced to ensure that the initial condition and the surface boundary condition are consistent at $\eta = 1$ and $\tau = 0$. Recall Example 5.13.

Case 1: $\theta(1,\tau) = 0$ $[T(L_o,\tau) = T_\infty]$ and $\theta(\eta,0) = F(\eta)$, where $F(\eta)$ can be any of the following three spatial distributions

$$F(\eta) = 1 - e^{-1000(\eta-1)^2} \quad \text{or} \quad 1-\eta \quad \text{or} \quad 1-\eta^2$$

Case 2: $\theta(\eta,0) = 0$ $[F(\eta) = 0]$ and $\theta(1,\tau) = b(\tau)$, where $b(\tau)$ can assume any of the following three temporal relations

$$b(\tau) = 1 - e^{-1000\tau} \quad \text{or} \quad 1 - e^{-p\tau} \quad \text{or} \quad 1 - \cos p\tau$$

where $p = aL_o^2/\alpha$ and a is a positive constant.

Case 3: $\theta(\eta,0) = 0$ $[F(\eta) = 0]$ and

$$\theta(1,\tau)e^{-1000\tau} + \left(\frac{\partial\theta(1,\tau)}{\partial\eta} - Q_o e^{-p\tau}\right)\left(1 - e^{-1000\tau}\right) = 0$$

Case 4: $\theta(\eta,0) = 0$ $[F(\eta) = 0]$ and

$$\theta(1,\tau) + \left(\frac{1}{Bi}\frac{\partial\theta(1,\tau)}{\partial\eta} - c(\tau)\right)\left(1 - e^{-1000\tau}\right) = 0$$

where $c(\tau)$ can assume either of the following two temporal relations

$$c(\tau) = e^{-p\tau} \quad \text{or} \quad \cos p\tau$$

The program that creates the interactive graphic that appears in Figure 11.1 is as follows.

```
Manipulate[fn=If[bcic==1,
    (* Specify selected boundary conditions and initial conditions *)
  Which[feta==1,1-Exp[-1000 (η-1)^2],
    feta==2,1-η,feta==3,1-η^2],0];
    bc1=Which[bcic==1,θ[1τ]==0,
     bcic==2,θ[1,τ]==Which[phibc==1,1-Exp[-1000. τ],
      phibc==2,1-Exp[-ptemp τ],
      phibc==3,1-Cos[ptemp τ]],
     bcic==3,θ[1,τ] Exp[-1000 τ]+((D[θ[η,τ],η]/.η->1)-
      fluxbc Exp[-pflux τ]) (1-Exp[-1000 τ])==0,
     bcic==4,Which[
      psibc==1,θ[1,τ]+((D[θ[η,τ],η]/.η->1)/biot-
         Exp[-pconv τ]) (1-Exp[-1000 τ])==0,
      psibc==2,θ[1,τ]+((D[θ[η,τ],η]/.η->1)/biot-
         Cos[pconv τ]) (1-Exp[-1000 τ])==0]];
                           (* Solve Eq. (11.1) *)
  temp=NDSolveValue[{
   D[θ[η,τ],τ]==D[η^n D[θ[η,τ],η],η]+go,
     (θ[η,0])==fn, (D[θ[η,τ],η]/.η->del)==0,bc1},
    θ,{η,del,1},{τ,0,3}];

                           (* Plot results *)
  Plot3D[temp[η,τ],{η,del,1},{τ,0,3},
   AxesLabel->{"η","τ","θ(η,τ)"}],
                    (* Create sliders and radio buttons *)
  Style["Heat Flow in Solids",Bold,12],
  "",
  Style["Type of solid",Bold,10],
  {{n,1,""},{1->"Slab",2->"Cylinder",3->"Sphere"},
   ControlType->RadioButton},
  "",
  Style["Constant heat generation",Bold,10],
  {{go,0,"Go"},0,5,0.1,Appearance->"Labeled",
   ControlType->Slider},
  "",
  Style["Select boundary and initial conditions",Bold,10],
```

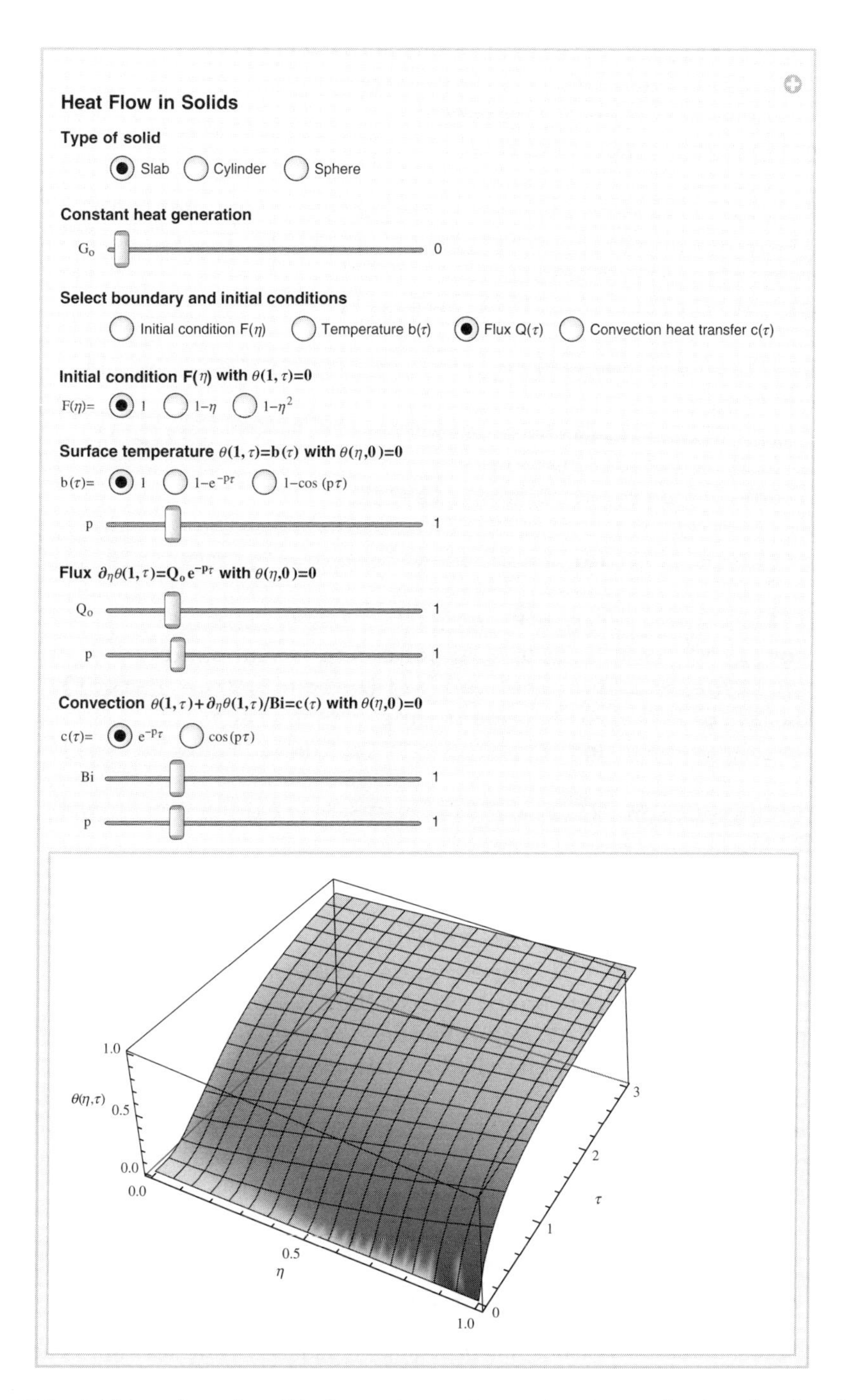

Figure 11.1 Initial configuration of the interactive graph to explore the transient heat flow in solids for different surface conditions and initial conditions

```
  {{bcic,3,""},{1->"Initial condition F(η)",
   2->"Temperature b(τ)",3->"Flux Q(τ)",
   4->"Convection heat transfer c(τ)"},
   ControlType->RadioButton},
  "",
  Style["Initial condition F(η) with θ(1,τ)=0",Bold,10],
  {{feta,1,"F(η)="},{1->"1",2->"1-η",3->"1-η^2"},
   ControlType->RadioButton,
   Enabled->If[bcic==1,True,False]},
  "",
  Style["Surface temperature θ(1,τ)=b(τ) with θ(1,τ)=0",
   Bold,10],
 {{phibc,1,"b(τ)="},{1->"1",2->"1-e^-pτ",3->"1-cos(pτ)"},
  ControlType->RadioButton,
  Enabled->If[bcic==2,True,False]},
 {{ptemp,1,"p"},0.1,5,0.1,Appearance->"Labeled",
  ControlType->Slider,Enabled->If[bcic==2,True,False]},
 "",
 Style["Flux ∂_ηθ(1,τ)=Q_o e^-pτ with θ(1,τ)=0",Bold,10],
 {{fluxbc,1,"Q_o"},0.1,5,0.1,Appearance->"Labeled",
  ControlType->Slider,Enabled->If[bcic==3,True,False]},
 {{pflux,1,"p"},0,5,0.1,Appearance->"Labeled",
  ControlType->Slider,Enabled->If[bcic==3,True,False]},
 "",
 Style["Convection θ(1,τ)+ ∂_ηθ(1,τ)/Bi=c(τ) with θ(1,τ)=0",
  Bold,10],
 {{psibc,1,"c(τ)="},{1->"e^-pτ",2->"cos(pτ)"},
  ControlType->RadioButton,
  Enabled->If[bcic==4,True,False]},
 {{biot,1,"Bi"},0,5,0.1,Appearance->"Labeled",
  ControlType->Slider,Enabled->If[bcic==4,True,False]},
 {{pconv,1,"p"},0,5,0.1,Appearance->"Labeled",
  ControlType->Slider,Enabled->If[bcic==4,True,False]},
 ControlPlacement->Top,
 TrackedSymbols:>{n,bcic,feta,phibc,fluxbc,psibc,biot,pconv,
  go,ptemp,pflux},
 Initialization:>(del=10^(-8))]
```

11.2.2 Heat Transfer in Concentric Spheres: Ablation of a Tumor

We shall consider the destruction of a spherical tumor of radius r_t by the insertion of an energy source of magnitude S and duration t_o. The tumor is surrounded by a concentric sphere of healthy tissue of outer radius r_b and temperature T_b. We shall further assume that the tumor and the healthy tissue have the same blood flow and, therefore, the same metabolic release rate S_m. In addition, the energy is carried away by the blood flow of mass flow rate per unit volume $\dot{m}_b$. The blood has a specific heat c_b. This type of biological system can be modeled as [1]

$$\frac{\partial \bar{T}}{\partial \tau} = \frac{1}{\eta^2}\frac{\partial}{\partial \eta}\left(\eta^2 \frac{\partial \bar{T}}{\partial \eta}\right) + \hat{S}U(\eta,\tau) + \hat{S}_m + \gamma\left(1 - \bar{T}\right) \tag{11.2}$$

where $\bar{T} = T(r,t)/T_b \rightarrow T(\eta,\tau)/T_b$ and

$$\begin{gathered} \tau = \frac{\alpha t}{r_b^2}, \quad \tau_o = \frac{\alpha t_o}{r_b^2}, \quad \gamma = \frac{\dot{m}_b c_b T_b}{\beta}, \quad \beta = \frac{\rho c \alpha T_b}{r_b^2} \quad \text{W}\cdot\text{m}^{-3}, \\ \hat{S} = \frac{S}{\beta}, \quad \hat{S}_m = \frac{S_m}{\beta}, \quad \eta = \frac{r}{r_b}, \quad \eta_t = \frac{r_t}{r_b} \\ U(\eta,\tau) = \left(u(\tau) - u(\tau_o)\right)\left(u(\eta) - u(\eta_t)\right) \end{gathered} \tag{11.3}$$

In these equations, $u(x)$ is the unit step function and α is the thermal diffusivity of the tissue, c its specific heat, and ρ its density. The function $U(\eta,\tau)$ defines the region over which the inserted energy S is applied. Also, T and T_b are expressed in degrees K. To express the results in °C, we use

$$T_c = T_b\bar{T} - 273.15$$

where T_c is the value of T in °C.

When solving these types of applications to the human body, the temperature of the region surrounding the tumor is often of interest. If the healthy tissue is subjected to a temperature exceeding a certain value, usually around 45 – 47 °C (318.15 – 320.15 K), the healthy tissue will die. The goal is to have only the tumor subjected to a temperature considerably higher than this value in order to kill the tumor. To identify the various temperature regions, which will subsequently be identified with two intersecting planes, we shall solve Eq. (11.2) with the following boundary conditions

$$\left.\frac{\partial \bar{T}}{\partial \eta}\right|_{\eta=\delta} = 0 \quad \text{and} \quad \left.\frac{\partial \bar{T}}{\partial \eta}\right|_{\eta=1} = 0$$

and with the following initial condition

$$\bar{T}(\eta, 0) = 1$$

Because the solution to Eq. (11.2) has a discontinuity at $\eta = 0$, we replace "zero" with $\delta = 10^{-8}$.

The following values for the system's parameters are assumed: $\dot{m}_b$= 0.18 kg·m^{-3}·s^{-1}, c_b = 3300 J·kg^{-1}·K^{-1}, S_m = 145 W·m^{-3}, T_b = 310.15 K, $\alpha = 10^{-7}$ m^2·s^{-1}, ρ = 850 kg·m^{-3},

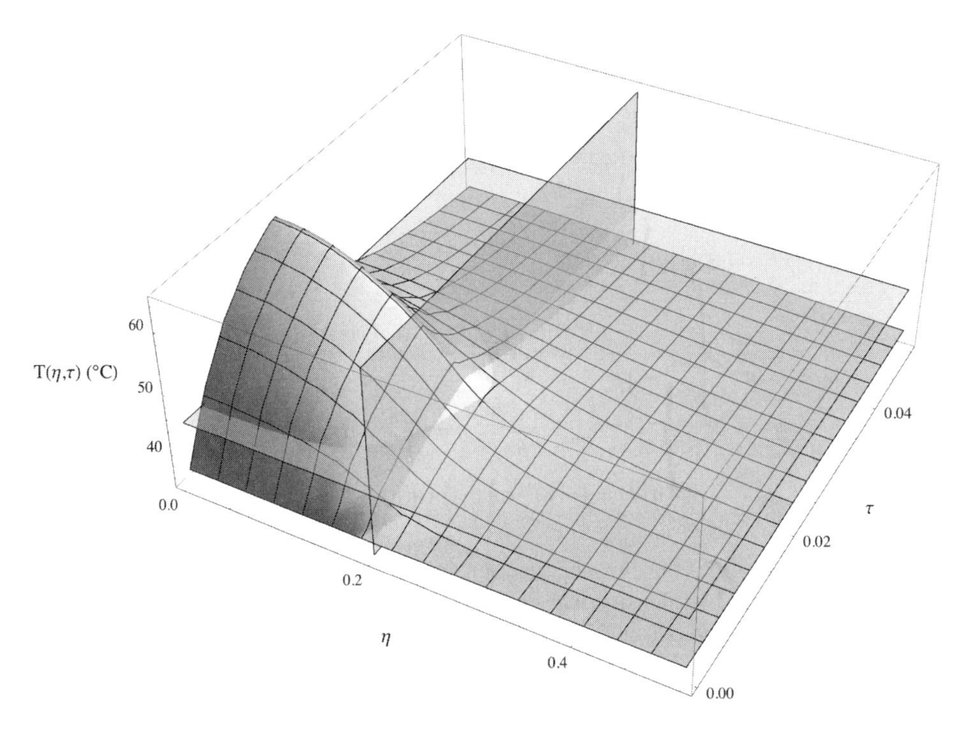

Figure 11.2 Temperature distribution within a sphere containing a concentric spherical tumor

$c = 3800\ \text{J}\cdot\text{kg}^{-1}\cdot\text{K}^{-1}$, $S = 4\times10^5\ \text{W}\cdot\text{m}^{-3}$, $r_t = 0.01$ m, $r_b = 0.05$ m, and $t_o = 400$ s. Substitution of these values into Eq. (11.3) yields

$$\tau_o = 0.016, \quad \gamma = 4.598, \quad \beta = 40{,}052\ \text{W}\cdot\text{m}^{-3},$$
$$\hat{S} = 9.987, \quad \hat{S}_m = 0.00362, \quad \eta_t = 0.2$$

The numerical results, which are shown in Figure 11.2, are obtained with the following program.

```
γ=4.598;  Ss=9.987;  Sm=0.00362;  τo=0.016;  ηt=0.2;
del=10^(-8); tb=310.15;
app=(UnitStep[τ]-UnitStep[τ-τo])*
  (UnitStep[η]-UnitStep[η-ηt]);

                    (* Solve Eq. (11.2) *)

temp=NDSolveValue[
  {D[T̄[η,τ],τ]== D[η^2 D[T̄[η,τ],η],η]/η^2+Ss app+
  γ (1-T̄[η,τ])+Sm,T̄[η,0]==1,(D[T̄[η,τ],η]/.η->1)==0,
  (D[T̄[η,τ],η]/.η->del)==0},T̄,{η,del,1},{τ,0,0.1}];
```

(* Plot results *)

```
Show[Plot3D[tb temp[η,τ]-273.15,{η,del,0.5},{τ,0,0.05},
   PlotRange->All,ImageSize->Large,
   AxesLabel->{Style["η",12],Style["τ",12],
   Style["T(η,τ) (°C)",12]}],
  Graphics3D[{Opacity[0.5],Polygon[{{0,0,45},
   {0.0,0.05,45},{0.5,0.05,45},{0.5,0.,45}}]}],
  Graphics3D[{Opacity[0.5],Polygon[{{0.2,0,35},{0.2,0,65},
   {0.2,0.05,65},{0.2,0.05,35}}]}]]
```

11.2.3 Heat Flow Through Fins

Fins are used to enhance heat convective heat transfer from surfaces by means of natural convection wherein the heat transfer coefficient h is relatively small. There are two types of fin geometry that we shall consider: the longitudinal fin, which is shown in Figure 11.3a, and the radial fin, which is shown in Figure 11.3b. Fins are typically longer than they are thick and therefore, it is common to assume that the temperature varies along the length direction and is constant through its thickness. The nondimensional form of the governing equation for steady-state heat flow and no volumetric heat generation in a longitudinal fin is [2]

$$\frac{d}{d\eta}\left(\bar{a}(\eta)\frac{d\theta}{d\eta}\right) - m_l^2\theta = 0 \tag{11.4}$$

where it has been assumed that $t/d_o \ll 1$. In Eq. (11.4), the following parameters have been introduced

$$\eta = \frac{x}{L}, \quad \theta = \frac{T - T_\infty}{T_b - T_\infty}, \quad m_l^2 = \frac{2hL^2}{kt}, \quad \bar{a}(\eta) = \frac{2}{t}y(\eta)$$

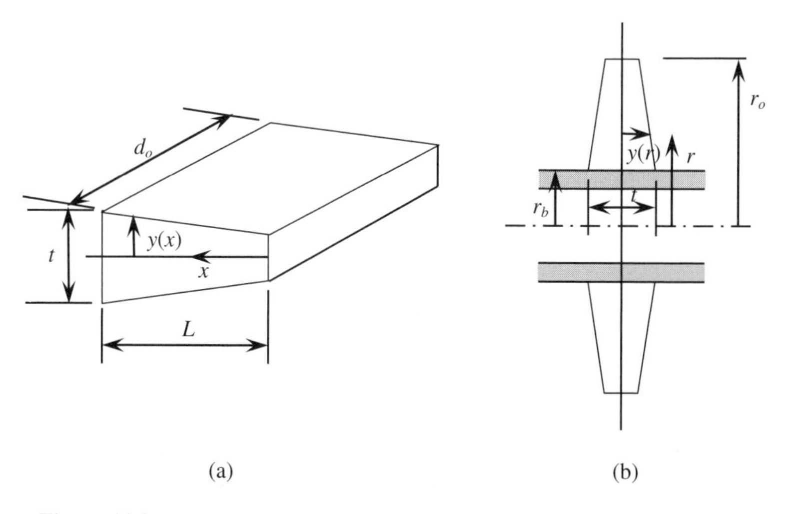

Figure 11.3 (a) Geometry of a longitudinal fin; (b) geometry of a radial fin

where $T = T(\eta)$ is the temperature along the length of the fin, T_b is the temperature at the base, T_∞ is the ambient temperature, and k is the thermal conductivity of the material.

It is assumed that the boundary conditions for this system are that at the base, which corresponds to $\eta = 1$,

$$\theta(1) = 1$$

and at the free end (tip), which corresponds to $\eta = 0$,

$$\frac{d\theta(0)}{d\eta} = 0$$

For a radial fin, the governing equation is written as

$$\frac{d}{d\xi}\left(\bar{a}(\xi)\frac{d\theta}{d\xi}\right) - \frac{m_r^2}{\alpha}\xi\theta = 0$$

where

$$\xi = \frac{r}{r_o}, \quad m_r^2 = \frac{2hr_o^2}{kt}, \quad \bar{a}(\xi) = \frac{2\alpha}{t}\xi y(\xi), \quad \alpha = \frac{r_b}{r_o} < 1$$

The boundary conditions for the radial fin are

$$\theta(\alpha) = 1$$

which is at the base and at the free end (tip), which corresponds to $\xi = 1$,

$$\frac{d\theta(1)}{d\xi} = 0$$

Thus, the solutions to these equations are obtained once $y(\eta)$ and $y(\xi)$ have been specified. The forms of these quantities are given in Table 11.1 for several different geometries.

A typical quantity of interest is the fin efficiency, which for the longitudinal fin is

$$\eta_l = \frac{1}{m_l^2}\frac{d\theta(1)}{d\eta} \tag{11.5}$$

and for the radial fin is

$$\eta_r = -\frac{2}{m_r^2}\frac{\alpha}{1-\alpha^2}\frac{d\theta(\alpha)}{d\xi} \tag{11.6}$$

To obtain a solution for each of the special cases, we use **DSolve** to obtain a symbolic solution for each case in Table 11.1 using the appropriate governing equation and corresponding boundary conditions. Then the derivative of the symbolic solution with respect to the appropriate spatial variable, either η or ξ, is taken and used in either Eq. (11.5) or Eq. (11.6) as the case may be. For cases 2, 3, and 4, however, there is a singularity at $\eta = 0$. To get around

Table 11.1 Nondimensional expressions for area functions and fin efficiencies for several longitudinal and radial fin shapes: definitions of m_l and m_r are given in the text

Case	Profile			
	Longitudinal	$y(\eta)$	$\bar{a}(\eta)$	Longitudinal fin efficiency η_l
1	Rectangular	$t/2$	1	$\frac{1}{m_l}\tanh(m_l)$
2	Triangular	$\eta t/2$	η	$\frac{I_1(2m_l)}{m_l I_0(2m_l)}$
3	Concave parabolic	$\eta^2 t/2$	η^2	$\frac{\sqrt{1+4m_l^2}-1}{2m_l^2}$
4	Convex parabolic	$\eta^{1/2} t/2$	$\eta^{1/2}$	$\frac{\sqrt{3}Ai'\left[(2m_l)^{2/3}\right]+Bi'\left[(2m_l)^{2/3}\right]}{\sqrt[3]{2}m_l^{4/3}\left(\sqrt{3}Ai\left[(2m_l)^{2/3}\right]+Bi\left[(2m_l)^{2/3}\right]\right)}$
	Radial	$y(\xi)$	$\bar{a}(\xi)$	Radial fin efficiency η_r
5	Rectangular	$t/2$	ξ/α	$\frac{2j\alpha(J_1(jm_r)Y_1(-jm_r\alpha)-Y_1(-jm_r)J_1(jm_r\alpha))}{(\alpha^2-1)m_r(J_1(jm_r)Y_0(-jm_r\alpha)+Y_1(-jm_r)J_0(jm_r\alpha))}$
6	Hyperbolic	$\alpha t/(2\xi)$	1	$\frac{2(Ai'(\beta)Bi'(\beta\alpha)-Ai'(\beta\alpha)Bi'(\beta))}{(\alpha^2-1)\beta^2(Ai'(\beta)Bi(\beta\alpha)-Ai(\beta\alpha)Bi'(\beta))}$, $\beta=\sqrt[3]{\frac{m_r^2}{\alpha}}$

In Cases 4 and 6, *Ai* and *Bi* are the Airy functions and the prime denotes the derivative of these functions.

this, we denote the boundary corresponding to 0 as δ and then take the limit as $\delta \to 0$. Thus, the form of the instructions to obtain the results for cases 2 to 4 are as follows

```
pp=θ[η]/.DSolve[{D[abar D[θ[η],η],η]-m_l^2 θ[η]==0,
  θ[1]==1,θ'[del]==0},θ,η][[1]];
pp1=FullSimplify[pp];
pps=Simplify[(D[pp1,η]/.η->1)/m_l^2];
result=Limit[pps,del->0]
```

where **abar** $= \bar{a}(\eta)$ and **result** is listed in the rightmost column of Table 11.1 for cases 2 to 4.

To obtain the solutions for cases 5 and 6, we use the following instructions

```
pp=θ[ξ]/.DSolve[{D[abar D[θ[ξ],ξ],ξ]-
  m_r^2 ξ θ[ξ]/α==0,θ[α]==1,θ'[1]==0},θ,ξ][[1]];
result=Simplify[-2 α (D[pp,ξ]/.ξ->α)/(m_r^2 (1-α^2))]
```

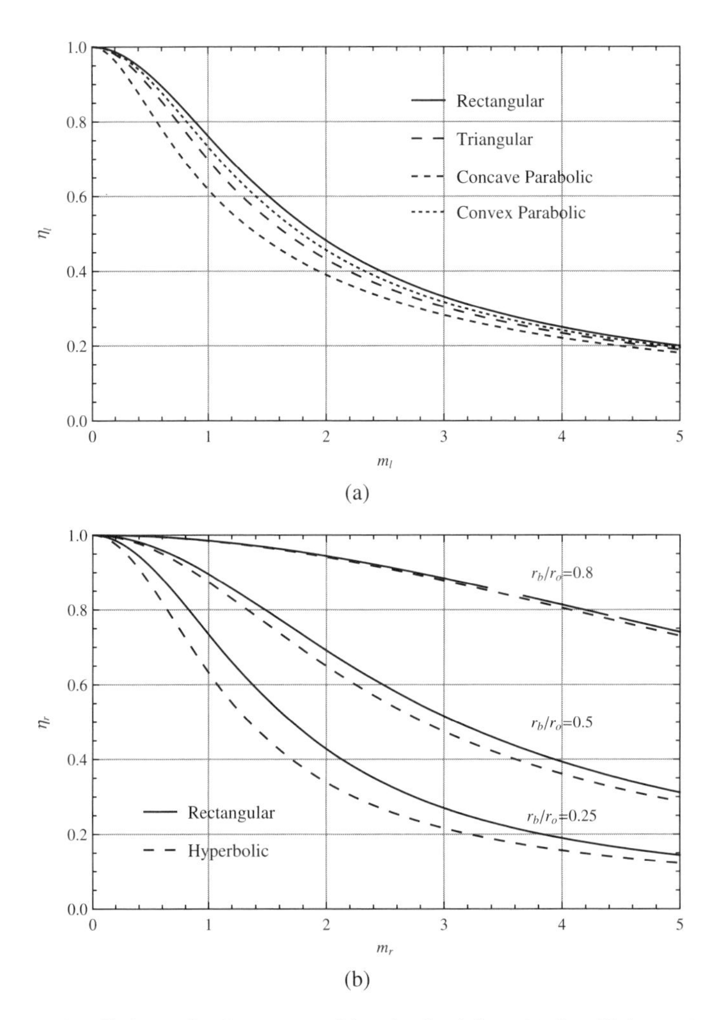

Figure 11.4 (a) Fin efficiency for four types of longitudinal fins; (b) fin efficiency for two types of radial fins

where **abar** = $\bar{a}(\xi)$, and **result** is listed in the rightmost column of Table 11.1 for cases 5 and 6.

We shall plot the results for cases 1 to 4 in Table 11.1 on one graph and the results for cases 5 and 6 for α equal to 0.25, 0.5, and 0.8 in another graph. The following program produces the results shown in Figure 11.4.

```
(* Create six functions, one for each case in Table 11.1 *)

case1[ml_]:=Tanh[ml]/ml
case2[ml_]:=BesselI[1,2 ml]/(ml BesselI[0,2 ml])
case3[ml_]:=(Sqrt[1+4 ml^2]-1)/(2 ml^2)
case4[ml_]:=(Sqrt[3] AiryAiPrime[2^(2/3) ml^(2/3)]+
```

```
  AiryBiPrime[2^(2/3) ml^(2/3)])/
  (2^(1/3) ml^(4/3) (Sqrt[3] AiryAi[2^(2/3) ml^(2/3)]+
  AiryBi[2^(2/3) ml^(2/3)]))
case5[mr_,α_]:=(2 I α (-BesselJ[1,I mr α]*
  BesselY[1,-I mr]+BesselJ[1,I mr] BesselY[1,-I mr α]))/
  (mr (-1+α^2) (BesselJ[1,I mr] BesselY[0,-I mr α]+
  BesselJ[0,I mr α] BesselY[1,-I mr]))
case6[mr_,α_]:=(β=(mr^2/α)^(1/3);
  (2 (-AiryAiPrime[β α] AiryBiPrime[β]+
   AiryAiPrime[β] AiryBiPrime[β α]))/
  (β^2 (-1+α^2) (AiryAiPrime[β] AiryBi[β α]-
   AiryAi[β α] AiryBiPrime[β])))
```

(* Define some options for Plot *)

```
lin={{Black},{Black,Dashing[Medium]},
  {Black,Dashing[Small]},{Black,Dashing[Tiny]}};
lin2=Flatten[{lin[[1;;2]],lin[[1;;2]],lin[[1;;2]]},1];
lab={Text["rb/ro=0.25",{4,0.25}],
  Text["rb/ro=0.5",{4,0.5}],
  Text["rb/ro =0.8",{4,0.9}]};
```

(* Plot results *)

```
Plot[{case1[ml],case2[ml],case3[ml],case4[ml]},{ml,0,5},
  PlotRange->{{0,5},{0,1}},PlotStyle->lin,
  Frame->True,FrameLabel->{"m1","η1]"},
  GridLines->Automatic,
  PlotLegends->Placed[{"Rectangular","Triangular",
   "Concave Parabolic","Convex Parabolic"},{0.7,0.7}]]
Plot[{case5[mr,0.25],case6[mr,0.25],case5[mr,0.5],
  case6[mr,0.5],case5[mr,0.8],case6[mr,0.8]},{mr,0,5},
  PlotRange->{{0,5},{0,1}},PlotStyle->lin2,Frame->True,
  FrameLabel->{"mr","ηr"},GridLines->Automatic,
  PlotLegends->Placed[{"Rectangular","Hyperbolic"},
  {0.2,0.2}],Epilog->lab]
```

11.3 Natural Convection Along Heated Plates

A heated vertical plate causes a buoyancy-driven flow when adjacent to a quiescent fluid. The temperature distribution and velocity of the fluid adjacent to the plate is given by [3, p. 503]

$$\begin{aligned} \frac{d^3f}{d\eta^3} + 3f\frac{d^2f}{d\eta^2} - 2\left(\frac{df}{d\eta}\right)^2 + \bar{T} = 0 \\ \frac{d^2\bar{T}}{d\eta^2} + 3\,\mathrm{Pr}\,f\frac{d\bar{T}}{d\eta} = 0 \end{aligned} \tag{11.7}$$

where

$$f = \frac{\psi}{4\nu_k \sqrt[4]{\mathrm{Gr}_x/4}}, \quad \eta = \frac{y}{x}\sqrt[4]{\mathrm{Gr}_x/4}$$

$$\mathrm{Gr}_x = \frac{g\beta x^3}{\nu_k}\left(T_s - T_\infty\right), \quad \bar{T} = \frac{T - T_\infty}{T_s - T_\infty}$$

In these relations, f is the modified stream function, ψ is the stream function, u and v, respectively, are the velocities in the x (vertical) and y (horizontal) directions such that $u = \partial\psi/\partial y$ and $v = \partial\psi/\partial x$, η is the similarity variable, $\mathrm{Pr} = \nu/\alpha$ is the Prandtl number, α is the thermal diffusivity, ν is the momentum diffusivity, Gr_x is the Grashof number, ν_k is the kinematic viscosity, β is the coefficient of thermal expansion, $T = T(y)$ is the temperature in the fluid, T_s is the surface temperature of the plate, T_∞ is the temperature at a large distance from the plate, and g is the gravity constant. The derivatives of f and $\bar{T}$ relate to different physical quantities as indicated in Figure 11.5.

The boundary conditions at $\eta = 0$ are

$$f = 0, \quad \frac{df}{d\eta} = 0, \quad \bar{T} = 1$$

and those at $\eta \to \infty$ are

$$\frac{df}{d\eta} \to 0, \quad \bar{T} \to 0$$

We shall create an interactive graphic that plots f and $\bar{T}$ and their appropriate derivatives as a function of Pr. After a little numerical experimentation, it is found that $\eta \to \infty$ can be approximated with $\eta_\infty = 5$ for $8 \geq \mathrm{Pr} \geq 1$ and with $\eta_\infty = 8$ for $\mathrm{Pr} < 1$. The program that creates the interactive graphic shown is Figure 11.5 is as follows.

```
Manipulate[If[Pr<1,xinf=8,xinf=5];
                    (* Solve Eq. (11.7) *)
  {ff,tts}=NDSolveValue[{f‴[η]+3 f[η] f″[η]-
    2 (f′[η])^2+T̄[η]==0,T̄″[η]+3 Pr f[η] T̄′[η]==0,
    f[0]==0,f′[0]==0,T̄[0]==1,f′[xinf]==0,T̄[xinf]==0},
    {f,T̄},{η,0,xinf}];
                    (* Plot results *)
  Plot[{ff[η],ff′[η],ff″[η],tts[η],tts′[η]},{η,0,xinf},
    PlotRange->All,PlotStyle->lin,ImageSize->Large,
    PlotLegends->Placed[leg,Above],Frame->True,
    FrameLabel->{Style["η",14],Style["Magnitudes",14]}],
                    (* Create sliders *)
  Style["Natural convection along a heated plate",
    Bold,12],
  "",
```

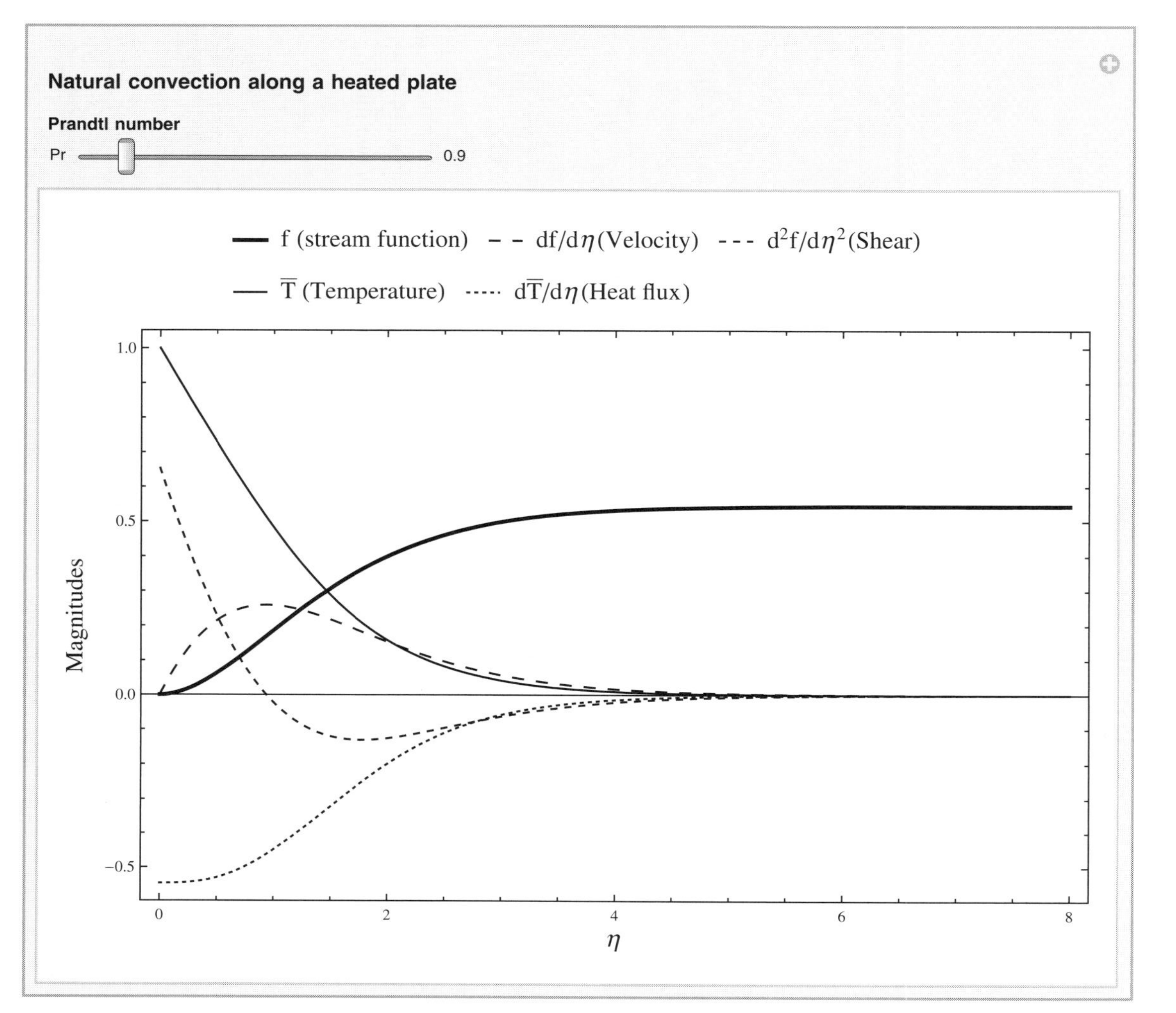

Figure 11.5 Initial configuration of the interactive graph to obtain the natural convection along a heated plated as a function of the Prandtl number

```
Style["Prandtl number",Bold,10],
{{Pr,0.9,"Pr"},0.1,8,0.1,Appearance->"Labeled",
 ControlType->Slider,ContinuousAction->False},
TrackedSymbols:>{Pr},
Initialization:>{
                (* Create labels and specify curve attributes *)
 leg={Style["f (stream function)",16],
   Style["df/dη (Velocity)",16],
   Style["d²f/dη² (Shear)",16],
   Style["T̄ (Temperature)",16],
   Style["dT̄/dη (Heat flux)",16]};
 lin={{Black,Thick},{Black,Dashing[Medium]},
  {Black,Dashing[Small]},{Black},
  {Black,Dashing[Tiny]}}}]
```

11.4 View Factor Between Two Parallel Rectangular Surfaces

The view factor for the two rectangular surfaces shown in Figure 11.6 is given by [3, p. 749]

$$F_{2-1} = \frac{1}{\pi(y_{2b} - y_{2a})(x_{2b} - x_{2a})} \int_{y_{1a}}^{y_{1b}} \int_{x_{1a}}^{x_{1b}} \int_{y_{2a}}^{y_{2b}} \int_{x_{2a}}^{x_{2b}} \frac{z_o^2}{\left((x_1 - x_2)^2 + (y_1 - y_2)^2 + z_o^2\right)^2} dx_2 dy_2 dx_1 dy_1 \tag{11.8}$$

We shall create an interactive graphic that determines the view factor for a wide range of surface sizes and separation distances. The corner of the surface with area A_1 is placed at that $x_{1a} = 0$ and $y_{1a} = 0$ and the remaining coordinates can vary for positive values of the coordinates. However, since the individual coordinates can vary independently, we must

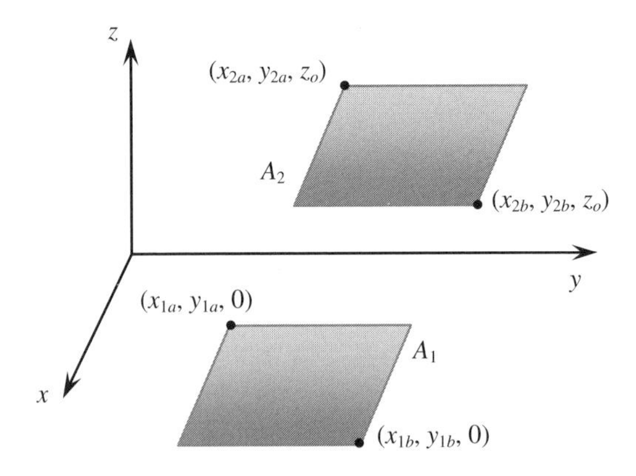

Figure 11.6 Geometry describing two parallel rectangular surfaces

ensure that $A_2 > 0$; that is, the calculation is not performed when this condition is not satisfied. Also, **ListLinePlot** is used to layout a top view of the overlapping rectangles by plotting a single point at the origin. Then the program to compute the view factor using the interactive graphic shown in Figure 11.7 is as follows.

```
Manipulate[A2=(y2b-y2a) (x2b-x2a);
                              (* Solve Eq. (11.8) *)
  If[Abs[A2]>0.001,
   VF=Quiet[NIntegrate[zo^2/((x1-x2)^2+
       (y1-y2)^2+zo^2)^2,{x2,x2a,x2b},{y2,y2a,y2b},
       {x1,0,x1b},{y1,0,y1b}]]/π/A2];
                        (* Plot rectangles and view factor *)
  ListLinePlot[{0,0},PlotRange->{{0,5},{0,5}},
    Epilog->{Red,Rectangle[{0,0},{x1b,y1b}],
     Blue,Opacity[0.5],Rectangle[{x2a,y2a},{x2b,y2b}]},
     PlotLabel->Row[{"View factor = ",
       NumberForm[VF,4]}]],
                              (* Create sliders *)
  Style["View factor between to parallel surfaces",
   Bold,14],
  " ",
  Style["Separation distance",Bold,12],
  {{zo,1.5,"zo"},1,5,0.1,ControlType->Slider,
   Appearance->"Labeled",ContinuousAction->False},
  Style["Lower plate",Bold,12],
  {{x1b,2,"x1b"},0.1,5,0.1,ControlType->Slider,
   Appearance->"Labeled",ContinuousAction->False},
  {{y1b,2.7,"y1b"},0.1,5,0.1,ControlType->Slider,
   Appearance->"Labeled",ContinuousAction->False},
  Style["Upper plate",Bold,12],
  {{x2a,0.9,"x2a"},0,5,0.1,ControlType->Slider,
   Appearance->"Labeled",ContinuousAction->False},
  {{y2a,1,"y2a"},0,5,0.1,ControlType->Slider,
   Appearance->"Labeled",ContinuousAction->False},
  {{x2b,3,"x2b"},0,5,0.1,ControlType->Slider,
   Appearance->"Labeled",ContinuousAction->False},
  {{y2b,4.1,"y2b"},0,5,0.1,ControlType->Slider,
   Appearance->"Labeled",ContinuousAction->False},
  TrackedSymbols:>{zo,x1b,y1b,y2a,x2a,y2b,x2b}]
```

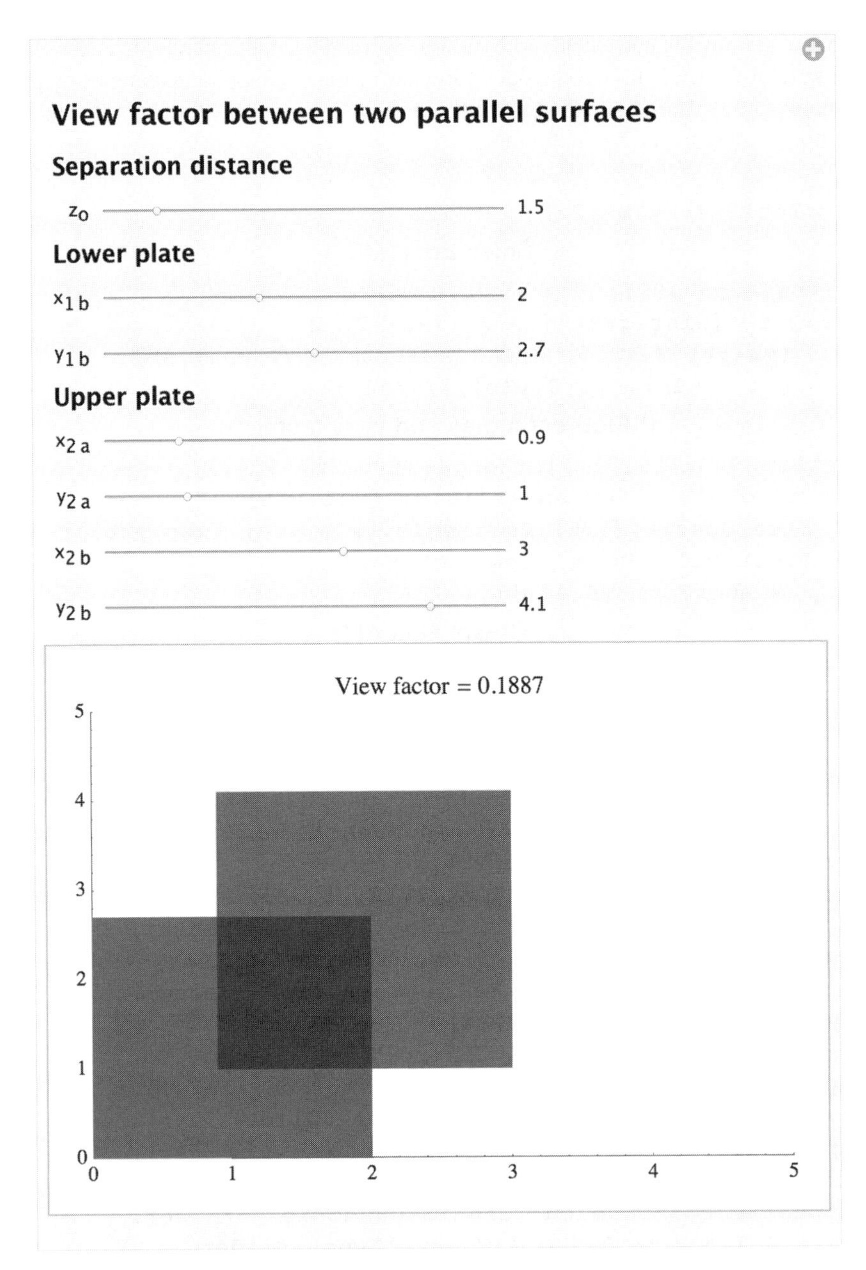

Figure 11.7 Initial configuration of the interactive graph to determine the view factor of two parallel rectangular surfaces separated by a distance z_o

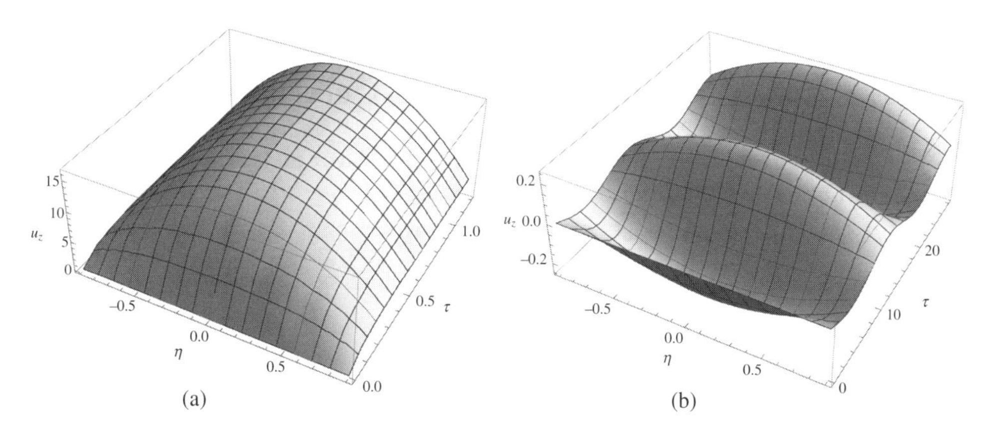

Figure 11.8 Laminar flow velocity in a horizontal pipe for (a) $P_G = -65.0$ and (b) $P_G = \sin(0.4\tau)$

11.5 Internal Viscous Flow

11.5.1 Laminar Flow in Horizontal Cylindrical Pipes

The governing equation for unsteady axisymmetric incompressible laminar flow in a circular tube of outer radius $r = a$ is given by [4]

$$\frac{\partial u_z}{\partial \tau} = -P_G(\tau) + \frac{1}{\eta}\frac{\partial}{\partial \eta}\left(\eta \frac{\partial u_z}{\partial \eta}\right) \tag{11.9}$$

where $u_z = u_z(\eta,\tau)$ (m·s^{-1}) is the velocity in the axial direction, $\eta = r/a$, $\tau = \nu t/a^2$, t (s) is time, ν (m^2·s^{-1}) is the kinematic viscosity, ρ (kg·m^{-3}) is the density, and P_G (m·s^{-1}) is the normalized axial pressure gradient; that is, if p_G (N·m^{-3}) is the pressure gradient, then $P_G = p_G a^2/(\rho\nu)$.

The boundary conditions for this system are that at $\eta = 0$, $\partial u_z/\partial\eta = 0$ and at $\eta = 1$, $u_z = 0$, which represents the no-slip condition. We shall evaluate Eq. (11.9) for two cases: (1) $P_G = -65.8$ and (2) $P_G = \sin(0.4\tau)$. In addition, the distribution of u_z across the pipe's diameter will be displayed; that is, we shall plot the results from $-1 \le \eta \le 1$. The program that evaluates Eq. (11.9) and creates the results shown in Figure 11.8 is as follows.

```
del=10^(-8); pg={65.8,-Sin[0.4 τ]}; tend={1.2,28.};
dt={0.1,1.}; plt={0,0};
                    (* Solve Eq. (11.9) for two cases *)
Do[u=NDSolveValue[{D[uz[η,τ],τ]==pg[[m]]+
  D[η D[uz[η,τ],η],η]/η,uz[η,0]==0,uz[1,τ]==0,
  (D[uz[η,τ],η]/.η->del)==0},uz,{η,del,1},
   {τ,0,tend[[m]]}];
                    (* Create plotting data from 0 to 1 *)
  up=Table[Table[{η,τ,u[η,τ]},{η,del,1,0.1}],
   {τ,0,tend[[m]],dt[[m]]}];
```

```
                    (* Create plotting data from 0 to -1 *)
    um=Table[Table[{-η,τ,u[η,τ]},{η,del,1,0.1}],
     {τ,0,tend[[m]],dt[[m]]}];
    plt[[m]]=ListPlot3D[{Flatten[up,1],Flatten[um,1]},
     AxesLabel->{"η","τ","u_z]"}],{m,1,2}]
                                    (* Plot results *)
plt[[1]]
plt[[2]]
```

11.5.2 Flow in Three Reservoirs

We shall create an interactive graphical interface that displays the flow rates from three reservoirs connected at a common junction as shown in the right-hand side of Figure 11.9. In the figure, h_j (m) is the elevation from the junction to the top surface of the water in the respective reservoir and each pipe has a diameter d_j (m), a roughness k_j (m), and a length L_j (m). If an open vertical tube were placed at the junction, then h_p (m) would be the height of the water column in that pipe.

The flow rates in each pipe Q_j ($\text{m}^3 \cdot \text{s}^{-1}$) must be such that

$$\sum_{j=1}^{3} Q_j = 0 \tag{11.10}$$

where a positive value indicates flow towards the junction and a negative value indicates flow away from the junction. In Eq. (11.10),

$$Q_j = \frac{\pi d_j^2}{4} V_j \text{sgn}(h_j - h_p) \quad j = 1, 2, 3 \tag{11.11}$$

where V_j is the velocity in the pipe and is given by

$$V_j = \sqrt{\frac{2gd_j \left|h_j - h_p\right|}{\lambda_j L_j}} \tag{11.12}$$

$g = 9.81$ m/s^2 is the gravity constant, and λ_j is the pipe friction coefficient determined from

$$\frac{1}{\sqrt{\lambda_j}} = -2\log_{10}\left(\frac{2.51}{\text{Re}_j\sqrt{\lambda_j}} + \frac{k_j/d_j}{3.7}\right) \tag{11.13}$$

In Eq. (11.13), Re_j is the Reynolds number given by

$$\text{Re}_j = \frac{V_j d_j}{\nu} \tag{11.14}$$

where $\nu = 10^{-6}$ $\text{m}^2 \cdot \text{s}^{-1}$ is the dynamic viscosity of water at 20 °C.

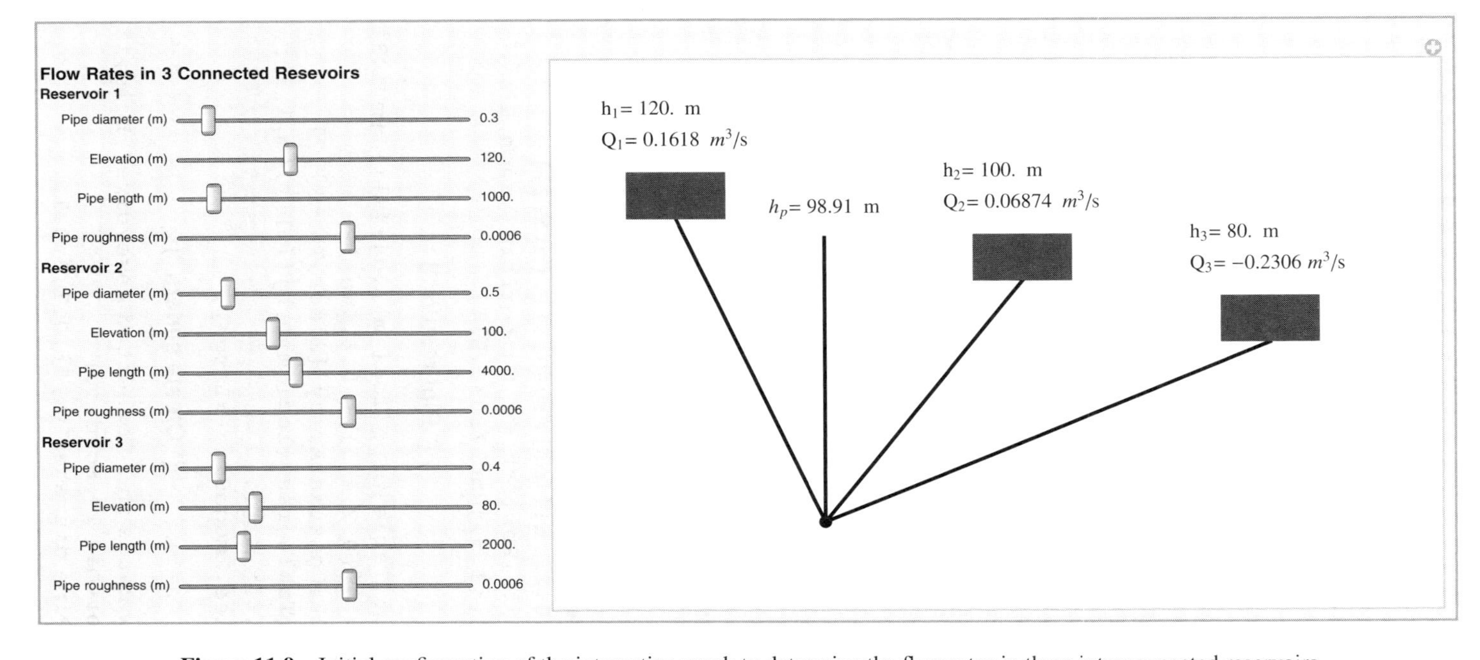

Figure 11.9 Initial configuration of the interactive graph to determine the flow rates in three interconnected reservoirs

The objective is to determine the value of h_p that satisfies Eq. (11.10), which will be done by using **FindRoot**. However, it is seen that several of the quantities used to determine the flow rates depend on V_j. Therefore, the following procedure is used. For a given set of h_j, d_j, k_j, and L_j, the iteration process is started by assuming that λ_j is

$$\lambda_j = \left[2\log_{10}\left(\frac{k_j/d_j}{3.7}\right)\right]^{-2}$$

which is obtained from Eq. (11.13) by neglecting the term containing Re_j. In addition, an initial guess $h_p = (\mathbf{Max}(h_j) + \mathbf{Min}(h_j))/2$ is selected. With these values, V_j can be determined from Eq. (11.12). For each value of V_j, **FindRoot** is used with Eqs. (11.13) and (11.14) to determine λ_j. Using these values of V_j and λ_j, Q_j is determined from Eq. (11.11) and then used in Eq. (11.10).

These equations and the method just described are used to create the interactive graphic shown in Figure 11.9, which is generated with the following program.

```
Manipulate[dj={d1,d2,d3}; kj={k1,k2,k3}; Lj={L1,L2,L3};
 hj={h1,h2,h3};  ymax=Ceiling[1.2 Max[hj]/30];
 lamgj=1/(2. Log10[3.7 dj/kj])^2;
 gues=(Max[hj]+Min[hj])/2.;
                    (* Determine Qj and hj *)
 junch=hp/.FindRoot[hfunc[hp,dj,hj,kj,Lj,lamgj,1]==0,
   {hp,gues}];
 qj=hfunc[junch,dj,hj,kj,Lj,lamgj,2];
                    (* Create labels *)
 linh={Thick,Line[{{1.5,0.25},{1.5,junch/30}}]};
 ttx=Table[Text[txt[qj[[m]],hj[[m]],m],
   {loc[[m]]+0.25,hj[[m]]/30+0.5}],{m,1,3}];
 txh=Text[Style[Row[{"hp= ",NumberForm[junch,4],
   " m"}],14],{1.5,junch/30+0.3}];
                    (* Plot results *)
 ListLinePlot[Table[pipe[loc[[m]],hj[[m]]],{m,1,3}],
   PlotRange->{{0.2,4.5},{-0.5,ymax}},
   PlotStyle->ConstantArray[{Black,Thick},2],
   Epilog->{Table[resv[loc[[m]],hj[[m]]],{m,1,3}],
    linh,pt,ttx,txh},
   Axes->False,ImageSize->Large],
                    (* Create sliders *)
 Style["Flow Rates in 3 Connected Resevoirs",Bold,12],
 Style["Reservoir 1",Bold,10],
 {{d1,0.3,"Pipe diameter (m)"},0.1,3,0.1,
  Appearance->"Labeled",ControlType->Slider },
```

```
{{h1,120.,"Elevation (m)"},10.,300.,10,
 Appearance->"Labeled",ControlType->Slider },
{{L1,1000.,"Pipe length (m)"},100.,10000.,50,
 Appearance->"Labeled",ControlType->Slider },
{{k1,0.0006,"Pipe roughness (m)"},0.00002,0.001,0.00002,
 Appearance->"Labeled",ControlType->Slider },
Style["Reservoir 2",Bold,10],
{{d2,0.5,"Pipe diameter (m)"},0.1,3,0.1,
 Appearance->"Labeled",ControlType->Slider },
{{h2,100.,"Elevation (m)"},10.,300.,10,
 Appearance->"Labeled",ControlType->Slider },
{{L2,4000.,"Pipe length (m)"},100.,10000.,50,
 Appearance->"Labeled",ControlType->Slider },
{{k2,0.0006,"Pipe roughness (m)"},0.00002,0.001,0.00002,
 Appearance->"Labeled",ControlType->Slider },
Style["Reservoir 3",Bold,10],
{{d3,0.4,"Pipe diameter (m)"},0.1,3,0.1,
 Appearance->"Labeled",ControlType->Slider },
{{h3,80.,"Elevation (m)"},10.,300.,10,
 Appearance->"Labeled",ControlType->Slider },
{{L3,2000.,"Pipe length (m)"},100.,10000.,50,
 Appearance->"Labeled",ControlType->Slider },
{{k3,0.0006,"Pipe roughness (m)"},0.00002,0.001,0.00002,
 Appearance->"Labeled",ControlType->Slider },
TrackedSymbols:>{h1,h2,h3,d1,d2,d3,L1,L2,L3,k1,k2,k3},
Initialization:>(
```

(* Function representing Eq. (11.13) *)

```
cole[λ_,Vj_,kj_,dj_]:=(Rej=Vj dj/10^(-6);
 1/Sqrt[λ]+2. Log10[2.51/(Rej Sqrt[λ])+
 kj/(3.7 dj)]);
```

(* Function representing Eqs. (11.11) and (11.12) *)

```
flowQ[hp_,dj_,hj_,kj_,Lj_,lambj_]:=
  Sign[hj-hp] π/4. Sqrt[19.62 dj Abs[hj-hp]/
    (lambj Lj)] dj^2;
```

(* Determine Q_j or sum of Q_j *)

```
hfunc[hp_?NumericQ,dj_,hj_,kj_,Lj_,lamgj_,opt_]:=
(qs={};Do[vj=Sqrt[19.62 dj[[m]] Abs[hj[[m]]-hp]/
 (lamgj[[m]] Lj[[m]])];
 If[vj==0,Qj=0,lamb=λ/.
  FindRoot[cole[λ,vj,kj[[m]],dj[[m]]],
   {λ,lamgj[[m]],0.01,0.35}];
```

```
Qj=flowQ[hp,dj[[m]],hj[[m]],kj[[m]],Lj[[m]],lamb]];
   AppendTo[qs,Qj],{m,1,3}];
If[opt==1,Total[qs],qs]);
```

(* Functions used for plotting results *)

```
pipe[loc_,hj_]:={{loc+0.25, hj/30-0.5},{1.5,0.25}};
resv[loc_,hj_]:={Blue,Rectangle[{loc,hj/30},
  {loc+0.5, hj/30-0.5}]};
pt={PointSize[Large],Point[{1.5,0.25}]};
loc={0.5,2.25,3.5};
```

(* Function for creating labels *)

```
txt[qj_,hj_,m_]:=Style[Column[{Row[
  {[{"h"ToString[m],"= "<>ToString[NumberForm[hj,4]]<>
   " m"}],
  Row[{"Q"ToString[m],"= "<>ToString[NumberForm[qj,4]]<>
   " m^3/s"}]}],14])]
```

It is pointed out that in the function **hfunc** the independent variable that is used by **FindRoot** has been restricted to numerical values only; that is, we have suppressed the use of any symbolic operations that are normally used by **FindRoot**. This restriction is invoked with the use of **?NumericQ** appended to the first argument of the function **hfunc**. If this limitation is not used, an internal symbolic operation is needed that can't be performed resulting in an error.

11.6 External Flow

11.6.1 Pressure Coefficient of a Joukowski Airfoil

The potential flow over a Joukowski airfoil is determined from a conformal transformation from the z-plane, where $z = x + jy$, to the ζ-plane, where $\zeta = \xi + j\eta$. Referring to Figure 11.10, the transformation is given by

$$z = \zeta + \lambda^2/\zeta$$

where λ is the real parameter given

$$\lambda = \xi_o + \sqrt{1 - \eta_o^2}$$

In other words, the transformation converts an airfoil shape in the z-plane to a unit circle whose center is located at (ξ_o, η_o) in the ζ-plane. In Figure 11.10, LE indicates the location of the leading edge and TE the location of the trailing edge. The region denoted by θ_{top} will map to the top of the airfoil; that is, $\zeta \to \zeta_{top} = e^{j\theta_{top}} + \zeta_o$, and the region denoted by θ_{bot} will map to the bottom of the airfoil; that is, $\zeta \to \zeta_{bot} = e^{j\theta_{bot}} + \zeta_o$.

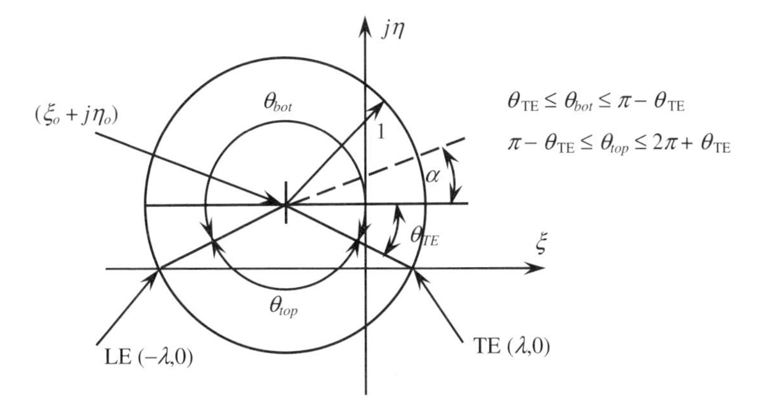

Figure 11.10 Geometry of the circle in the ζ-plane

The pressure coefficient for the airfoil is given by [5]

$$C_p = 1 - |w(\zeta)|^2 \tag{11.15}$$

where

$$w(\zeta) = \frac{1}{1 - \lambda^2/\zeta^2}\left[e^{-j\alpha} + \frac{e^{j\alpha}}{(\zeta - \zeta_o)^2} + \frac{2j\sin(\alpha - \theta_{TE})}{\zeta - \zeta_o}\right] \tag{11.16}$$

and α is an angle relative to the ξ-axis, $\zeta_o = \xi_o + j\eta_o$, and $\theta_{TE} = -\sin^{-1}(\eta_o)$. The angle α is called the angle of attack.

The shape of the airfoil is determined from

$$y = 4\lambda\left(\sqrt{0.25\left(1 + \frac{1}{16H^2}\right) - X^2} - \frac{1}{8H} \pm 0.385T(1 - 2X)\sqrt{1 - 4X^2}\right) \tag{11.17}$$

where

$$X = \frac{x}{L}, \quad H = \frac{h}{L}, \quad T = \frac{t}{L}$$

and $L = 4\lambda$ is the chord length of the airfoil, h is the maximum value of the camber of the midline of the airfoil, and t is the maximum thickness of the airfoil. In addition, the plus sign corresponds to the top of the airfoil and the minus sign to its bottom. When $H \ll 1$ and $T \ll 1$, these parameters can be approximated by

$$H \cong 0.5\sin^{-1}\eta_o \quad T \cong 3\sqrt{3}\left(\frac{1}{L} - \frac{1}{4}\right)$$

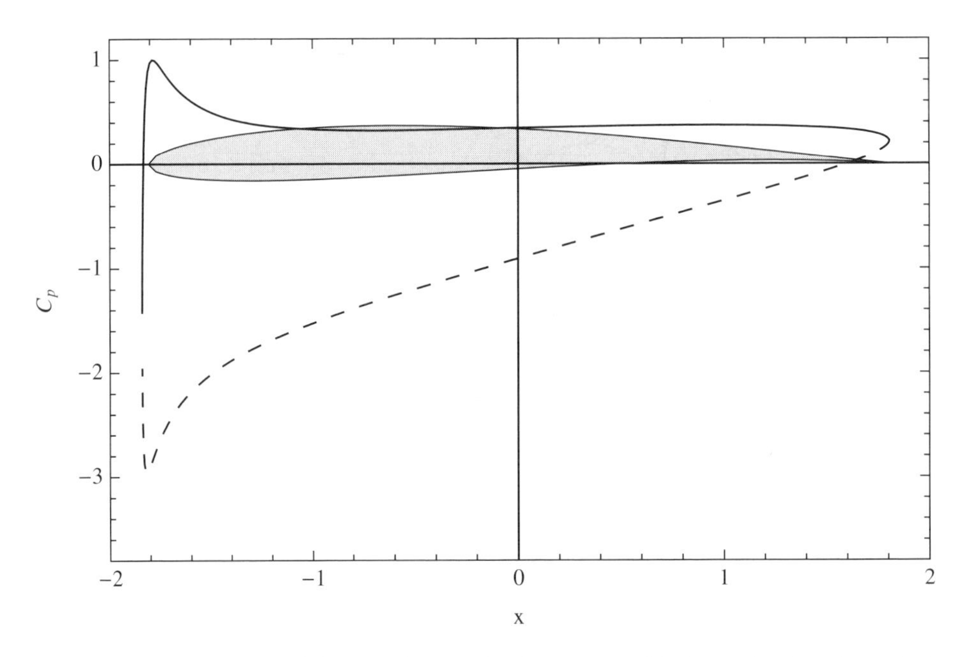

Figure 11.11 Pressure coefficient of a Joukowski airfoil for $\alpha = 7°$, $\xi_o = -0.093$, and $\eta_o = 0.08$

The values of C_p are plotted as a function of x along with the shape of the airfoil. The program that evaluates C_p and plots the results shown in Figure 11.11 for the case where $\alpha = 7°$, $\xi_o = -0.093$, and $\eta_o = 0.08$ is as follows.

```
(* Function representing Eq. (11.16) *)

w[ζ_,ζo_,λ_,α_]:= (Exp[-I α]-Exp[I α]/
   (ζ-ζo)^2+I 2.Sin[α+ArcSin[ηo]]/(ζ-ζo))/(1-λ^2/ζ^2)

(* Function representing Eq. (11.17) *)

jow[Xx_,Hh_,Tt_,sign_]:=Sqrt[0.25 (1+1/(16. Hh^2))-Xx^2]-
  1/(8. Hh)+0.385 sign Tt (1-2 Xx) Sqrt[1-4 Xx^2]

(* Generate parameter values *)

ξo=-0.093;  ηo=0.08;  α=7. Degree;  num=101;
ζo=ξo+I ηo;  θTE=-ArcSin[ηo];  λ=ξo+Sqrt[1-ηo^2];
thetop=Range[θTE+0.001,π-θTE,(π-2 θTE)/num];
zetatop=Exp[I thetop]+ζo;
ztop=zetatop+λ^2/zetatop;

(* Evaluate Eq. (11.15) *)

Cptop=Table[{Re[ztop[[n]]],1-Abs[w[zetatop[[n]],
  ζo,λ,α]]^2},{n,1,num}];
thebot=Range[π-θTE,2. π+θTE,(π+2 θTE)/num];
zetabot=Exp[I thebot]+ζo;
```

```
zbot=zetabot+λ^2/zetabot;
Cpbot=Table[{Re[zbot[[n]]],1-Abs[w[zetabot[[n]],
  ζo,λ,α]]^2},{n,1,num}];
                    (* Create airfoil coordinates *)
ptsp=Table[{4 λ Xx,4 λ jow[Xx,0.5 ArcSin[ηo],
  3. Sqrt[3] (1/(4 λ)-0.25),1]},{Xx,-0.5,0.5,0.01}];
ptsm=Table[{4 λ Xx,4 λ jow[Xx,0.5 ArcSin[ηo],
  3. Sqrt[3] (1/(4 λ)-0.25),-1]},{Xx,-0.5,0.5,0.01}];
                    (* Plot results *)
Show[ListLinePlot[{ptsp,ptsm},PlotStyle->{Black,Black},
  Filling->{1->{{2},LightGray}},Frame->True,
  FrameLabel->{"x","Cp"},PlotRange->{{-2,2},{-3.8,1.2}}],
  ListLinePlot[{Cptop,Cpbot},PlotStyle->{{Black,
   Dashing[Medium]},{Black}}]]
```

11.6.2 Surface Profile in Nonuniform Flow in Open Channels

We shall determine the surface profile of the flow of water in an open channel with a trapezoidal shape as shown in Figure 11.12 when the volume flow rate Q is known. It is assumed that the channel ends in an abrupt drop-off. The profile is determined from [6]

$$x(\eta) = \frac{h_n}{S_f}\left[\eta - \int_0^{\eta} \frac{d\xi}{1-\xi^{N(\xi)}} + \int_0^{\eta} \left(\frac{h_c}{h_n}\right)^{M(\xi)} \frac{\xi^{N(\xi)-M(\xi)}}{1-\xi^{N(\xi)}} d\xi\right] \tag{11.18}$$

where

$$\begin{aligned} N(\xi) &= \frac{10\left(1+2mh_b\xi\right)}{3\left(1+mh_b\xi\right)} - \frac{8h_b\xi\sqrt{1+m^2}}{3\left(1+2h_b\xi\sqrt{1+m^2}\right)} \\ M(\xi) &= \frac{3\left(1+2mh_b\xi\right)^2 - 2mh_b\xi\left(1+mh_b\xi\right)}{\left(1+2mh_b\xi\right)\left(1+mh_b\xi\right)} \end{aligned} \tag{11.19}$$

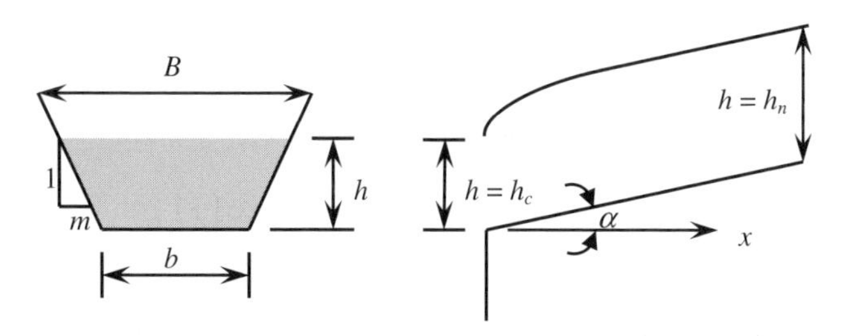

Figure 11.12 Geometry and definitions defining the flow in an open channel of trapezoidal shape

and $h_b = h_n/b$, $\eta = h/h_n$, and $S_f = \sin\alpha$. For a given Q, the quantity h_n is determined from

$$Q = \frac{h_n(b + mh_n)\sqrt{S_f}}{n}\left(\frac{h_n(b + mh_n)}{b + 2h_n\sqrt{1 + m^2}}\right)^{2/3} \tag{11.20}$$

where n is the Manning roughness coefficient ($\mathrm{m}^{-1/3}\mathrm{s}$). The quantity h_c is determined from

$$Q = \sqrt{\frac{g\left(h_c(b + mh_c)\right)^3}{b + 2mh_c}} \tag{11.21}$$

where g = 9.81 $\mathrm{m{\cdot}s^{-2}}$ is the gravity constant. It is noted that when $m = 0$, the channel has a rectangular shape ($B = b$) and when $b = 0$, the channel has a triangular shape.

It is noted that in Eq. (11.18) the range of η is $\eta_c \le \eta \le 1 - \varepsilon$, where $\eta_c = h_c/h_n$ and $0 < \varepsilon \ll 1$. The quantity ε has been introduced to avoid the singularity in Eq. (11.18) at $\xi = 1$. The starting value of x, that is, $x(\eta_c)$, will vary depending on the parameters Q, m and b. Therefore, to start each surface profile at $x = 0$, the profile's coordinates are taken as $(x(\eta_c) - x(\eta), h_n\eta)$. However, we shall display the profile relative to the sloping channel floor whose height is $y_f(x) = x(\eta)\tan\alpha$. In this case, the coordinates of the profile become $(x(\eta_c) - x(\eta), h_n\eta + x(\eta)\tan\alpha)$. Lastly, it is noted that the automatic symbolic manipulation of **NIntegrate** has been suppressed as discussed in Section 5.2. This suppression is done with the option **Method** as shown in the function **Xeta**.

The program that generates the interactive graphic shown in Figure 11.13 is as follows. (Note: There may be combinations of the parameters that yield unrealistic results.)

```
Manipulate[Sf=Sin[α Degree];
                    (* Determine hn and hc *)
  hn=xx/.FindRoot[Hn[xx,m,Sf,mann,b,Qq],{xx,1.5}];
  hc=zz/.FindRoot[Hc[zz,m,b,Qq],{zz,1.5}];
                    (* Create data to plot *)
  hcn=hc/hn;
  xshift=hn Xeta[hcn,m,hn/b,Sf,hcn];
  xxc=Join[Range[hcn,0.7,(0.7-hcn)/25],
    Range[0.71,0.999,0.289/10]];
  ptsx=Table[xshift-hn Xeta[cc,m,hn/b,Sf,hcn],{cc,xxc}];
  xmax=Max[ptsx];
  ymax=0.999 hn+Last[ptsx] Tan[α Degree];
  bed={{0,0},{xmax,0},{xmax,0.999 hn}};
  pts=Table[{ptsx[[n]],hn xxc[[n]]+
    ptsx[[n]] Tan[α Degree]},{n,Length[xxc]}];
                    (* Plot results *)
  ListLinePlot[pts,PlotRange->{{0,Ceiling[xmax]},
    {0,Ceiling[ymax]}},Epilog->{Gray,Polygon[bed]}],
    AxesLabel->{"x","h"}],
```

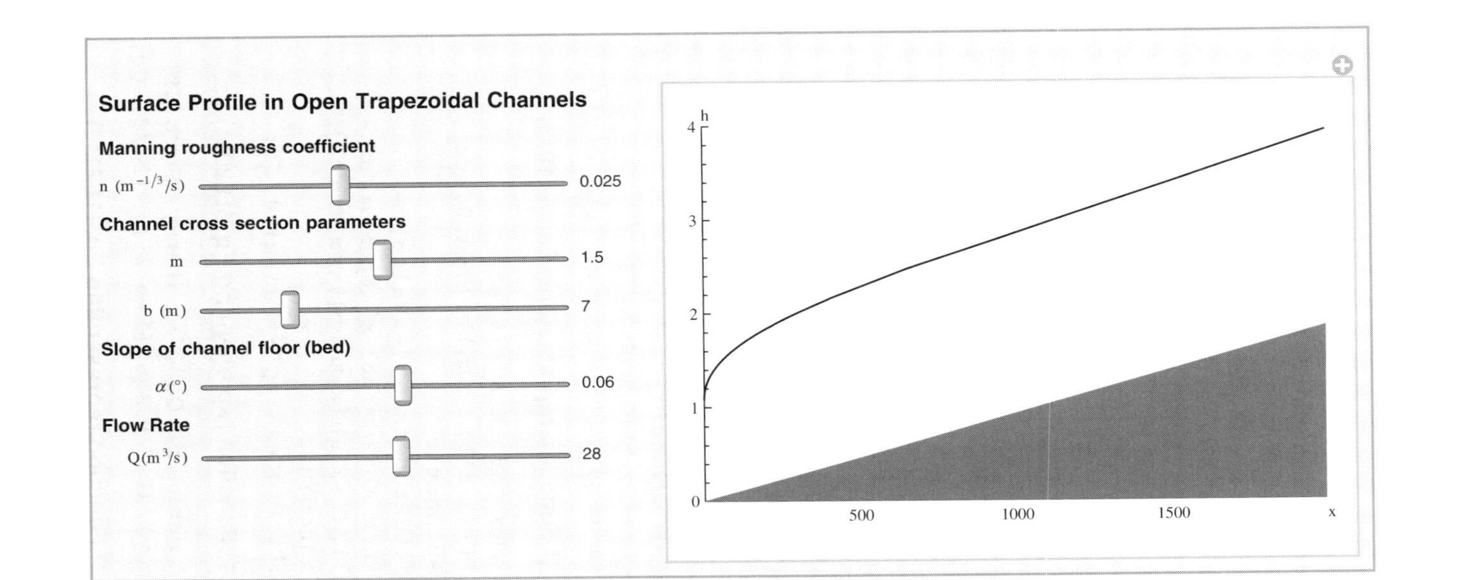

Figure 11.13 Initial configuration of the interactive graph to display the surface profile of nonuniform flow in open trapezoidal channels

(* Create sliders *)

```
Style["Surface Profile in Open Trapezoidal Channels",
 Bold,12],
"",
Style["Manning roughness coefficient",Bold,10],
{{mann,0.025,"n (m^-1/3/s)"},0.01,.05,.001,
  Appearance->"Labeled",ControlType->Slider},
Style["Channel cross section parameters",Bold,10],
{{m,1.5,"m"},0,3,.1,Appearance->"Labeled",
  ControlType->Slider},
{{b,7,"b (m)"},0.5,30,.5,Appearance->"Labeled",
  ControlType->Slider},
Style["Slope of channel floor (bed)",Bold,10],
{{α,0.06,"α (°)"},0.01,0.1,.01,Appearance->"Labeled",
  ControlType->Slider},
Style["Flow Rate",Bold,10],
{{Qq,28,"Q (m^3/s)"},1,50,0.5,Appearance->"Labeled",
  ControlType->Slider},
TrackedSymbols:>{mann,m,b,α,Qq},
Initialization:>{
```

(* Functions representing Eq. (11.19) *)

```
 Nh[z_,m_,hb_]:=10./3 (1+2 m hb z)/
   (1+m hb z)-8/3 Sqrt[1+m^2] hb z/
   (1+2 Sqrt[1+m^2]hb z);
 Mh[z_,m_,hb_]:=(3. (1+2 m hb z)^2-
   2 m hb z (1+m hb z))/((1+2 m hb z) (1+m hb z));
```

(* Function representing Eq. (11.18) *)

```
 Xeta[eta_,m_,hb_,Sf_,hcn_]:=(eta-
  NIntegrate[1/(1-x^Nh[x,m,hb]),{x,0,eta},
   Method->{Automatic,
        "SymbolicProcessing"->False}]+
  NIntegrate[hcn^Mh[x,m,hb] x^(Nh[x,m,hb]-
   Mh[x,m,hb])/(1-x^Nh[x,m,hb]),{x,0,eta},
   Method->{Automatic,
        "SymbolicProcessing"->False}])/Sf;
```

(* Function representing Eq. (11.20) *)

```
 Hn[y_,m_,Sf_,mann_,b_,Qq_]:=y (b+m y) Sqrt[Sf]/
  mann (y (b+m y)/(b+2 y Sqrt[1+m^2]))^(2./3)-Qq;
```

(* Function representing Eq. (11.21) *)

```
 Hc[y_,m_,b_,Qq_]:=
  Sqrt[9.81 (y (b+m y))^3/(b+2 y m)]-Qq}]
```

References

[1] H. H. Pennes, "Analysis of tissue and arterial blood temperature in the resting human foreman," *Journal of Applied Physiology*, 1948, 1, pp. 93–102.

[2] M. D. Mikhailov and M. N. Özişik, *Unified Analysis and Solutions of Heat and Mass Diffusion*, John Wiley and Sons, New York, 1984, Section 6.2.

[3] F. P. Incropera and D. P. Dewitt, *Introduction to Heat Transfer*, 4th edn, John Wiley & Sons, New York, 2002.

[4] J. H. Duncan, Fluid Mechanics, Section 11.2.1, in E. B. Magrab, et al., *An Engineer's Guide to MATLAB®*, Prentice Hall, Upper Saddle River, New Jersey, 2011, p. 621.

[5] R. L. Panton, *Incompressible Flow*, 4th edn, John Wiley & Sons, Chichester, United Kingdom, 2013, p. 477.

[6] W. H. Graf, *Fluvial Hydraulics*, John Wiley & Sons, Chichester, United Kingdom, 1998, pp. 158 and 194–6.

Index

An Engineer's Guide to Mathematica®, First Edition. Edward B. Magrab.

Companion Website: www.wiley.com/go/magrab